Lattice gas hydrodynamics is the name used to describe the approach to fluid dynamics starting from a simple micro-world constructed as an automaton universe. In this universe, the microscopic dynamics is based not on a realistic description of interacting particles, but merely on the laws of symmetry and of invariance of macroscopic physics.

We imagine point-like particles residing on a regular lattice, where they move from node to node and undergo collisions when their trajectories meet at the same node. If the collisions occur according to some simple logical rules, and if the lattice has the proper symmetry, then the automaton shows global behavior very similar to that of real fluids. This book carries two important messages. First, it shows how an automaton universe with very simple microscopic dynamics – the lattice gas – can exhibit macroscopic behavior in accordance with the phenomenological laws of classical physics. Second, it demonstrates that lattice gases have spontaneous microscopic fluctuations which capture the essentials of actual fluctuations in real fluids. The two aspects are closely related, offering a microscopic approach to the description of fluid systems. Accordingly, the book follows the philosophy of classical statistical mechanics, and introduces the reader to the world of the discrete analogue of continuous molecular hydrodynamics.

This book will be of interest to graduate students and researchers in statistical physics, computational physics, hydrodynamics, applied mathematics and engineering.

Jean Pierre Boon studied at the University of Brussels and at the University of Chicago. He has been FNRS Research Fellow and Professor at the University of Brussels since 1975. He is a former Visiting Professor at Bell Telephone Laboratories, at the Kernforschungsanlage Jülich, at the Massachusetts Institute of Technology and at the University of Nice. J.P. Boon's main field of research is statistical physics. He is the co-author of *Molecular Hydrodynamics* by J.P. Boon and S. Yip.

Jean-Pierre Rivet received his PhD in 1988 from the University of Nice for work on theoretical and numerical aspects of lattice gas hydrodynamics. Since then, he has been working on applications of the lattice gas method to computational fluid dynamics. Since 1996 he has worked on relativistic stochastic processes and statistical physics, in collaboration with researchers at the Laboratory of Radio-Astronomy of the Ecole Normale Supérieure (Paris). In addition to his research activity, he has given several courses in Physics and Astronomy at the University of Paris VI, and at the University of Nice. He is presently CNRS researcher at the Observatoire de la Côte d'Azur (Nice).

Cambridge Nonlinear Science Series 11

EDITORS

Professor Boris Chirikov
Budker Institute of Nuclear Physics, Novosibirsk

Professor Predrag Cvitanović
Niels Bohr Institute, Copenhagen

Professor Frank Moss
University of Missouri, St Louis

Professor Harry Swinney
Center for Nonlinear Dynamics,
The University of Texas at Austin

Lattice Gas Hydrodynamics

TITLES IN PRINT IN THIS SERIES

G. M. Zaslavsky, R. Z. Sagdeev, D. A. Usikov and A. A. Chernikov

1. Weak chaos and quasi-regular patterns

J. L. McCauley

2. Chaos, dynamics and fractals: an algorithmic approach to deterministic chaos

K. Nakamura

3. Quantum chaos: a new paradigm for nonlinear dynamics

C. Beck and F. Schlögl

4. Thermodynamics of chaotic systems: an introduction

P. Meakin

5. Fractals, scaling and growth far from equilibrium

R. Badii and A. Politi

6. Complexity – hierarchical structures and scaling in physics

H. Kantz and T. Schreiber

7. Nonlinear time series analysis

T. Bohr, M. H. Jensen, G. Paladin and A. Vulpiani

8. Dynamical systems approach to turbulence

P. Gaspard

9. Chaos, scattering and statistical mechanics

E. Schöll

10. Nonlinear Spatio-Temporal Dynamics and Chaos in Semiconductors

J.-P. Rivet and J. P. Boon

11. Lattice Gas Hydrodynamics

Lattice Gas Hydrodynamics

J.-P. Rivet
Observatoire de la Côte d'Azur, France

and J. P. Boon
Université de Bruxelles, Belgium

CAMBRIDGE
UNIVERSITY PRESS

CAMBRIDGE UNIVERSITY PRESS
Cambridge, New York, Melbourne, Madrid, Cape Town, Singapore, São Paulo

Cambridge University Press
The Edinburgh Building, Cambridge CB2 2RU, UK

Published in the United States of America by Cambridge University Press, New York

www.cambridge.org
Information on this title: www.cambridge.org/9780521419444

First published 2001
This digitally printed first paperback version 2005

A catalogue record for this publication is available from the British Library

ISBN-13 978-0-521-41944-4 hardback
ISBN-10 0-521-41944-1 hardback

ISBN-13 978-0-521-01971-2 paperback
ISBN-10 0-521-01971-0 paperback

Contents

Preface

... Feynman told us to explain it like this: We have noticed in nature that the behavior of a fluid depends very little on the nature of the individual particles in that fluid. [...] We have therefore taken advantage of this fact to invent a type of imaginary particle that is especially simple for us to simulate. This particle is a perfect ball bearing that can move at a single speed in one of six directions. The flow of these particles on a large enough scale is very similar to the flow of natural fluids.

W.D. Hillis, *Physics Today*, February 1989

The story of lattice gas automata started around 1985 when pioneering studies established theoretically and computationally the feasibility of simulating fluid dynamics via a microscopic approach based on a new paradigm : a fictitious oversimplified micro-world is constructed as an automaton universe based not on a realistic description of interacting particles (as in molecular dynamics), but merely on the laws of symmetry and of invariance of macroscopic physics. Imagine point-like particles residing on a regular lattice where they move from node to node and undergo collisions when their trajectories meet at the same node. The remarkable fact is that, if the collisions occur according to some simple logical rules and if the lattice has the proper symmetry, this automaton shows global behavior very similar to that of real fluids. Then, to observe the fluid-like behavior of the automaton, we look at the lattice from a distance and, to measure its properties, we perform measurements over averaged quantities to obtain for instance the fluid velocity field or to visualize its stream lines (see Figure 1).

At the beginning, most of the effort was invested into lattice gas *hydrodynamics* through theoretical developments recovering Navier–Stokes equations, simulations showing wakes and eddy formation, and hardware elaboration of dedicated computers. By the late eighties, the field of lattice gas automata had flourished into a broad range of theoretical studies and applications : statistical mechanics theories had gone deeper into the basics of lattice gases and models had been proposed for a variety of phenomena. The concomitant consequence was the sophistication of the new models and of the methods developed for their implementation. As a result, the field was also branching out into specialized sub-areas, such as statistical mechanics of lattice gases, three-dimensional

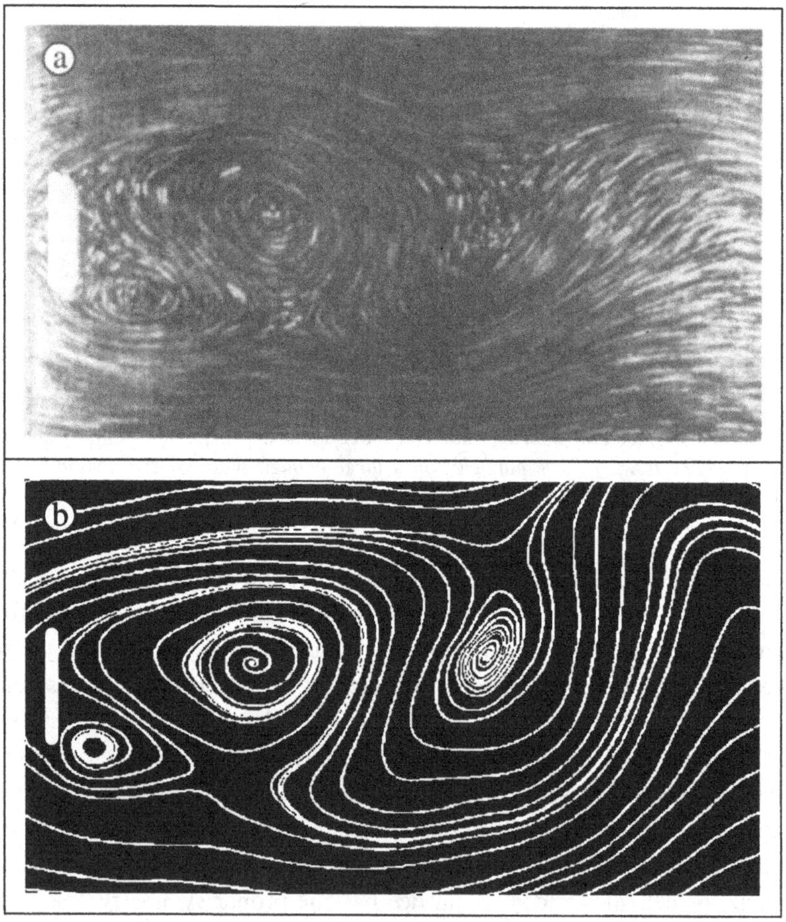

Figure 1 Beyond some threshold value of the Reynolds number, the flow behind an obstacle produces a non-stationary wake, the Bénard–von Kármán vortex street illustrated in (a) by a picture taken by Prandtl and Tietjens in 1934; (b) shows the result of a lattice gas simulation (performed on the dedicated machine 'RAP-1', see Chapter 10) under conditions similar to those of the laboratory experiment. In both (a) and (b), the Reynolds number is of the order of 200. ((a) In *An Introduction to Fluid Dynamics*, G.K. Batchelor, Cambridge University Press, 1967; (b) J.P. Rivet, 1987.)

models for relatively high Reynolds numbers and transition to turbulence, multi-species and multi-phase models for low Reynolds number complex flows (with or without surface tension), boundary layers, lattice Boltzmann approach to turbulence, thermal lattice gases, reactive systems. Despite an apparent diversity, lattice gas automata developed for generic or specific physical problems all share a common basis, conceptually and operationally. The reason can be found in the following considerations.

Often systems with a large number of degrees of freedom exhibit macroscopic behavior where the details of the microscopic dynamics are relatively unimportant. This feature opens the way to a generic description: systems with different microscopic characteristics can be described at the macroscopic scale with a generic set of equations where the specific nature of the constituents of the system enter only a restricted number of coefficients at the level of constitutive relations. Since the macroscopic description does not provide any content to these coefficients, their specificity can be (at least partly) eliminated when the equations are properly recast into non-dimensional form. Then the global behavior of the system depends on a limited number of universal control parameters where the microscopic nature of the elementary constituents does not appear explicitly. The macroscopic level of analysis therefore provides a phenomenological description where (i) the complexity of identifying the microscopic degrees of freedom is bypassed, and (ii) many different physical phenomena are recognized to belong to a limited number of classes. Classical fluid mechanics offers a striking example through Reynolds's dynamical similarity law (1883). Nevertheless, it is obvious that in this process of reasoning the connection between the phenomenology and the underlying microscopic mechanisms has been lost. Consequently, the macroscopic level of description which uses average quantities prevents analysis of, for instance, how large-scale phenomena can be triggered by local and/or transient deviations from averages, that is, fluctuations. This is one of the objects of the statistical mechanical approach which establishes the microscopic basis of the properties of many-body systems. But we have observed that it is not necessary to have a complete knowledge of the details of the microscopic interactions to understand how macroscopic phenomena emerge. It will appear as logical conduct to develop a microscopic approach to macroscopic phenomena by modeling the microscopic dynamics by means of a *simplified description* provided the basic requirements of fundamental physics are correctly incorporated, essentially – although further requirements may be in order – the conservation laws and the symmetry properties.

Our description uses a class of simple *discrete models*, where space and time are incremented by finite elementary units, as in the physical picture that we presented where point particles move step by step on a regular lattice. A logical question to ask is: why are we interested in such discrete models? The fact is that despite their simplicity at small scale (at the microscopic level), these discrete systems can show extremely complex behavior at larger scales (the mesoscopic and macroscopic levels), and this global behavior can be very similar to what is observed in natural phenomena (see Figure 1). Now many of these natural phenomena are very complicated (for instance the diversity of forms that a fluid can exhibit in the turbulent regime), and often the best one can do is to offer a phenomenological analysis of what is observed. Yet the objective is to try to understand how things come about by providing an analysis in terms of the elementary processes that are responsible for the emergence of large-scale phenomena, a formidable task if one starts with a realistic description of all the basic interactions that govern the underlying microscopic dynamics.

The philosophy that constitutes the conceptual basis of the lattice gas automaton (LGA) approach is that a microscopically simple system which can exhibit complex macroscopic behavior in accordance with observed phenomena should contain, at the elementary level, the essentials which are responsible for the emergence of complexity, and thereby can teach us something about the basic mechanisms from which complexity builds up.

As a matter of fact, when LGAs were initially developed around 1985, the goal was to produce simplified models that could provide new computational accessibility to complicated problems in fluid dynamics such as 2-D and 3-D turbulence. While earlier studies (in the seventies) had considered even simpler lattice models for basic problems in statistical mechanics, subsequently a variety of models were constructed for more involved physical applications: hydrodynamic instabilities, flows in porous media, phase transitions, reactive systems, etc. All these models share a common structure where point particles move along the links of a regular lattice and interact on the nodes through collisions. The key point is that these simple basic ingredients (along with some symmetry requirement for the lattice structure) suffice for the emergence of complex global behavior from elementary processes. Furthermore, the lattice gas automaton exhibits two other important features: (i) it usually resides on a large lattice, and so possesses a large number of degrees of freedom; (ii) its microscopic Boolean nature (which we shall describe in detail in the next two chapters) combined with the (generally) stochastic rules which govern its microscopic dynamics, result in intrinsic fluctuations. Because of these spontaneous fluctuations and because of its large number of degrees of freedom, the lattice gas can be considered as a 'reservoir of thermal excitations' in much the same way as an actual fluid (see Figure 2).

The principal goal of this book is to carry two messages. The first is to show how an automaton universe with simple microscopic dynamics can exhibit macroscopic behavior in accordance with the phenomenological description of classical physics. The second is to establish that the correlations of the lattice gas intrinsic fluctuations capture the essentials of actual fluctuations in real fluids.

The two aspects are of course closely connected. Therefore our emphasis is on the statistical mechanics of lattice gas automata as a microscopic approach to the description of fluid systems, and in this sense we proceed very much along the lines of the general philosophy of classical statistical mechanics.

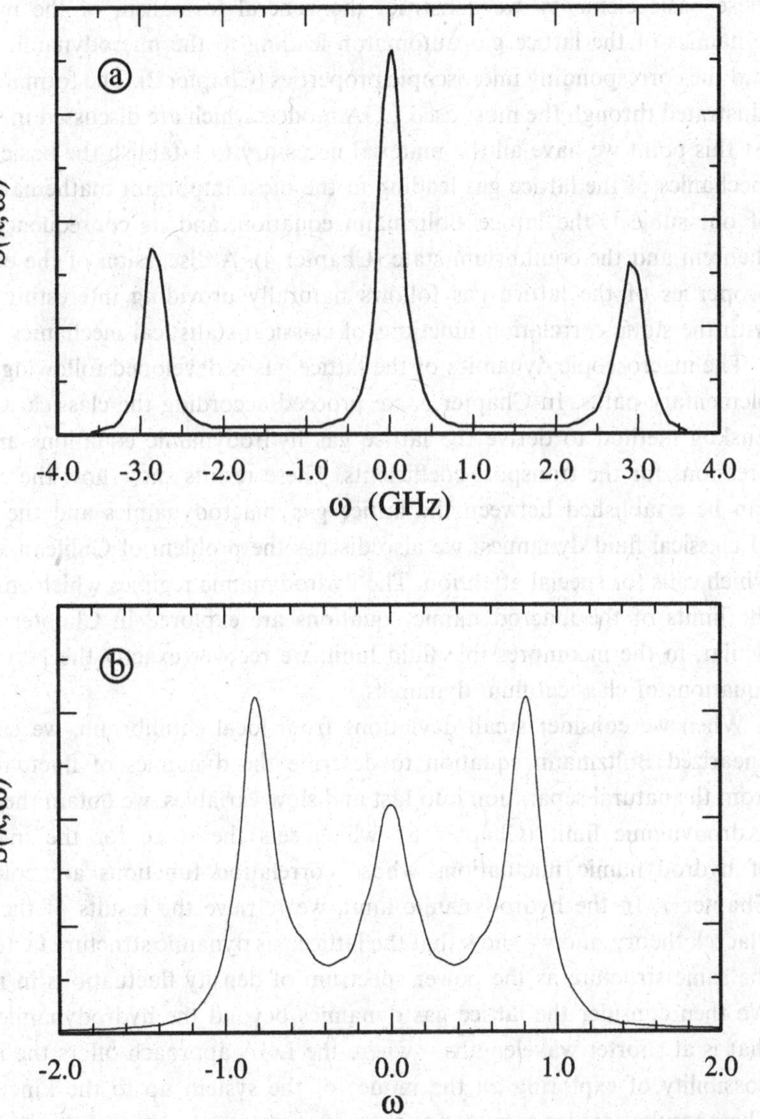

Figure 2 The spectrum of density fluctuations in the hydrodynamic regime. (a) The Rayleigh–Brillouin–Mandelstahm spectrum measured by light scattering spectroscopy in liquid argon (Fleury and Boon, 1969); (b) the dynamic structure factor $S(k, \omega)$ for the GBL thermal lattice gas (see Section 3.6) (Grosfils, Boon, Lallemand, 1992). In both (a) and (b), $S(k, \omega)$ is in arbitrary units, and in (b) ω is in reciprocal simulation time units.

After this simple presentation of the lattice gas picture and of the ideas underlying the philosophy of the lattice gas automaton approach, we introduce the basic notions of cellular automata and of lattice gas automata from the physical point of view and from the mathematical point of view, and we discuss the complementary aspects of the two viewpoints (Chapter 1). With these basic elements, we construct the general formalism of the microscopic dynamics of the lattice gas automaton leading to the microdynamic equations and the corresponding microscopic properties (Chapter 2). The formalism is then illustrated through the most used LGA models which are discussed in Chapter 3. At this point we have all the material necessary to establish the basic statistical mechanics of the lattice gas leading to the most important mathematical object of our subject: the lattice Boltzmann equation, and its consequences, the H-theorem and the equilibrium state (Chapter 4). A discussion of the equilibrium properties of the lattice gas follows naturally providing interesting analogies with the static correlation functions of classical statistical mechanics.

The macroscopic dynamics of the lattice gas is developed following two complementary paths. In Chapter 5, we proceed according the classical Chapman–Enskog method to derive the lattice gas hydrodynamic equations and the expressions for the transport coefficients. These results show how the connection can be established between the lattice gas macrodynamics and the equations of classical fluid dynamics; we also discuss the problem of Galilean invariance, which calls for special attention. The hydrodynamic regimes which emerge from the limits of the macrodynamic equations are explored in Chapter 9; in particular, in the incompressible fluid limit, we recover exactly the Navier–Stokes equations of classical fluid dynamics.

When we consider small deviations from local equilibrium, we can use the linearized Boltzmann equation to describe the dynamics of fluctuations, and from the natural separation into fast and slow variables, we obtain the linearized hydrodynamic limit (Chapter 6), which sets the stage for the investigation of hydrodynamic fluctuations whose correlation functions are computed in Chapter 7. In the hydrodynamic limit, we retrieve the results of the Landau–Placzek theory, and we show that the lattice gas dynamic structure factor exhibits the same structure as the power spectrum of density fluctuations in real fluids. We then consider the lattice gas dynamics beyond the hydrodynamic regime – that is at shorter wavelengths – where the LGA approach offers the interesting possibility of exploring all the modes of the system up to the kinetic regime. These results are also extended to the two-component system (the 'colored' lattice gas) to incorporate mass diffusion. Larger deviations for systems arbitrarily far from equilibrium are considered in Chapter 8 where, using a projector technique, we derive the full non-linear hydrodynamic equations and the general form of the Green–Kubo expression for the transport coefficients.

Numerical simulations are an important contribution to our subject: they provide tests of validity of the LGA theory and, in cases such as those illustrated in Figures 1 and 2, the results of lattice gas simulations can be compared with laboratory experimental results; but most importantly they can also offer access – theoretically or numerically – to regimes which cannot be easily reached by other methods. Chapter 10 presents examples of lattice gas simulations which are illustrated and discussed for various hydrodynamic flows, along with the essentials of the procedure for computer implementation. Finally, we close with an epilogue in the form of a guide for further reading (Chapter 11) indicating topics and areas of research where lattice gas automata offer, technically or conceptually, a particularly interesting and useful approach.[1]

The ideas expressed in this book and the material presented here reflect our own choices and views. The materialization of these ideas has been influenced in various ways by the interaction, the stimulation and the discussions we had over the past fifteen years with our colleagues and friends. To all of them we express our appreciation, with special thanks to David Dab, Matthieu Ernst, Uriel Frisch, Patrick Grosfils, David Hanon, Michel Hénon, Dominique d'Humières, Ray Kapral, Pierre Lallemand, Alain Noullez, Dan Rothman, Alberto Suarez, Sauro Succi, Olivier Tribel, Jörg Weimar, and Stéphane Zaleski. JPR acknowledges support by the *Centre National de la Recherche Scientifique* (CNRS, France) and JPB acknowledges support by the *Fonds National de la Recherche Scientifique* (FNRS, Belgium).

[1] As much as possible, we avoided disrupting the course of the main text by deferring mathematical details and technical developments to the appendix.

Chapter 1

Basic ideas

When the lattice gas was introduced in statistical physics around 1985 (see Frisch, Hasslacher and Pomeau, 1986), it was originally constructed as a physical model for hydrodynamics. In fact, the concept of the lattice gas is as much a physical concept – and we shall indeed start with intuitive physical ideas – as it is a mathematical concept, as a more formal definition can also be given. We first present the point of view of the physicist (Section 1.1), then we describe the lattice gas automaton from the mathematical viewpoint (Section 1.2), and in Section 1.3 we discuss the two aspects.

1.1 The physicist's point of view

A lattice gas can be viewed as a simple, fully discrete microscopic model of a fluid, where fictitious particles reside on a finite region of a regular Bravais lattice.[1] These fictitious particles move at regular time intervals from node to node, and can be scattered by local collisions according to a node-independent rule that may be deterministic or non-deterministic. Thus, time, space coordinates and velocities are discrete at the microscopic scale, that is, at the scale of particles, lattice nodes and lattice links.

The stationary states in statistical equilibrium and thus the large-scale dynamics of a lattice gas will crucially depend on its conservation properties, that is, on the quantities preserved by the microscopic evolution rule of the system. Suitable

[1] A Bravais lattice is a special case of regular lattice; see Section 1.3.4 for details.

conservations are necessary, but not sufficient conditions in order that a lattice gas exhibit physically realistic collective large-scale dynamics: an appropriate *crystallographic space group* for the lattice[2] is another necessary condition for macroscopic dynamics with correct continuous invariances. These points will be emphasized in later chapters.

Another important feature is the 'exclusion principle': two absolutely indistinguishable particles cannot reside simultaneously on the same node. However, two particles with different velocities can coexist at a node as well as particles with the same velocity as far as they can be distinguished by some other property (mass, color, label, etc.). A consequence of the exclusion principle is the following: provided there exists only a finite number b of distinguishable kinds of particles, the occupation state of a node is completely determined by b Boolean variables, each variable encoding the presence or the absence of a particle of each kind on the node. This feature is of considerable importance for the implementation of lattice gases on special-purpose machines as well as on general purpose computers.

So the physicist would view a lattice gas as a universe of fictitious zero-dimensional particles undergoing one-dimensional displacements on the links and colliding on the nodes of a regular Bravais lattice embedded in a D-dimensional space. But the Boolean encoding of the state of a lattice gas leads naturally to a mathematically more formal point of view.

1.2 The mathematician's point of view

We shall first introduce the concepts of 'finite automata' and 'cellular automata'.

1.2.1 Finite automata

A 'finite automaton' is a mathematical model of a processor with a finite number of possible internal states, which evolves and produces output data according to a rule depending on an input symbol belonging to a finite alphabet (for more details about finite automata and related notions, see Hopcroft and Ullman, 1979). Notice that a processor with a finite number of states, whose evolution depends on *several* input symbols belonging to several finite alphabets, also qualifies as a finite automaton, provided the number of inputs is finite. Indeed, a finite number of input symbols, where each one belongs to a finite alphabet, can be viewed as a single symbol belonging to a larger, yet finite

[2] The 'crystallographic space group' or 'global invariance group' of a Bravais lattice is the group of all rigid (i.e. distance-preserving) transformations, including translations, rotations, symmetries and their combinations, that leave the lattice globally unchanged (see Ashcroft and Mermin, 1976).

alphabet. The above definition prescribes a deterministic evolution rule for the internal state of the automaton. Nevertheless, a non-deterministic automaton also fits the above definition if its non-deterministic evolution rule can be considered as a deterministic rule depending on a finite number of additional input symbols belonging to finite alphabets, produced by 'external' (pseudo-) random generators.

Since the internal state of a finite automaton can change, the automaton undergoes some kind of evolution, and there must be an underlying notion of 'before' and 'after': a 'past' and a 'future'. However, these primitive notions do not really imply a temporal structure for the automaton, since the concept of 'time interval between events' is not included in the definition.

1.2.2 Cellular automata

A 'cellular automaton' (see Wolfram, 1986) is a set of synchronized identical finite automata with identical evolution rules, which exchange data according to a regular finite connection scheme. Each individual automaton exchanges information (input and output symbols) with a finite sub-set of automata in some prescribed neighborhood, according to a connection rule which is the same for all individual automata. For example, if the automata are indexed by a signed integer i, the connection rule could be 'automaton number i receives input symbols from automata number $i-1$, i and $i+1$'.

This definition does not imply any geometrical structure. Individual automata must be labeled, for example by signed integers, in order to define a connection rule, but the automata are not supposed to be attached to geometrical points of a metric space; there is no notion of distance or angle contained in the definition. However, the key notion of *homogeneity* is already present: homogeneity, in this context, simply refers to the fact that all individual finite automata are identical (same internal structure, same evolution rule, and same connection rule).

Another key notion in cellular automata is their *synchronization*, which implicitly requires that the concept of 'past' and 'future' be universal, that is, independent of the individual automaton: an event which is in the past (future) for *one* automaton must be in the past (future) for *all* individual automata. So, there must be a single clock for all individual automata, which justifies the notion of simultaneity and thus of synchronization.

1.2.3 Lattice gases

From the mathematical point of view, a lattice gas is a particular class of cellular automata which verify the following additional constraints:

(i) The individual automata are tied geometrically to the nodes of a regular Bravais lattice embedded in a Euclidean space of dimension D. This relation between individual automata and lattice nodes yields a one-to-one mapping between the set of all the individual automata of the cellular automaton and a connected part of the Bravais lattice. Because of this one-to-one mapping, the individual automata can also be called 'nodes'. The nodes are labeled by their position vector $\mathbf{r}_\star$; the star subscript indicates that the vector takes only discrete values.

(ii) The elementary evolution process of the cellular automaton is repeated at regularly spaced discrete times separated by a time increment Δt (the *time step*), which is usually taken equal to unity by a proper choice of the time scale.

(iii) The number of possible internal states of any individual automaton is 2^b, where b is an integer. Any possible internal state of an individual automaton can thus be represented by b Boolean variables, which will be called 'channels'[3] because they correspond to 'communication channels' between nodes, as we shall see below. The b channels on any node are labeled by an integer i ranging from 1 to b or from 0 to $b-1$. Since all the individual automata are taken identical, the number b of channels per node and their labeling are node-independent.

(iv) The elementary evolution process of the cellular automaton that occurs at each time step is a sequence of two distinct phases:

(a) *The collision phase:* during this phase, the internal state of each individual automaton evolves according to a *purely local*, node-independent rule which can be non-reversible and non-deterministic. More precisely, the post-collision state of an individual automaton depends only on its own pre-collision state and on a finite number of random input symbols for non-deterministic models (see comments on local versus non-local collisions in Section 1.3).

(b) *The propagation phase:* during this phase, the information present in the automata's 'memory' (the post-collision state) is shifted from node to node without loss or gain, according to a node-independent, *reversible* and *deterministic* rule. In practice, each channel index i between 0 and $b-1$ is associated to a unique lattice vector $\mathbf{c}_i$, such that the Boolean value of channel i on any node $\mathbf{r}_\star$ is transferred during the propagation phase to channel i of node $\mathbf{r}_\star + \mathbf{c}_i$. The vectors $\mathbf{c}_i$ are called the 'velocity vectors', and must verify the following conditions: (i) the set of velocity vectors remains globally invariant under the action of the

[3] Channels are sometimes called 'cells', but the word 'cell' may be misleading because of possible confusion with the crystallographic notion of (primitive) cell of a lattice.

crystallographic point group G of the lattice,[4] (ii) they include at least D generating vectors of the Bravais lattice, that is, D lattice vectors that generate the whole lattice by linear combinations with signed integer coefficients.

We shall consider the set of properties listed above as *our* definition of lattice gases.[5]

1.3 Comments

1.3.1 The velocity vectors

We first justify the name 'velocity vectors'. If the time scale is chosen in such a way that the time step Δt is equal to unity, then the information (particle) present in channel i at node $\mathbf{r}_\star$ goes to node $\mathbf{r}_\star + \mathbf{c}_i$ in one unit of time. The $\mathbf{c}_i$s can thus be viewed as the microscopic velocity vectors of fictitious particles occupying channels i.

All velocity vectors are not necessarily distinct: lattice gas models with multiple channels can be designed to model, for instance, gas mixtures or reaction-diffusion systems where distinguishable particles representing different 'chemical species' have the same kinematics (i.e. the same propagation rules). In addition, some of the velocity vectors can be zero: this is the case for models with rest particles which stay on a node during the propagation phase.

The velocity vectors, which must have the same local symmetries as the lattice, are also constrained to include D generating vectors of the Bravais lattice. This guarantees that any two nodes of the lattice can be connected by a finite chain of velocity vectors, and thus, that the lattice cannot separate into sub-lattices with independent evolutions. Note that for a given Bravais lattice, there may exist several sets of D lattice vectors that generate the whole lattice, and several sets of velocity vectors compatible with the symmetries of the lattice. To illustrate this point, take for instance the three-dimensional body-centered lattice ('bcc' lattice) with unit lattice constant (length of the spatial periodicity). This Bravais lattice can be generated by the set $\mathscr{S}_1$ containing the three lattice vectors $(\frac{1}{2}, -\frac{1}{2}, -\frac{1}{2})$, $(-\frac{1}{2}, \frac{1}{2}, -\frac{1}{2})$, and $(\frac{1}{2}, \frac{1}{2}, \frac{1}{2})$. The smallest set of velocity vectors that can be built around $\mathscr{S}_1$ contains *eight* vectors which can be constructed by applying all isometries of the local invariance group of the bcc lattice, to all three vectors of $\mathscr{S}_1$. These eight vectors with modulus $\frac{\sqrt{3}}{2}$ have the form $(\epsilon_1 \frac{1}{2}, \epsilon_2 \frac{1}{2}, \epsilon_3 \frac{1}{2})$. Here, $(\epsilon_1, \epsilon_2, \epsilon_3)$ are any of the eight possible combinations of '+'

[4] The point group or local invariance group of a lattice is the subgroup of the space group of the lattice which leaves individual nodes unchanged; see Ashcroft and Mermin (1976).

[5] This definition should not be considered as universal: with the same basic ideas, one can relax, modify or add constraints to construct different definitions of lattice gases.

and '–'. Now, the bcc lattice can *also* be generated by the set $\mathscr{S}_2$ containing the three vectors (1,0,0), (0,1,0) and $(\frac{1}{2}, \frac{1}{2}, \frac{1}{2})$. The smallest set of velocity vectors that can be built around $\mathscr{S}_2$ contains *fourteen* vectors that include the eight vectors with modulus $\frac{\sqrt{3}}{2}$, plus six vectors with unit modulus: $(\pm 1, 0, 0)$, $(0, \pm 1, 0)$ and $(0, 0, \pm 1)$. This example shows that with one single Bravais lattice, one can construct several lattice gas models with different propagation rules and thus different complexities. It motivates the introduction of the concept of 'minimal model' for a given Bravais lattice. A minimal model built on a given Bravais lattice is a model which has the smallest possible number b of channels per node compatible with the local symmetries of the lattice. There may be several minimal models associated to the same Bravais lattice, since they can differ by their collision rules, but also by the 'physical' properties attached to the channels.

1.3.2 Space and time

One of the most important features that distinguish lattice gas automata from cellular automata is that lattice gases have a spatial structure and geometry. Distances, angles, surfaces, volumes, and thus densities and fluxes are meaningful quantities which can be defined. The same holds for time: it makes sense to speak of time intervals for lattice gases. This feature renders possible the comparison of collective motions in lattice gases with space- and time-dependent behaviors of physical objects such as real fluids.

1.3.3 The exclusion principle

The constraint that two indistinguishable particles cannot reside at the same node at the same time (see Section 1.1), has been imposed to enable a natural Boolean description and encoding of the states of lattice gases. However, models permitting the coexistence of up to n indistinguishable particles are also admissible, provided this number n is of the form $2^p - 1$, with p integer. Indeed, such models can be recast in a way that does not violate the exclusion principle, by simply using the binary expression of the number of identical particles residing on a node. Any number of indistinguishable particles less than or equal to $n = 2^p - 1$ can be viewed as the sum of zero or one single particle, plus zero or one pair of particles, plus zero or one quadruplet of particles, and so on, up to zero or one 2^{p-1}-multiplet of particles. To get an intuitive picture, one can think of a pair of particles as a single particle with double mass, a quadruplet as a single particle with quadruple mass, etc. These 'super-particles' with multiple mass are now distinguishable and their presence on a node can thus be fully described by p Boolean quantities. This picture in terms of multiple particles is correct provided the collision phase can 'break' particles with multiple mass in such a way that it produces no spurious conservations.

Note that n *distinguishable* particles require n Boolean quantities, whereas n *indistinguishable* particles only require $p \simeq \log_2 n$ Boolean quantities, provided n is of the form $2^p - 1$. The problem of distinguishable versus indistinguishable particles occurs, for example, for some three-dimensional lattice gas models (see Section 3.7).[6]

1.3.4 Bravais lattices

Because of the homogeneity principle inherent in the concept of cellular automata, the connection scheme between the nodes must be node-independent. As a consequence, the set of velocity vectors of the lattice gas, and thus the local neighborhood of a node, must be identical for all nodes. The only regular lattices which verify this condition are, by definition, Bravais lattices.[7]

To illustrate the difference between an ordinary regular lattice and a Bravais lattice consider two regular lattices: the *triangular lattice* with hexagonal symmetry, and the *hexagonal lattice* with triangular symmetry (also called the *'honeycomb lattice'*). From Figure 1.1, it is clear that the triangular regular lattice is a Bravais lattice, but the hexagonal one, while regular, is not a Bravais lattice.

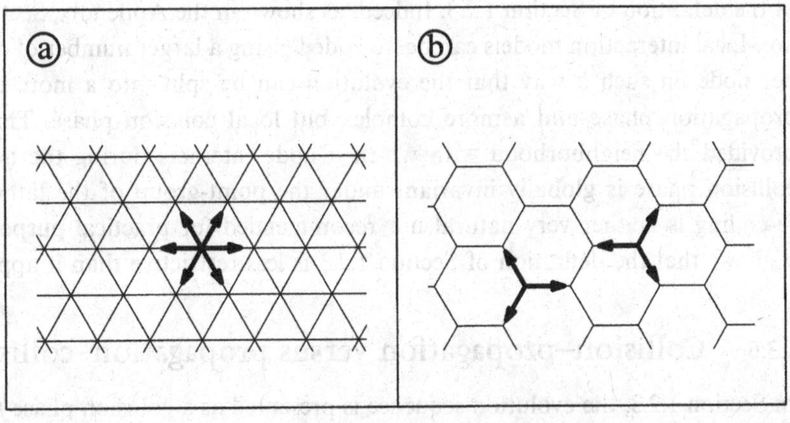

Figure 1.1 (a) An example of a two-dimensional Bravais lattice: the triangular lattice. (b) An example of a two-dimensional regular but not Bravais lattice: the hexagonal 'honeycomb' lattice. For both lattices, the arrows represent the neighborhood of a node. The hexagonal lattice has two different kinds of neighborhoods, depending on the parity of the node number.

6 One of these three-dimensional models, the FCHC-8 model, admits up to three indistinguishable particles with zero-velocity (rest particles). Only two Boolean quantities are used for them, one encoding the presence of a particle of mass 1, the other encoding the presence of a particle with mass 2 (that is, a pair of particles of mass 1).

7 Detailed definitions and discussions of Bravais lattices are given in books on solid state physics; see for example Ashcroft and Mermin (1976).

Indeed, for the latter, the set of nearest neighbors is different for 'odd' and 'even' nodes. No node-independent connection scheme involving nearest neighbors can be designed on the honeycomb lattice.

One can classify exhaustively and enumerate all possible Bravais lattices in given dimensions, according to their symmetry groups: there exist only *five* different Bravais lattices in two dimensions, and *fourteen* in three dimensions, as first observed by Auguste Bravais in 1845 (see Cauchy, 1851). Notice that the constraint of using a Bravais lattice is not too severely restrictive, since several lattice gas models can be built on one single Bravais lattice by changing the connection scheme and the collision rules. Moreover, discrete models within the spirit of lattice gases, but involving non-regular or non-Bravais or quasi-lattices can exist, but will not be considered here, as we keep with the definition of lattice gases given in Section 1.2.3.

1.3.5 Local versus non-local collisions

According to the definition of lattice gases given in Section 1.2.3, the collision phase should be local, that is, the post-collision state at any node $\mathbf{r}_\star$ depends only on the pre-collision state at the same node $\mathbf{r}_\star$. This constraint seems to preclude all models with finite-range interactions which require non-local collisions (see for example Appert and Zaleski, 1990). Such models can actually fit the definition of Section 1.2.3. Indeed, as shown in the Appendix, Section A.1, non-local interaction models can be 're-coded' using a larger number of channels per node, in such a way that the evolution can be split into a more complex propagation phase and a more complex but local collision phase. This holds provided the neighborhood with which a node interacts during the non-local collision phase is globally invariant under the point-group of the lattice. This re-coding is neither very natural nor recommended for practical purposes, but it shows that the definition of Section 1.2.3 is less restrictive than it appears.

1.3.6 Collision–propagation versus propagation–collision

In Section 1.2.3, the evolution sequence is presented as a collision phase *followed* by a propagation phase. This order was chosen by anticipating its convenience for writing the microdynamic equation (see Chapter 2). In fact, this choice is arbitrary and not restrictive: models with the reverse order can be re-coded to fit the chosen order. This re-coding is operationally similar to the re-coding discussed above in Section 1.3.5.

In general, the order in the sequence is unimportant when long-time behaviors and large-scale properties are considered. However, if one considers fluctuating processes which are measured and analyzed via fluctuation correlations, the order in the sequence is not irrelevant: the result of the correlation measurements may depend on the phase where the iteration sequence is stopped (see Chapter 4). Care must be taken in using proper averaging procedure.

1.3.7 Mathematical versus physical

The mathematical point of view, where lattice gases are considered as a special class of cellular automata,[8] is most useful when a formal definition is needed. But this more abstract view may result in some loss in physical intuition. For example, the crucial role of the conservation laws appears more clearly by analogy with usual statistical mechanics when considering the lattice gas as a gas of point particles. The picture which shows particles moving and colliding, rather than information exchanges and local evolution, often yields physical insight. We shall use both languages, but with some preference for the physicist's language, as we want to put the emphasis on the physical properties of lattice gases.

[8] A different point of view, discrediting the concept of cellular automata for lattice gases, is presented by Hénon (Hénon, 1989).

Chapter 2

Microdynamics: general formalism

In this chapter, we develop the 'microdynamic formalism', which describes the instantaneous microscopic configuration of a lattice gas and its discrete-time evolution. This exact description of the microscopic structure of a lattice gas is the basis for all further theoretical developments, in particular for the prediction of large-scale continuum-like behavior of LGAs.

We first introduce the basic tools and concepts for a general instantaneous description of the microscopic configurations (Section 2.1). The time evolution of the lattice gas is then given in terms of the 'microdynamic equations' (Section 2.2). Thereafter, we define microscopic characteristics (e.g. various forms of reversibility), which have a crucial incidence on the macroscopic behavior of the gas, and therefore on its suitability to simulate real physical situations (Section 2.3). The last section is devoted to special rules needed to handle boundary problems (obstacles, particle injections, etc.).

This chapter deals with rather abstract concepts which will find their application in Chapter 3, where lattice gas models are described at the microscopic level.

2.1 Basic concepts and notation

2.1.1 The lattice and the velocity vectors

One of the most important features of lattice gases is the underlying Bravais lattice structure which gives a geometrical support to the abstract notion of a cellular automaton. Strictly, a Bravais lattice is by definition infinite. We consider that the cellular automaton only occupies a connected subset $\mathscr{L}$ of the D-dimensional underlying Bravais lattice. In practical applications, this subset contains only a finite number $\mathscr{N}$ of lattice nodes. However, for theoretical developments, $\mathscr{L}$ will be assumed to be infinite when necessary.

We label any given node of the Bravais lattice by its position vector $\mathbf{r}_\star$ with respect to some arbitrary origin. The vectors $\mathbf{r}_\star$ belong to the D-dimensional space $\mathbb{R}^D$ in which the Bravais lattice is embedded. The subscript '$\star$' will be systematically used to distinguish discrete-valued quantities such as discrete positions ($\mathbf{r}_\star$) or discrete times ($t_\star$) from the corresponding continuous quantities. The components of position vectors, and more generally of any D-dimensional vector will be labeled by Greek indices ranging from 1 to D. For example, the components of the position vector $\mathbf{r}_\star$ are denoted $\mathbf{r}_{\star\alpha}$, $\alpha = 1, \ldots, D$.

To each node of $\mathscr{L}$ corresponds a finite automaton (see Section 1.2); so the cellular automaton contains $\mathscr{N}$ identical automata, and each of them has a finite number of internal states coded with b binary digits. In more physical terms, each node, i.e. each individual automaton, can be viewed as a set of b 'channels' each of which can be 'occupied' by a fictitious particle (binary value 1) or 'unoccupied' (binary value 0). The channels and, more generally, any quantity related to channels will be labeled with a Roman index ranging from 1 to b. For example, the velocity vector corresponding to channel i is denoted $\mathbf{c}_i$, $i = 1, \ldots, b$. For convenience, the channel index can also run from 0 to $b - 1$ (e.g. for models with one rest particle, see Section 3.3).

From the standard definition of a finite automaton, the internal state evolves according to a rule that depends on some external inputs. In the case of a lattice gas, the external input data come from a fixed finite number B ($B \leq b$) of neighboring nodes, and, in the case of non-deterministic models, from a certain number of random variable generators associated to each automaton. The number B, which is called the 'connectivity' or 'connection number' of the lattice gas model, is the maximum number of *different non-zero* vectors in the set of b velocity vectors $\mathbf{c}_i$ (see Section 1.2.3). These B distinct non-zero vectors are called the 'connection vectors' of the model.

If the evolution rule involves exchanges of information with all nearest neighbors, and only those, then the connectivity B of the lattice gas model is identical to the coordination number of the Bravais lattice, as defined in solid state physics (see e.g. Ashcroft and Mermin, 1976). If, in addition, all the velocity vectors are non-zero, then the model is said to be 'homokinetic', all velocity vectors having the same modulus.

Notice that the connectivity B of a lattice gas model may be equal to, or may differ from, the number b of channels per node; it depends on whether all $\mathbf{c}_i$s are identical and non-zero, or not (see the models described in Chapter 3).

2.1.2 The Boolean field

Each channel has only two possible states: 0 or 1, which means physically that each channel is either empty or occupied by at most one particle (exclusion principle). For any given node, we denote by $s = (s_i,\ i = 1, \ldots, b)$ the b-bit word that describes the instantaneous internal state of the individual automaton, or, equivalently, the occupation of the b channels of the node. This b-bit word is called the 'Boolean state' of the node, and the set γ of the 2^b possible Boolean states of one node defines what we call the 'single-node phase space' or the 'local phase space'.

Similarly, the instantaneous state of the whole lattice will be encoded by a collection of $\mathcal{N}$ b-bit words, one word for each node. This collection of $\mathcal{N} \times b$ Boolean quantities $S = (S_i(\mathbf{r}_\star),\ i = 1, \ldots, b,\ \mathbf{r}_\star \in \mathscr{L})$ is called the 'Boolean configuration' of the lattice, and the set Γ of all possible Boolean configurations defines the 'global phase space' or simply 'phase space'.

In order to simplify the algebraic treatment of Boolean configurations, we use algebraic binary quantities whose values are '1' or '0', instead of logical quantities with values 'TRUE' or 'FALSE'. The equation governing the time evolution of the lattice will then be an algebraic relation with the same content as the corresponding logical equation.

For any discrete time $t_\star$, we define the 'Boolean field' n as the function mapping the set of all channels of all nodes onto the set $\{0, 1\}$, associating with each channel of each node, its occupation level (0 or 1):

$$\begin{array}{rrcl} n: & \mathscr{L} \times \{1, \ldots, b\} & \rightarrow & \{0, 1\} \\ & (\mathbf{r}_\star, i) & \mapsto & n_i(\mathbf{r}_\star), \end{array} \tag{2.1}$$

where $n_i(\mathbf{r}_\star)$ is the occupation of channel i at node $\mathbf{r}_\star$. There is a semantic distinction between the Boolean configuration $S_i(\mathbf{r}_\star)$ and the Boolean field $n_i(\mathbf{r}_\star)$: the Boolean field is a generic notion, whereas a Boolean configuration is a particular realization of the Boolean field.

At this stage, the Boolean field is a function of only two discrete-valued variables: the channel index i and the node position $\mathbf{r}_\star$. For the dynamic description to be presented in Section 2.2, it is useful to include a third variable: the discrete time $t_\star$, and so to work with the time-dependent Boolean field $n_i(\mathbf{r}_\star, t_\star)$.

2.1.3 Observables

In order to establish the connection between purely abstract automata and concrete gas models, physical properties (mass, momentum, energy, etc.) must be associated with each channel of each node of the lattice. We emphasize that such physical quantities (observables) are not associated with the particles themselves, but with the channels: particles loose their identity during collisions, whereas channels do not. Nevertheless, for simplicity and intuition, we shall use terms such as 'the mass of particle i' or 'the momentum of particle i' rather than 'the mass associated with channel number i if occupied' or 'the momentum associated with channel number i if occupied'. Mass, momentum and energy are examples of observables with physical content; yet, the concept of observable, as we define it below, is in principle not restricted to quantities with an equivalent in the physical world.

It is systematically assumed that the amount of a given observable assigned to a channel depends only on the channel index, and not on the position of the node: if occupied, channel i of any node always carries the same amount of the observable.

Under the above assumption, an observable is defined as a collection of b values (q_i, $i = 1, \ldots, b$), indicating the amount of the underlying observable quantity which is present in the channel labeled by the index i, when this channel is occupied. The q_is can be scalars, vectors or more generally tensors of arbitrary order.

The 'value of the observable measured at node $\mathbf{r}_\star$' is the total amount of the observable quantity present at node $\mathbf{r}_\star$: $\sum_{i=1}^{b} q_i n_i(\mathbf{r}_\star)$. It is called the 'microscopic density per node' of the observable or simply its 'microscopic density', and is denoted by $N_q^\star(\mathbf{r}_\star)$ in the most general case:

$$N_q^\star(\mathbf{r}_\star) = \sum_{i=1}^{b} q_i n_i(\mathbf{r}_\star), \tag{2.2}$$

where the superscript '$\star$' indicates that it is a *non-averaged* microscopic quantity. Now, when divided by v_0, the volume[1] of the primitive unit cell of the Bravais lattice, this density becomes a 'microscopic density per unit volume', and when divided by b, the number of channels per node, it becomes a 'microscopic density per channel'. Similarly, the 'microscopic flux' of the observable q is defined by

$$\mathbf{\Phi}_q^\star(\mathbf{r}_\star) = \sum_{i=1}^{b} q_i \mathbf{c}_i n_i(\mathbf{r}_\star). \tag{2.3}$$

Ensemble- and/or space- and/or time-averaging of the microscopic densities $N_q^\star$ and of the microscopic fluxes $\mathbf{\Phi}_q^\star$ yields the 'macroscopic densities' N_q and 'macroscopic fluxes' $\mathbf{\Phi}_q$ (without '$\star$' superscript). These 'densities' and 'fluxes' (without further specification) are the physically relevant fields in the macroscopic theory of lattice gases (see Chapters 4 and 5). In order to shed some light on the above definitions, we now give examples of some useful observables, frequently encountered in lattice gas theory.

[1] Length, surface, volume or hyper-volume, depending on the dimensionality D.

2.1.3.1 Example 1

The collection $q = (1, 1, 1, \ldots, 1)$ of b quantities equal to 1 is a scalar observable called the 'particle number observable', as its value measured at node $\mathbf{r}_\star$ is $\sum n_i(\mathbf{r}_\star)$ which is clearly the total number of particles present at node $\mathbf{r}_\star$.

2.1.3.2 Example 2

Denoting by m_i the individual mass assigned to any particle present in channel i, the collection $q = (m_1, \ldots, m_b)$ is a scalar observable called the 'mass observable'. Its value $\sum m_i n_i(\mathbf{r}_\star)$, measured at node $\mathbf{r}_\star$, is the total mass present at node $\mathbf{r}_\star$. If the individual masses m_i, $(i = 1, \ldots, b)$ are all equal to a single value m (which can be taken equal to 1 by a proper choice of the mass scale), then the mass observable is equal to the particle number observable. This is the most frequent case.

2.1.3.3 Example 3

Denoting by $p_{i\alpha}(\alpha = 1, \ldots, D)$ the components of the individual momentum assigned to any particle present in channel i, the collection $q_\alpha = (p_{1\alpha}, \ldots, p_{b\alpha})$, $(\alpha = 1, \ldots, D)$ is a vector observable called the 'momentum observable'. Its value $q_\alpha = \sum p_{i\alpha} n_i(\mathbf{r}_\star)$,[2] measured at node $\mathbf{r}_\star$, is the total momentum present at node $\mathbf{r}_\star$. For many models, the individual momentum $\mathbf{p}_i$ of channel i is chosen equal to $m_i \mathbf{c}_i$, which is physically consistent with the interpretation of the $\mathbf{c}_i$s as microscopic velocities. This choice also guarantees that the microscopic density of momentum is just the microscopic flux of mass.

2.1.3.4 Example 4

Denoting by e_i the individual energy assigned to any particle present in channel i, the collection $q = (e_1, \ldots, e_b)$ is a scalar observable called the 'energy observable'. Its value $\sum e_i n_i(\mathbf{r}_\star)$, measured at node $\mathbf{r}_\star$, is the total energy present at node $\mathbf{r}_\star$. If particles carry only kinetic energy, it is reasonable to take e_i equal to $\mathbf{p}_i^2/(2m_i)$.

2.1.4 Generalized observables

The observables, as defined above, are linear, local and homogeneous since their value, measured at node $\mathbf{r}_\star$, is a linear combination of the n_is at the *same* node, with node-independent coefficients (the q_is). Relaxing these constraints, we can build the abstract notion of 'generalized observable'. A generalized observable q is any physical quantity whose value measured at node $\mathbf{r}_\star$ is a function $q(n, \mathbf{r}_\star)$ of the Boolean field n and of the space variable $\mathbf{r}_\star$.

When the value $q(n, \mathbf{r}_\star)$ of a generalized observable q, measured at node $\mathbf{r}_\star$,

[2] Remember that the Greek subscript α refers to the components of the vectors, and the Roman subscript i refers to the channel index.

does not *explicitly*[3] depend on $\mathbf{r}_\star$, then the generalized observable is said to be 'homogeneous'. When the value of the observable measured at node $\mathbf{r}_\star$ only depends on the n_is at this very node $\mathbf{r}_\star$, then the generalized observable is said to be 'local'.

The following simple examples should provide a more physical picture of generalized observables.

2.1.4.1 Example 1: space-averaged quantities

Let $\mathscr{I}$ be a connected subset of $\mathscr{L}$. Consider the generalized observable whose value, measured at node $\mathbf{r}_\star$ is:

$$q(n, \mathbf{r}_\star) = \sum_{\mathbf{r}_\star' \in \mathscr{I}} n_1(\mathbf{r}_\star + \mathbf{r}_\star'). \tag{2.4}$$

This generalized observable is linear and homogeneous, but non-local, since its value, measured at node $\mathbf{r}_\star$, depends on the Boolean field at nodes $\mathbf{r}_\star + \mathbf{r}_\star'$, with $\mathbf{r}_\star' \in \mathscr{I}$. The resulting value can be understood as the population of channel number 1, space-averaged over the domain $\mathscr{I}$ surrounding node $\mathbf{r}_\star$.

2.1.4.2 Example 2: potential energy

Suppose the fictitious particles are in interaction with a space-dependent force field that derives from a scalar potential $U(\mathbf{r}_\star)$. The potential energy measured at node $\mathbf{r}_\star$ is a generalized observable whose value is:

$$q(n, \mathbf{r}_\star) = U(\mathbf{r}_\star) \sum_{i=1}^{b} n_i(\mathbf{r}_\star). \tag{2.5}$$

This generalized observable is linear and local, but non-homogeneous, since its value depends explicitly on $\mathbf{r}_\star$ through the scalar potential $U(\mathbf{r}_\star)$.

2.2 The microdynamic equation

The cellular automaton rule governing the evolution[4] of a lattice gas can be viewed as a logical equation expressing the Boolean field at time $t_\star + 1$ from the Boolean field at time $t_\star$. When written in algebraic form, the resulting equation is called the 'microdynamic equation' of the lattice gas, as it governs the microscopic evolution (microdynamics) of the lattice (Frisch, d'Humières, Hasslacher, Lallemand, Pomeau and Rivet, 1987). The microdynamic equation is an *exact* equation which contains the complete dynamics of the lattice gas.

[3] The value of q depends *implicitly* on the space variable $\mathbf{r}_\star$, through the dependence in the Boolean field.

[4] The time scale is chosen such that the time increment of the cellular automaton is one.

2.2.1 Formal expression

The cellular automaton rule can be fully deterministic or not. For the deterministic case, the Boolean field at time $t_\star + 1$ is a well-defined function of the Boolean field at time $t_\star$. This function, mapping Γ (the cellular automaton phase space) onto itself, is the 'evolution operator' $\mathscr{E}$:

$$n(t_\star + 1) = \mathscr{E}\big(n(t_\star)\big). \tag{2.6}$$

For the non-deterministic case, the transition between times $t_\star$ and $t_\star+1$ involves random choices at each node. The microdynamic equation then reads:

$$n(t_\star + 1) = \mathscr{E}\big(n(t_\star),\, \xi(t_\star)\big), \tag{2.7}$$

where $\xi(t_\star)$ is a collection of $\mathscr{N}$ random functions of the discrete time variable $t_\star$ (one function for each node).

This formalism is quite general and involves few assumptions about the discrete gas model. In the following, we give a more explicit expression for the evolution operator, using the fundamental property that the evolution rule of a lattice gas, as defined in Section 1.2, is a sequence of two phases: (i) a *collision phase* during which the Boolean state of each node is modified locally (that is, without exchanging any information between nodes), and (ii) a *propagation phase* during which different nodes can exchange information in a deterministic and reversible way. The evolution operator $\mathscr{E}$ can thus be written as the product $\mathscr{S} \circ \mathscr{C}$ where the 'collision operator' $\mathscr{C}$ is local but may be non-deterministic and non-invertible, and the 'propagation operator' or 'streaming operator' $\mathscr{S}$ is non-local, but deterministic and invertible.

2.2.2 The propagation operator

According to the definition of the propagation phase given in Section 1.2, the action of the propagation operator on the Boolean field is such that the state of channel i at node $\mathbf{r}_\star - \mathbf{c}_i$ before propagation becomes, after propagation, the state of channel i at node $\mathbf{r}_\star$. The propagation operator carries the information from channel i of node $\mathbf{r}_\star$ to channel i of node $\mathbf{r}_\star + \mathbf{c}_i$. So the expression for the propagation operator reads:

$$\mathscr{S}\big(n\big)_i(\mathbf{r}_\star) = n_i(\mathbf{r}_\star - \mathbf{c}_i). \tag{2.8}$$

This definition of the propagation operator raises the problem of finite size lattices. Indeed, if the subset $\mathscr{L}$ of the Bravais lattice under consideration is finite, the node $\mathbf{r}_\star + \mathbf{c}_i$ may be outside $\mathscr{L}$, even if the node $\mathbf{r}_\star$ is inside.

One solution to this problem is to introduce a periodic wrapping in such a way that any particle driven out of $\mathscr{L}$ be re-injected elsewhere in $\mathscr{L}$. More precisely, we need a finite domain $\mathscr{L}'$ of the underlying Bravais lattice which contains all nodes of $\mathcal{L}$ and which can perfectly tile the full Bravais lattice by successive lattice-preserving translations. So there must exist a subgroup of the lattice-preserving translations group such that the nodes[5] in $\mathcal{L}'$ and their images by the elements of this subgroup cover the full infinite Bravais lattice without any gap or overlap (see Figure 2.1(a)). It is then possible to wrap $\mathcal{L}'$ on itself and to 'connect' opposite boundaries in such a way that $\mathcal{L}'$ become a topological torus (see Figure 2.1(b)). In terms of particle motions, this wrapping leads to *periodic boundary conditions*[6] at the boundaries of $\mathcal{L}'$.

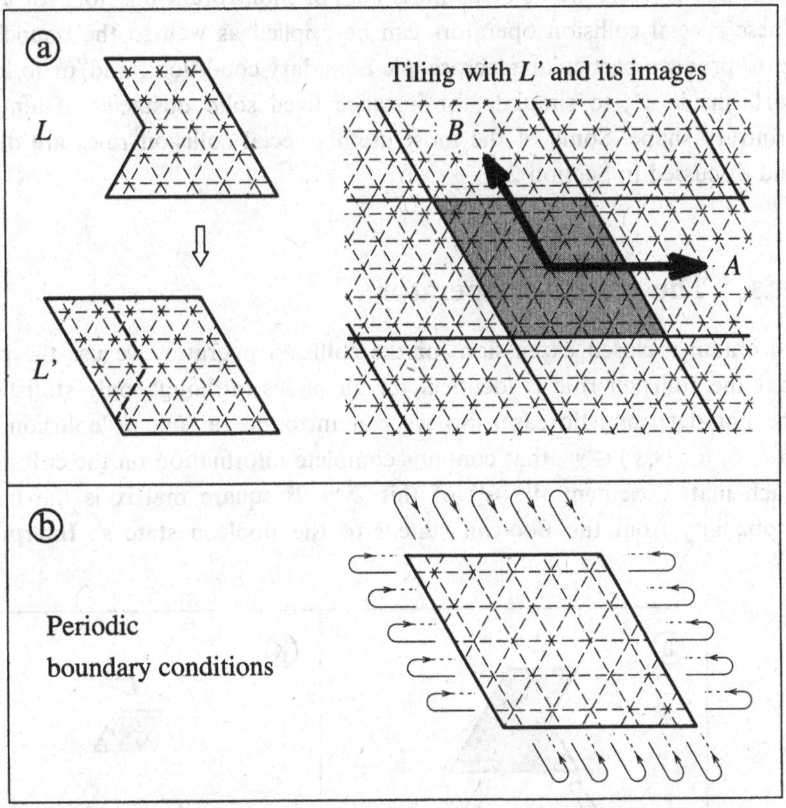

Figure 2.1 The construction of periodic boundary conditions is illustrated for the two-dimensional triangular lattice. (a) Starting from a lattice domain $\mathcal{L}$ with arbitrary shape, one constructs $\mathcal{L}'$ which tiles perfectly the full Bravais lattice. This tiling is realized by applying to $\mathcal{L}'$ all the translations generated by the two lattice-preserving translations with vectors **A** and **B**. (b) The sides of $\mathcal{L}'$ are reconnected in such a way that it becomes a topological torus.

[5] Note that $\mathcal{L}$ and $\mathcal{L}'$ must be thought of as sets of lattice nodes and not as geometrical surfaces or volumes.

[6] In solid state physics, periodic boundary conditions are also called 'Born–von Kármán boundary conditions'.

For a given lattice gas model and a given domain $\mathscr{L}$, there can exist several different periodic boundary conditions. Indeed, the same domain $\mathscr{L}$ can lead to several perfect tilings with different tile shapes, even if one restricts the choice to tiles which contain a minimal number of nodes (see Figure 2.2(a)). Moreover, the same tile shape can lead to several different tilings with different translation subgroups (see Figure 2.2(b)).

We emphasize that this periodic wrapping is applied *only* to the propagation operation; it *does not imply* that the macroscopic variables satisfy periodic conditions in the usual sense. Indeed, special collision operators can be applied to the boundary nodes in such a way that the boundaries of $\mathscr{L}$ behave macroscopically as solid walls, or as mass and/or momentum injectors for example. These special collision operators can be applied as well to the boundaries of $\mathscr{L}$ to produce particular macroscopic boundary conditions, and/or to localized parts inside $\mathscr{L}$, to simulate for instance fixed solid obstacles of almost any arbitrary shape. Some of the most useful special collision rules are described and discussed in Section 2.4.

2.2.3 The collision operator

To obtain a closed expression for the collision operator, we use the property that the collision rule is identical for all nodes (although only statistically in the non-deterministic case). So we can introduce a unique 'collision matrix' $A(s \to s')$ for $(s, s') \in \gamma^2$, that contains complete information on the collision rule. Each matrix element $A(s \to s')$ of this $2^b \times 2^b$ square matrix is the transition probability from the Boolean state s to the Boolean state s'. Interpreted as

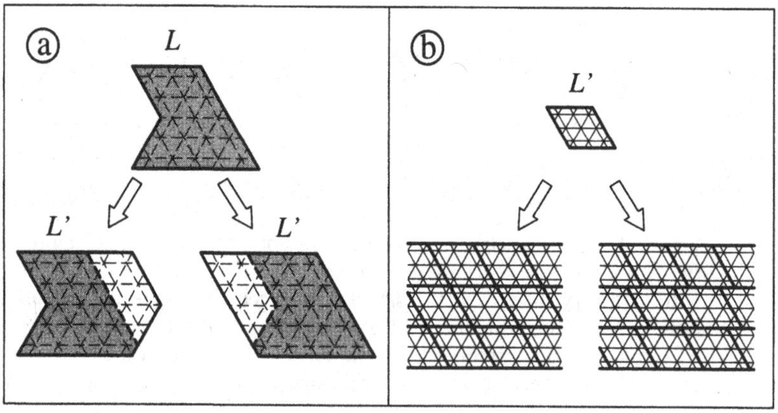

Figure 2.2 The same domain $\mathscr{L}$ can lead to several different tilings with (a) different tile shapes and/or (b) different translation subgroups. The two-dimensional triangular Bravais lattice is used to illustrate the point.

transition probabilities, the matrix elements are real numbers between zero and one, and satisfy the normalization condition:

$$\sum_{s' \in \gamma} A(s \to s') = 1, \quad \forall s \in \gamma. \tag{2.9}$$

Under the above assumption, one can write the collision operator in an explicit form that involves the collision matrix indirectly:

$$\mathscr{C}(n)_i(\mathbf{r}_\star) = \sum_{(s,s') \in \gamma^2} s'_i \xi(s \to s') \prod_{j=1}^{b} n_j^{s_j} \overline{n}_j^{\overline{s}_j}, \tag{2.10}$$

where $\overline{s}_j \equiv 1 - s_j$ and $\overline{n}_j \equiv 1 - n_j$. The n_js and $\overline{n}_j$s are taken at node $\mathbf{r}_\star$. The $\xi(s \to s')$ s are the elements of a $2^b \times 2^b$ square matrix of Bernoulli random variables (values 0 or 1) whose averages are the $A(s \to s')$s, and such that

$$\sum_{s' \in \gamma} \xi(s \to s') = 1, \quad \forall s \in \gamma. \tag{2.11}$$

Equation (2.10) expresses, in a generic way, the post-collision Boolean field $\mathscr{C}(n)$ as an explicit function of the pre-collision Boolean field n. It is an analytic transcription of the cellular automaton collision rule. There are two important points which justify the analytic structure of (2.10):

(i) For a given statistical realization of the Bernoulli variables $\xi(s \to s')$, and for a given Boolean state s, there exists one and only one Boolean state s' such that $\xi(s \to s') = 1$, all $\xi(s \to s'')$s for $s'' \neq s'$ being zero. This is a consequence of (2.11) and of the fact that the $\xi(s \to s')$s are Bernoulli random variables. So, the $\xi(s \to s')$ factor on the r.h.s. of (2.10) selects, among all possible pairs of Boolean states (s, s'), those which are pre- and post-collision states of the collision that takes place.

(ii) The product $\prod n_j^{s_j} \overline{n}_j^{\overline{s}_j}$ is a 'matching operator' which is one if $n_j(\mathbf{r}_\star) = s_j$ (for all $j = 1, \ldots, b$), and zero otherwise. Operationally, it is equivalent to a 'Kronecker symbol' $\delta_{kr}(n_.(\mathbf{r}_\star) - s_.)$. So, it selects, among all possible Boolean states s, the one which is the actual value of the pre-collision Boolean field n at node $\mathbf{r}_\star$.

As an obvious consequence, the product

$$\xi(s \to s') \prod n_j^{s_j} \overline{n}_j^{\overline{s}_j}$$

selects, among all possible pairs of Boolean states (s, s'), the one for which s is the pre-collision state of node $\mathbf{r}_\star$, and s' is the actually realized post-collision state. Equation (2.10) assigns to $\mathscr{C}(n)(\mathbf{r}_\star)$ the value of this selected post-collision state s'.

The above formalism is applicable to both deterministic and non-deterministic models. For deterministic models, the matrix elements $A(s \to s')$ are either 0 or 1, and the $\xi(s \to s')$s are deterministic and thus simply equal to the $A(s \to s')$s.

2.2.4 Analytic expressions of the microdynamic equation

Using the expressions (2.8) and (2.10) for the propagation and collision operators, the complete microdynamic equation can be written as:

$$n_i(\mathbf{r}_\star + \mathbf{c}_i, t_\star + 1) = \sum_{(s,s')\in\gamma^2} s'_i \xi(s \to s') \prod_{j=1}^{b} n_j^{s_j} \bar{n}_j^{\bar{s}_j}, \tag{2.12}$$

where the n_js on the r.h.s. are implicitly taken at node $\mathbf{r}_\star$ and at time $t_\star$. An alternative form, strictly equivalent to (2.12), reads:

$$n_i(\mathbf{r}_\star + \mathbf{c}_i, t_\star + 1) = n_i + \sum_{(s,s')\in\gamma^2} (s'_i - s_i) \xi(s \to s') \prod_{j=1}^{b} n_j^{s_j} \bar{n}_j^{\bar{s}_j}. \tag{2.13}$$

The equivalence between (2.12) and (2.13) follows immediately from the identity:

$$n_i = \sum_{(s,s')\in\gamma^2} s_i \xi(s \to s') \prod_{j=1}^{b} n_j^{s_j} \bar{n}_j^{\bar{s}_j}, \tag{2.14}$$

which is a consequence of (2.11). The virtue of the alternative form (2.13) is that it distinguishes 'effective collisions' with $s' \neq s$ from 'passive collisions' which leave the Boolean state unchanged.

The microdynamic equation plays for lattice gases the role of Hamilton's equations for a system of interacting particles; it involves no approximation and constitutes the starting point of all further theoretical developments.

2.3 Microscopic properties of a lattice gas

As for real gases, the macroscopic (large-scale) dynamics of lattice gases depends crucially on their microscopic properties. We now define some of the most important microscopic characteristics that a given lattice gas model may or may not possess. The consequences of these microscopic characteristic properties on the large-scale macroscopic behavior will be discussed in Chapters 4 and 5.

2.3.1 Detailed and semi-detailed balance

The collision process is said to be 'micro-reversible' if any collision has the same probability as the reverse collision, that is, if and only if:

$$\forall (s, s') \in \gamma^2, \quad A(s \to s') = A(s' \to s). \tag{2.15}$$

This symmetry condition for the matrix $A(s \to s')$ is usually called 'detailed balance'. A weaker form, called 'semi-detailed balance', reads:

$$\forall s' \in \gamma, \quad \sum_{s\in\gamma} A(s \to s') = \sum_{s\in\gamma} A(s' \to s), \tag{2.16}$$

and may be viewed as a state-averaged form of the detailed balance (2.15). The physical interpretation of semi-detailed balance is that for any state s', the 'number of collisions'[7] with output state s' is the same as the number of collisions with input state s': when semi-detailed balance is satisfied, collisions 'create' any given state s' statistically as often as they 'destroy' it.

Using the normalization relation (2.9), semi-detailed balance can be written as:

$$\forall s' \in \gamma, \quad \sum_{s\in\gamma} A(s\to s') = 1, \tag{2.17}$$

which is the most commonly used expression for the semi-detailed balance condition.

The semi-detailed balance will be crucial in deriving the macroscopic theory of lattice gases. In particular, it affects deeply the nature of equilibrium states and the validity of the Boltzmann approximation.

2.3.2 Duality

For any Boolean state s, the dual state $\bar{s}$ is obtained by exchanging particles and 'holes' (occupied and unoccupied channels) or 0s and 1s in s. If s is considered as a collection of b logical variables, then the dual state is $\bar{s}$, the logical bitwise negation of s.

A lattice gas model is said to be 'self-dual' if the dynamics of holes is, at least statistically, identical to the dynamics of particles. The propagation operator moves information without any loss or gain, and is thus self-dual. Consequently, a necessary and sufficient condition for a model to be self-dual is that the collision matrix $A(s\to s')$ satisfy the relation:

$$\forall (s,s') \in \gamma^2, \quad A(s\to s') = A(\bar{s}\to\bar{s}'). \tag{2.18}$$

2.3.3 Conservation laws

As we emphasized earlier, it is crucial to know the set of the configuration-dependent quantities which are conserved by the evolution operator, for any initial configuration of the lattice. Indeed, the nature and the number of these conserved quantities have a direct influence on the existence and the nature of statistical equilibrium states and on the large-scale collective behavior of the lattice gas. Therefore, the conserved quantities are determinant elements for the validity of the lattice gas as a model system for the simulation of physical phenomena.

The accurate description of the conservation laws in a lattice gas model calls

[7] weighted by their probability.

on the notions of 'collisional invariants' and of 'geometrical invariants', where one distinguishes 'static' and 'moving' geometrical invariants (d'Humières, Qian and Lallemand, 1989, 1990). We now introduce these notions.

2.3.3.1 Collisional invariant

A collisional invariant, also called 'microscopic invariant' or 'local invariant' is an observable quantity whose value, measured at a node in any state, remains unchanged under the action of the sole *collision* operator. In other words, if an observable is a collisional invariant, the transition probability between two Boolean states corresponding to different values of the observable must be zero. So, the observable q is a collisional invariant if, and only if, the collision matrix $A(s \to s')$ satisfies the condition:

$$\forall (s, s') \in \gamma^2, \qquad \sum_{i=1}^{b} q_i (s'_i - s_i) A(s \to s') = 0. \tag{2.19}$$

Of course, during the propagation phase, the value at *individual* nodes of an observable which is a collisional invariant can change, since it is redistributed over neighboring nodes. However, the sum over the whole lattice is conserved if the standard collision operator is applied *everywhere* (no obstacles, no mass or momentum creation or annihilation).

Now, considering the complete evolution operator, we express that an observable q is a collisional invariant, in terms of the Boolean field:

$$\sum_{i=1}^{b} q_i n_i(\mathbf{r}_\star + \mathbf{c}_i, t_\star + 1) = \sum_{i=1}^{b} q_i n_i(\mathbf{r}_\star, t_\star). \tag{2.20}$$

This equation, which must hold for all discrete times $t_\star$ and for all nodes in $\mathscr{L}$, expresses that the amount of the collisional invariant q present at time $t_\star$ at node $\mathbf{r}_\star$ is conserved by the collision operator, and distributed over the b neighboring nodes $\mathbf{r}_\star + \mathbf{c}_i$ by the propagation operator.[8] Equation (2.20) displaying collisional invariance will be explicitly used as a starting point for the macroscopic theory of lattice gases.

2.3.3.2 Static geometrical invariant

A static geometrical invariant is a triplet $(q, \mathscr{I}, k)$ where q is an observable, $\mathscr{I}$ a sub-lattice of $\mathscr{L}$ and k an integer independent of the size of $\mathscr{L}$, such that for any initial configuration of the lattice, the value of the observable, summed

[8] If a special treatment (reflection, injection, forcing, etc., see Section 2.4) is applied to the boundaries or to localized parts of $\mathscr{L}$, then Equation (2.20) holds only in the bulk of $\mathscr{L}$.

over all nodes in $\mathcal{I}$ remains unchanged after k successive applications of the *full* evolution operator. In terms of the Boolean field, this statement reads:

$$\sum_{\mathbf{r}_\star \in \mathcal{I}} \sum_{i=1}^{b} q_i n_i(\mathbf{r}_\star, t_\star + k) = \sum_{\mathbf{r}_\star \in \mathcal{I}} \sum_{i=1}^{b} q_i n_i(\mathbf{r}_\star, t_\star). \tag{2.21}$$

For instance, the triplet formed by any collisional invariant, the whole lattice $\mathcal{L}$, and any integer k is a static geometrical invariant, since the propagation operator only redistributes the total amount of the observable initially present in $\mathcal{L}$, without any loss or gain. Of course, this only holds when periodic conditions are applied (with no boundaries elsewhere).

2.3.3.3 Moving geometrical invariant

In order to define a moving geometrical invariant, we need a triplet $(q, \mathcal{I}, \tau)$, where q is an observable, $\mathcal{I}$ a sub-lattice of $\mathcal{L}$, and τ a lattice-preserving translation which does not leave $\mathcal{I}$ globally invariant. It is then assumed that, for any initial configuration of the lattice, the amount of observable present in $\mathcal{I}$ be transferred completely to $\tau(\mathcal{I})$. Re-expressed in terms of the Boolean field, this reads:

$$\sum_{\mathbf{r}_\star \in \tau(\mathcal{I})} \sum_{i=1}^{b} q_i n_i(\mathbf{r}_\star, t_\star + 1) = \sum_{\mathbf{r}_\star \in \mathcal{I}} \sum_{i=1}^{b} q_i n_i(\mathbf{r}_\star, t_\star). \tag{2.22}$$

2.3.3.4 Conserved quantities

A conserved quantity is simply a static geometrical invariant with $k = 1$. It is a doublet formed by an observable q and a sub-lattice $\mathcal{I}$ of $\mathcal{L}$, such that the value of the observable summed over all nodes in $\mathcal{I}$ remains unchanged under the action of the full evolution operator, for any initial configuration of the lattice. A collisional invariant, summed over the whole lattice is always a conserved quantity (except if special conditions are applied, e.g. local injection or removal of particles, biased velocity field).

2.3.4 G-invariance

The property of G-invariance is quite intuitive. It characterizes lattice gas models whose microscopic structure (propagation rules, collision rules, observables) is compatible with the local symmetries of their underlying Bravais lattice. G-invariant lattice gases are those whose physical structure has the same symmetries as the Bravais lattice.

2.3.4.1 Definition

A lattice gas model is said to be G-invariant whenever, for any isometry $\mathbf{g}$ in the crystallographic point group (local invariance group) G of the Bravais lattice, there exists a unique index permutation g acting on $\{1, 2, \ldots, b\}$ such that:

- the set of velocity vectors $\mathbf{c}_i$, $i = 1, \ldots, b$ of the model satisfies:

$$\mathbf{g}(\mathbf{c}_i) = \mathbf{c}_{g(i)}, \quad i = 1, \ldots, b,$$

- any relevant observable q_i, $i = 1, \ldots, b$ of the model satisfies:

$$\mathbf{g}(q_i) = q_{g(i)}, \quad i = 1, \ldots, b,$$

- the collision matrix satisfies:

$$A\big(\mathscr{G}(s) \to \mathscr{G}(s')\big) = A\big(s \to s'\big), \quad \forall (s, s') \in \gamma^2,$$

where $\mathscr{G}$ is the permutation acting on γ, such that

$$\mathscr{G}(s)_i = s_{g^{-1}(i)}, \quad \forall (s) \in \gamma, \quad i = 1, \ldots, b.$$

Note that the second condition must hold for all relevant observables, whether they are scalars, vectors, or higher order tensors. This definition of the G-invariance gives a formal frame to the intuitive notion of compatibility between the lattice symmetries and the 'physical' microscopic properties of the lattice gas.

2.3.4.2 The classes of channels

When G-invariance is satisfied, or at least when its restricted version omitting the third condition about collisions is satisfied, then it becomes possible to define an equivalence relation within the indices $i = 1, \ldots, b$, leading to a classification of the b channels. Indeed, consider that channel i and channel j are equivalent whenever there exists an index permutation g associated to an isometry $\mathbf{g}$ in G such that $j = g(i)$. This relation is clearly an equivalence relation. This comes directly from the fundamental properties of the group structure of G. This equivalence relation naturally leads to a partition of the set of all channels into one or more classes of equivalent channels. Conventionally, each class is labeled by an integer index that will be denoted (when needed) by capital letters such as I or J. The symbol $Cl_{(i)}$ denotes the index of the class to which channel i belongs. In the case of the most simple lattice gas models, only one class exists, and this notation is not necessary.

2.3.4.3 Examples of models with one or several classes

To illustrate the notion of class, we use the simple geometry of the two-dimensional square lattice. The point group G related to this lattice is usually denoted 'C_{4v}' according to the standard Schoenflies convention (see Eliott and Dawber, 1979). It contains the symmetries S_y, S_{x-y}, S_x, S_{x+y} with respect to the directions $(1,0)$, $(1,1)$, $(0,1)$, $(-1,1)$ and the rotations R_0, $R_{\pi/2}$, R_π, $R_{3\pi/2}$ around an axis normal to the plane of the lattice. We present two possible sets of velocity vectors compatible with the symmetries of this lattice. The first set, defined in Figure 2.3(a), corresponds to the simple HPP model which will be described in Chapter 3. It contains only four channels with four different velocity vectors $\mathbf{c}_1$ to $\mathbf{c}_4$. Each of these vectors has unit modulus and corresponds to one of the four natural directions of the lattice. In this simple case, there exists only one class, since any of these four vectors can be changed into any other by an isometry of G. The second set of velocity vectors corresponds to an extended square model as shown in Figure 2.3(b). It contains 24 channels with 24 different velocity vectors. These 24 channels can be collected into five classes of equivalent channels, as listed in Table 2.3.4.3. It is clear that if a lattice gas model is G-invariant, any associated scalar observable must have the same value for all channels in the same class.

We must keep in mind that G-invariance is a *microscopic* property, that does

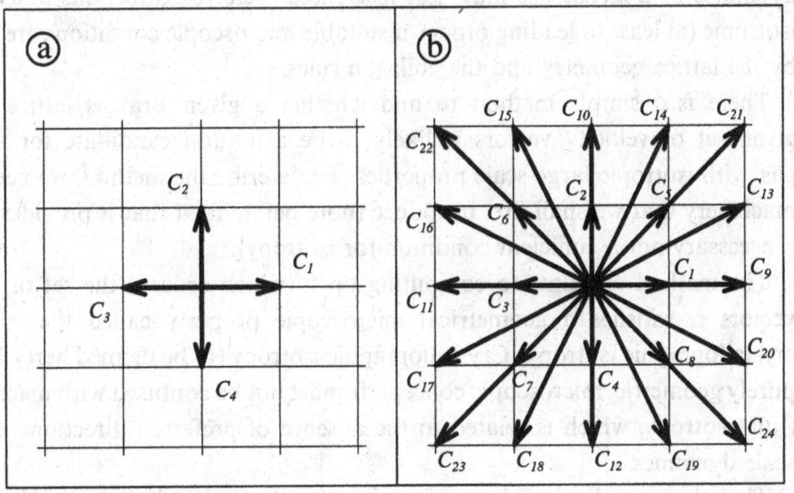

Figure 2.3 (a) A lattice gas model (the HPP model) with only one class of channels. (b) A more complex model with five classes.

Table 2.1 The classes of channels for the extended HPP model of Figure 2.3. The first column gives the class index I. The second column describes the channels in each class. The third and fourth columns give the number of elements in each class and the moduli of the corresponding vectors respectively.

Class index	Channels	No. of elements	Modulus
1	1 to 4	4	1
2	5 to 8	4	$\sqrt{2}$
3	9 to 12	4	2
4	13 to 20	8	$\sqrt{5}$
5	21 to 24	4	$2\sqrt{2}$

not imply directly the isotropy of *macroscopic* collective (large-scale) dynamics. An example is the HPP model (see Section 3.1), which is G-invariant, but whose large-scale dynamics is not isotropic.

2.3.5 Crystallographic isotropy

In lattice gases, the 'symmetries' (invariance properties not restricted to mathematical symmetries) of large-scale dynamics depend crucially on the microscopic symmetries of the underlying lattice. Although the microscopic structure of lattice gases only displays discrete symmetries, the resulting macroscopic behavior may possess continuous invariances. Obviously, a lattice cannot have the full isotropy of the continuum since it has preferred directions. Still, the large-scale dynamics of a lattice gas may 'not feel' these preferred directions and be fully isotropic (at least to leading order), if suitable microscopic conditions are fulfilled by the lattice geometry and the collision rules.

There is a simple method to find whether a given Bravais lattice with a given set of velocity vectors is likely to be a natural candidate for a lattice gas with isotropic large-scale properties. To describe the method, we need some machinery that we shall first introduce (note beforehand that it provides neither a necessary nor a sufficient condition for isotropy).

The method amounts to computing up to which order n the set of velocity vectors $\mathbf{c}_i$ satisfies a geometrical microscopic property called the 'n^{th} order crystallographic isotropy'. Crystallographic isotropy (to be defined hereafter), is a purely geometric, microscopic concept. It must not be confused with macroscopic (full) isotropy, which is related to the absence of preferred directions in large-scale dynamics.

That the crystallographic isotropy property be satisfied up to the desired order is not sufficient to guarantee macroscopic isotropy for large-scale dynamics: indeed, non-G-invariant collision rules can introduce some 'artificial' anisotropic contributions to the large-scale equations of motion. Moreover, the requirement of crystallographic isotropy may even be bypassed as a proper bias in the collision rule may compensate for the lack of crystallographic isotropy. However, for simple G-invariant homokinetic models with collisions satisfying semi-detailed balance, crystallographic isotropy up to desired order will be a necessary and sufficient condition for macroscopic isotropy. This point will be discussed in Chapter 5.

In order to describe the crystallographic isotropy property, we first define some geometrical properties of lattices and tensors.

2.3.5.1 Isotropic tensors

A tensor of order n defined in the D-dimensional Euclidean space $\mathbb{R}^D$ is said to be 'isotropic' if, and only if, it remains invariant under the action of any element of the orthogonal group O_D in dimension D, that is, the group of all distance-preserving linear transformations acting on $\mathbb{R}^D$ (see Tung, 1985). There exists an equivalent and more intuitive criterion to characterize the isotropy of a tensor: a tensor of order n is isotropic if, and only if, its contraction with n arbitrary 'test' vectors of $\mathbb{R}^D$ can be expressed exclusively in terms of inner products of these vectors. With this criterion, it is obvious that odd-order tensors cannot be isotropic, unless they are null. It is also easy to convince oneself that a second order tensor $T_{\alpha_1\alpha_2}$ is isotropic if, and only if, it is a multiple of the unit tensor $\delta_{\alpha_1\alpha_2}$ (Kronecker symbol). In the same way, a fourth order isotropic tensor $T_{\alpha_1\alpha_2\alpha_3\alpha_4}$ must be a linear combination of $\delta_{\alpha_1\alpha_2}\delta_{\alpha_3\alpha_4}$, $\delta_{\alpha_1\alpha_3}\delta_{\alpha_2\alpha_4}$, and $\delta_{\alpha_1\alpha_4}\delta_{\alpha_2\alpha_3}$. More generally, an nth rank isotropic tensor (with n even) must be a linear combination of all products of the form $\delta_{\alpha_{\sigma(1)}\alpha_{\sigma(2)}}\cdots\delta_{\alpha_{\sigma(n-1)}\alpha_{\sigma(n)}}$, where σ is any permutation on $(1,\ldots,n)$.

If, in addition, the tensor is fully symmetric (i.e. invariant under any permutation of its indices), then there exists a second isotropy criterion which is equivalent to the previous one, but technically much simpler to use: A fully symmetric n^{th} order tensor $T^{(n)}$ is isotropic if, and only if, its n-fold contraction with *one single* arbitrary test vector $\mathbf{x}$ of $\mathbb{R}^D$ is proportional to the nth power of the modulus $\|\mathbf{x}\|$ of the test vector, that is, if, and only if, there exists a real number a such that:

$$\forall \mathbf{x} \in \mathbb{R}^D, \quad \sum_{\alpha_1=1}^{D}\cdots\sum_{\alpha_n=1}^{D} T^{(n)}_{\alpha_1\ldots\alpha_n} x_{\alpha_1}\ldots x_{\alpha_n} = a\|\mathbf{x}\|^n. \tag{2.23}$$

2.3.5.2 Crystallographic tensor of order n

Consider a tensor $\mathbf{T}^{(n)}$ of order $n \geq 1$ constructed from the $\mathbf{c}_i$ vectors by summing over the channel index i, the n-fold tensor product[9] of $\mathbf{c}_i$ with itself, weighted by a set of coefficients $\theta(Cl(i))$ which depend only on the class $Cl(i)$ to which channel i belongs (see Section 2.3.4):

$$T^{(n)}_{\alpha_1\ldots\alpha_n} = \sum_{i=1}^{b} \theta(Cl(i)) c_{i\alpha_1}\ldots c_{i\alpha_n}. \tag{2.24}$$

Such a tensor will be called a 'crystallographic tensor' of the lattice gas.

2.3.5.3 Crystallographic isotropy of order n

A lattice gas model has the 'n^{th} order crystallographic isotropy' property if all crystallographic tensors with order less than or equal to n are isotropic. We emphasize again that this property of crystallographic isotropy must not be confused with ordinary isotropy: a lattice cannot be isotropic in the ordinary

[9] The underlying metrics is supposed to be Euclidean. We thus make no distinction between covariant and contravariant subscripts.

sense, but its local geometry (the $\mathbf{c}_i$s) may however yield crystallographic isotropy up to some order.

Crystallographic tensors built on the $\mathbf{c}_i$ vectors of lattice gas models are fully symmetric by construction. Hence, the second isotropy criterion provides a simple algebraic method to check the crystallographic isotropy properties of a given lattice gas model. Indeed, consider the nth order crystallographic tensor $T^{(n)} = \sum_{i=1}^{b} c_{i\alpha_1} \dots c_{i\alpha_n}$ of a lattice gas model; if n is odd, the isotropy test is simple: isotropy holds if, and only if, $T^{(n)}$ is zero. If n is even, then the full contraction of $T^{(n)}$ with an arbitrary vector $\mathbf{x}$ reduces to $\sum_{i=1}^{b}(\mathbf{c}_i \cdot \mathbf{x})^n$ which is a polynomial function with even order of the coordinates $(x_1, \dots, x_D)$ of the vector $\mathbf{x}$. The test for crystallographic isotropy is then to check whether this function is proportional to $(\sum_{\alpha=1}^{D} x_\alpha^2)^{\frac{n}{2}}$. Such algebraic manipulations, involving only the use of multinomial expansions, can be carried out by hand at least for orders lower than or equal to 6 or 8. (The use of a symbolic manipulator can simplify the task.)

2.3.6 Irreducibility

Irreducibility is a symmetry-related property which is not as natural as the notions defined previously. However, some of the calculations presented in Chapters 4 and following, are dramatically simplified if this property is verified. We thus wish to spend some time to define and visualize clearly this notion, and to explain its consequences.[10]

2.3.6.1 Definition

A lattice gas model is said to be 'irreducible' if, and only if, for any channel index $i = 1, \dots, b$, the invariant subspace Σ_i of the subgroup G_i of all lattice-preserving isometries that also preserve $\mathbf{c}_i$, is linearly generated by $\mathbf{c}_i$. In other words, any vector $\mathbf{u}$ preserved by all the isometries in G which also leave $\mathbf{c}_i$ unchanged, must be a multiple of $\mathbf{c}_i$.

We shall try to justify the name 'irreducibility' given to this property. Let us consider the subgroup G_i and the hyper-plane Π_i of dimension $D-1$, orthogonal to $\mathbf{c}_i$. We construct the set G_i' collecting the restrictions to Π_i of all elements in G_i. The irreducibility property simply states that G_i' is an irreducible representation of G_i on Π_i.[11]

2.3.6.2 Examples

To illustrate the notion of irreducibility for a lattice gas, we use once more the simple HPP and extended HPP models presented in Section 2.3.4. Let us

[10] On first reading, the reader can omit the details of the present section and just keep in mind its conclusions and consequences.

[11] For basics on irreducible representations of groups, the reader is referred to text books on group theory, for example Eliott and Dawber (1979), or Tung (1985).

consider first the simple HPP model (see Figure 2.3(a) for notations). We pick channel 1 corresponding to the velocity vector $\mathbf{c}_1 = (1,0)$. The subgroup G_1 of G leaving $\mathbf{c}_1$ unchanged contains only the identity (R_0) and the mirror symmetry S_y. The only vectors in $\mathbb{R}^2$ that remain invariant under the action of G_1 are clearly proportional to $\mathbf{c}_1$. Since all four vectors $\mathbf{c}_1$ to $\mathbf{c}_4$ are equivalent, what is true for $\mathbf{c}_1$ is true for all $\mathbf{c}_i$ vectors. This completes the proof that the simple HPP model possesses the property of irreducibility. Consider now the extended HPP model defined in Figure 2.3(b). We pick channel 13 corresponding to the velocity vector $\mathbf{c}_{13} = (2,1)$. The subgroup G_{13} of G leaving $\mathbf{c}_{13}$ unchanged only contains the identity (R_0). All vectors in $\mathbb{R}^D$ are invariant under the action of G_{13}, not only those proportional to $\mathbf{c}_{13}$. This shows that the extended HPP model does not possess the irreducibility property.

2.3.6.3 Consequences of the irreducibility property

As already stressed earlier, the irreducibility property can be used to greatly simplify algebraic calculations on the statistical and macroscopic properties of a lattice gas model. Indeed, suppose that, in some algebraic calculation, we encounter a set of vectors $\mathbf{T}_i$ such that, for all $\mathbf{g}$ in G and for all $i = 1,\ldots,b$, we have $\mathbf{g}(\mathbf{T}_i) = \mathbf{T}_{g(i)}$ (for the definitions of $\mathbf{g}$ and g, see Section 2.3.4). If the model considered is G-invariant and verifies the irreducibility condition, then we can immediately state that there exists a collection of coefficients $\zeta(I)$ where $I = Cl(i)$, such that $\mathbf{T}_i = \zeta(I)\mathbf{c}_i$, for all $i = 1,\ldots,b$. This considerably simplifies the algebra.

Moreover, the same kind of simplification also holds for second order tensors. Consider a set of symmetric[12] second order tensors $\mathbf{T}_i$ such that, for all $\mathbf{g}$ in G and for all $i = 1,\ldots,b$, we have $\mathbf{g}(\mathbf{T}_i) = \mathbf{T}_{g(i)}$.[13] If the lattice gas model is G-invariant and verifies the irreducibility condition, then the tensors $\mathbf{T}_i$ must have the form $\mathbf{T}_i = \lambda(I)\mathbf{c}_i \otimes \mathbf{c}_i + \mu(I)\delta$, where δ is the Kronecker tensor, and $\lambda(I)$ and $\mu(I)$ are two collections of coefficients that depend only on the class $I = Cl(i)$ of the channel.

The presentation of these technically very useful properties closes this section on the microscopic properties of lattice gases.

2.4 Special rules

If the part of lattice $\mathscr{L}$ actually occupied by the lattice gas were infinite in all directions, then the cellular automaton rule described by the microdynamic

[12] In this context, 'symmetry' means that $T_{i\alpha\beta} = T_{i\beta\alpha}$, where $T_{i\alpha\beta}$ are the Cartesian components of the tensor $\mathbf{T}_i$.

[13] For simplicity, we denote by the same symbol $\mathbf{g}$ the isometry acting on vectors and on second order tensors.

equation (2.12) could be applied everywhere. In practice however, that part of the lattice is finite and the propagation phase requires periodic wrapping such that no particle be lost through the boundaries of $\mathscr{L}$. As explained in Section 2.2.2, this is realized (i) by embedding $\mathscr{L}$ into a wider domain $\mathscr{L}'$, that can perfectly tile the full Bravais lattice by successive translations, and (ii) by reconnecting opposite sides in such a way that $\mathscr{L}'$ becomes a topological torus. In most cases, $\mathscr{L}$ can itself perfectly tile the Bravais lattice, and is thus identical to $\mathscr{L}'$. From now on, we assume this to be the case, and we use the notation $\mathscr{L}$ only.

If the collision operator described in Section 2.2.3 were applied everywhere in $\mathscr{L}$, then the macroscopic quantities would verify periodic boundary conditions in the usual sense. When different macroscopic boundary conditions are imposed, one must apply to the boundary nodes of $\mathscr{L}$ special collision rules that differ from those applied in the bulk. Note that special collision rules are not necessarily restricted to boundary nodes: they can also be applied to localized zones in the bulk, to simulate physical situations such as, the presence of solid obstacles in a flow.

Several special collision rules are designed to simulate various macroscopic boundary conditions. In the following subsections, we describe some of those which are most frequently used in lattice gas simulations of fluid dynamics. They can be used plainly or in a statistically mixed manner to obtain particular effects. For example, a solid obstacle collision rule statistically mixed with the normal collision rule can produce a semi-permeability effect, with a permeability adjusted by tuning the respective probabilities for the two collision rules.

2.4.1 Solid impermeable obstacles

A fixed solid obstacle inside, or at the boundaries of, a fluid flow can be implemented by applying a special collision operator to all the obstacle nodes. This special collision operator, called the 'reflection operator', must act on the Boolean state of obstacle nodes so as to send back any incoming particle to regular bulk nodes. In other words, the channels connecting an obstacle node to any other obstacle node must be empty after the application of the reflection operator. This guarantees that the nodes *inside* the obstacle (those which cannot be connected to any node of the bulk by one of the $\mathbf{c}_i$ vectors) will remain empty if initially empty. Only the nodes of the surface *layers* of the obstacle (those which can be connected to a bulk node by one of the $\mathbf{c}_i$s) will not remain empty. This raises the question of the position of the effective surface of the obstacle, that is, the location where the normal mean velocity actually vanishes. Cornubert, d'Humières and Levermore (1991) have shown, for some two-dimensional models, that the actual position of the effective surface may depend on the orientation of the surface with respect to the lattice symmetry axes, at least for specular reflections. If solid obstacle conditions are applied to the boundaries of $\mathscr{L}$, for example to simulate wall-bounded flows, then some care must be taken to avoid leakage. The layer of boundary nodes where the obstacle collision operator is applied must be sufficiently *thick* to guarantee that during the propagation phase, no particle be sent directly outside the layer of boundary nodes. This point is discussed in Chapter 10.

There exist several ways to realize particle reflection, leading to several different reflection operators, and to different macroscopic effects. In the next paragraphs, we describe some of the simplest reflections: the 'bounce-back', 'specular', and 'diffusive' reflection operators.

2.4.1.1 Bounce-back reflection

The bounce-back reflection operator sends back all the particles impinging on an obstacle node to where they come from (see Figure 2.4(a)). This operation

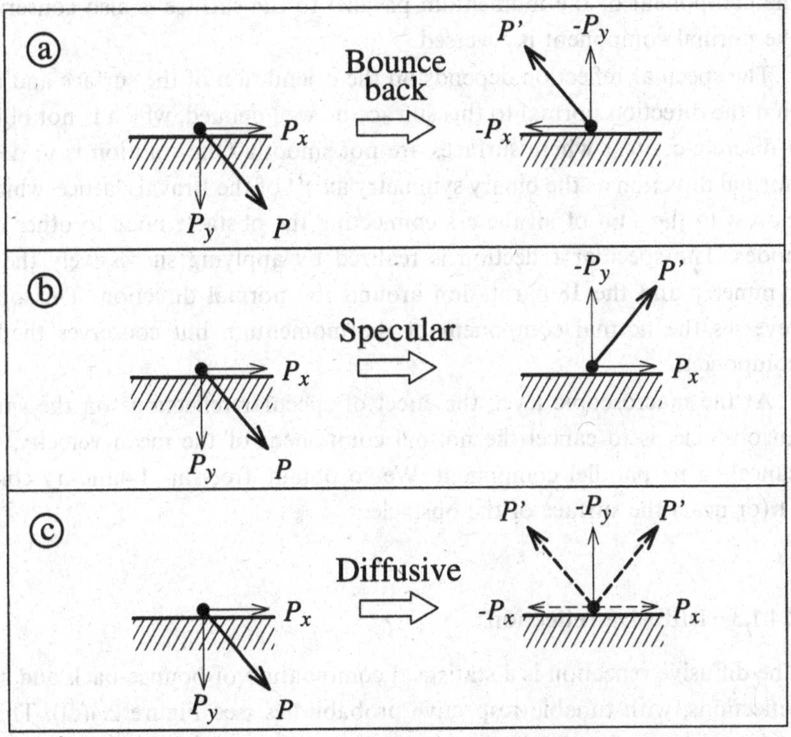

Figure 2.4 The figure illustrates, in a two-dimensional case, the effect on the total momentum of (a) the bounce-back reflection, (b) the specular reflection and (c) the diffusive reflection. The solid arrows represent total momenta at surface nodes. They do not represent the individual velocity vectors $\mathbf{c}_i$ as in the figures of Chapter 3. P_x and P_y denote respectively the parallel and orthogonal components of the total momentum $\mathbf{P}$.

conserves the total mass and energy observables, but the momentum observable is reversed.

The bounce-back reflection mode is by far the simplest one and does not depend on the orientation of the surface. The easiest realization is to apply the central symmetry to the state of obstacle nodes. A quite efficient numerical strategy to implement this central symmetry is described in Chapter 10. At the macroscopic level, the global effect of bounce-back reflections is that all components of the mean velocity are canceled at the surface of the obstacle, where 'no-slip' boundary conditions are thus realized.

2.4.1.2 Specular reflection

The specular reflection operator makes the particles bounce elastically on the surface of the obstacle, like a light beam reflected on a perfect mirror (see Figure 2.4(b)). This operation conserves the total mass and energy observables; the component of the momentum parallel to the surface is also conserved, but the normal component is reversed.

The specular reflection depends on the orientation of the surface and requires that the direction normal to this surface be well defined, which is not obvious in a discrete context where surfaces are not smooth. One solution is to define the normal direction as the binary symmetry axis[14] of the Bravais lattice which is the closest to the sum of all the $\mathbf{c}_i$ s connecting the obstacle node to other obstacle nodes. The specular reflection is realized by applying successively the central symmetry and the 180° rotation around the normal direction. This operation reverses the normal component of the momentum but conserves the parallel component.

At the macroscopic level, the effect of specular reflections on the surface of an obstacle, is to cancel the normal component of the mean velocity, without canceling its parallel component. We so obtain 'free-slip' boundary conditions at (or near) the surface of the obstacle.

2.4.1.3 Diffusive reflection

The diffusive reflection is a statistical combination of bounce-back and specular reflections, with tunable respective probabilities (see Figure 2.4(c)). This operation conserves the total mass and energy observables; the component of the momentum normal to the surface is reversed, and the parallel component is conserved or reversed according to a random choice.

[14] A binary symmetry axis is such that a 180° rotation around the axis leaves the Bravais lattice globally invariant.

2.4.2 Sources and sinks of observable quantities

It is often useful to simulate localized zones at the boundaries or in the bulk of $\mathcal{L}$, where some observable quantity, which would be conserved otherwise, is being injected or absorbed (sources or sinks). As an example, consider a thermal lattice gas model designed to simulate fluid flow with thermal exchanges: local heating or cooling can be simulated by locally injecting or absorbing kinetic energy (otherwise conserved), while conserving mass and momentum. This kind of effect can be achieved by applying suitable combinations of the usual collision operator with some special collision operators. The latter have almost the same conservation rules as the usual operator, except that each of these special collision operators maximizes the increase or decrease of one of the usually conserved observable quantities. A suitable tuning of the relative statistical weights in the combination produces the desired rate of injection/absorption for each of the usually conserved quantities. In the example of tunable local heating (cooling), one combines the usual collision operator with a 'heating operator' ('cooling operator'), with adjustable statistical weights. The heating operator (cooling operator) must be designed so as to maximize (minimize) kinetic energy while conserving mass and momentum. Another example is the incorporation of body forces in a lattice gas model for fluid flow. The realization proceeds along the same lines. Consider a body force in the direction x_1, acting over a more or less extended region of $\mathcal{L}$. Two special operators are needed that maximize respectively the increase and decrease of the component p_1 of the momentum, while conserving all other usually conserved quantities (mass, momenta p_2 to p_D, energy, etc.). The sign and intensity of the body force is adjusted by a proper tuning of the statistical weights of the normal and special collision operators.

2.5 Comments

We have presented the framework and the tools for the description and classification of lattice gases at the microscopic level. These microdynamic concepts and properties have a crucial importance for the theory to be developed in later chapters. The material presented so far may be perceived as rather abstract. Therefore, before engaging in the systematic study of the statistical properties of lattice gases at equilibrium and out of equilibrium, we devote the next chapter to the description of various examples of lattice gas models in order to illustrate the microdynamic formalism.[15]

[15] The reader familiar with the microdynamic formalism and with lattice gas models may want to move directly to Chapter 4.

Chapter 3

Microdynamics: various examples

We now illustrate the abstract microdynamic notions of Chapter 2, with a presentation of lattice gas models in terms of the microdynamic tools. The models are chosen to illustrate the various microdynamical concepts; further models will be considered briefly in Chapter 11.

We start with the simplest two-dimensional model based on the square lattice, the earliest lattice gas model (1973) labeled HPP according to the initials of the authors: Hardy, de Pazzis and Pomeau. Sections 3.2 to 3.4 are devoted to models constructed on the triangular lattice and based on the FHP model initially introduced by Frisch, Hasslacher and Pomeau (1986). A 'colored' version of the FHP model, developed as a two-components lattice gas is presented in Section 3.5. A slightly more complex model, also based on the triangular lattice, but with thermal properties (Grosfils, Boon and Lallemand, 1992) is described in Section 3.6. We then move to three-dimensional systems in Section 3.7, as we introduce the basic (pseudo-four-dimensional) lattice gas model of d'Humières, Lallemand and Frisch (1986).

Except for the HPP model, all the models presented in this chapter, have been designed to exhibit large-scale dynamics in accordance with the Newtonian viscous behavior of isotropic fluids.

3.1 The HPP model

Historically, the first lattice gas model was introduced in the early seventies by Hardy, de Pazzis and Pomeau (1973) with motivations focusing on fundamental aspects of statistical physics (see also Hardy *et al*, 1972, 1976 and 1977). Their model (HPP) is based on the two-dimensional square lattice with unit lattice constant (see Figure 3.1).

The HPP model is a 4-bit model, where each node has $b = 4$ channels, corresponding to the four directions of the square lattice (east, north, west, south) labeled by integers from 1 to 4. The $\mathbf{c}_i$s have unit modulus and their Cartesian components are:

$$\mathbf{c}_1 = \begin{pmatrix} 1 \\ 0 \end{pmatrix}, \quad \mathbf{c}_2 = \begin{pmatrix} 0 \\ 1 \end{pmatrix}, \quad \mathbf{c}_3 = \begin{pmatrix} -1 \\ 0 \end{pmatrix}, \quad \mathbf{c}_4 = \begin{pmatrix} 0 \\ -1 \end{pmatrix}. \tag{3.1}$$

These vectors connect a node to its four nearest neighbors. So, the model's connectivity B and the number b of channels per node are both equal to the coordination number of the two-dimensional square lattice, namely four. Furthermore, the four connection vectors are identical to the four velocity vectors $\mathbf{c}_i$, which are all different and non-zero (see Section 2.1.1).

The masses m_i of particles are equal, and by a suitable choice of the mass scale, their common value may be taken as one unit mass.[1] The momentum $\mathbf{p}_i$ of each particle is equal to $\mathbf{c}_i$, which is physically consistent with unit mass and unit time step. The kinetic energy of each particle is taken equal to $\frac{1}{2}$, again with physical consistency. The choice of unitary quantities for lattice constant, time step, and particle mass completely defines a natural system of units for microscopic mechanical quantities.

As prescribed by the general definition of lattice gases (see Section 1.2.3),

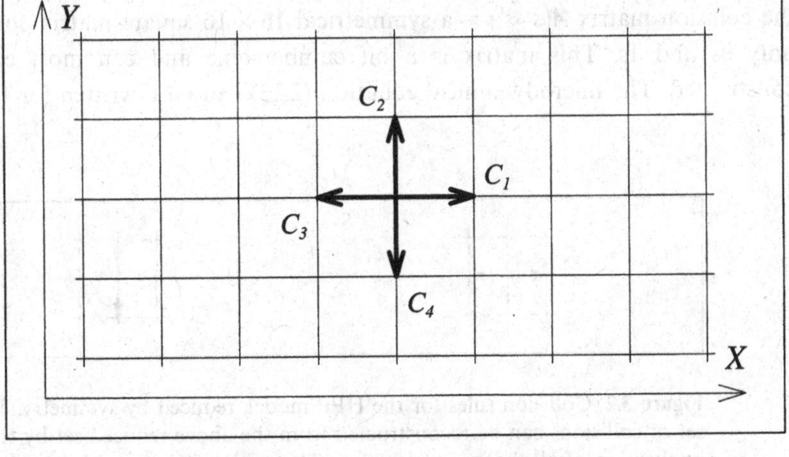

Figure 3.1 The HPP lattice with the corresponding $\mathbf{c}_i$s.

[1] Most often, the mass will therefore not appear explicitly in the equations.

the evolution rule of the HPP model is a two-phase sequence: a propagation phase and a collision phase. During the propagation phase, a particle present at node $\mathbf{r}_\star$ in channel i moves to node $\mathbf{r}_\star + \mathbf{c}_i$, its nearest neighbor in the direction i. During the collision phase, a pair of particles, arriving at the same node from opposite directions, for example *north–south*, is turned into a pair of particles heading in the two other directions, here *east–west* (see Figure 3.2). The reverse collision transforming east–west into north–south also has probability one, but is not shown on Figure 3.2 because it can be obtained from the collision (north–south → east–west) by the application of an isometry ($+\pi/2$ rotation) belonging to the local invariance group of the lattice. All other states which are not strictly 'north–south' ($s = 0101$) or 'east–west' ($s = 1010$) remain unchanged in the collision step. Notice that only two among the $2^4 = 16$ possible states, can undergo an 'effective collision', that is, with a post-collisional state different from the pre-collisional state. The 'collisional efficiency' (the percentage $\sum_{s \neq s'} A(s \to s') / \sum_{s,s'} A(s \to s')$ of efficient collisions) is 12.5 %. In other words, when the Boolean state of a given node $\mathbf{r}_\star$ is randomly chosen among the 16 possible states of the single-node phase-space γ (with equal probabilities), effective collisions occur at node $\mathbf{r}_\star$ with a probability of 0.125.

3.1.1 The microdynamical equation

The number b of channels per node being quite small for the HPP model, (namely four), the number of possible states per node is only $2^b = 16$. In addition, the HPP model is deterministic and verifies detailed balance (2.15). So, the collision matrix $A(s \to s')$ is a symmetrical 16×16 square matrix containing only 0s and 1s. This matrix is a bit cumbersome and can most easily be constructed. The microdynamical equation (2.13) can be rewritten for the HPP

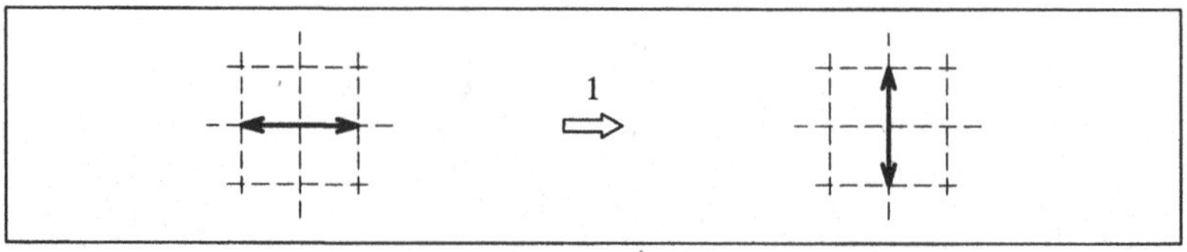

Figure 3.2 Collision rules for the HPP model, reduced by symmetry. The full set of collisions can be reconstructed from the above reduced set by the application of all the isometries of $\mathcal{G}$. The number '1' above the open arrow indicates that the collisional change of configuration occurs with probability 1.

model by taking all the $\xi(s \to s')$s equal to the $A(s \to s')$s (since the model is deterministic). It reads:

$$\begin{aligned} &\text{for } i = 1, \ldots, 4: \\ n_i(\mathbf{r}_\star + \mathbf{c}_i, t_\star + 1) &= n_i \\ &- n_i \overline{n}_{i+1} n_{i+2} \overline{n}_{i+3} \\ &+ \overline{n}_i n_{i+1} \overline{n}_{i+2} n_{i+3}. \end{aligned} \tag{3.2}$$

Here, $\overline{n}_i$ stands for $(1 - n_i)$ and the subscript i (channel index) is cyclic (modulo 4). All variables on the right hand side are taken implicitly at time $t_\star$ and node $\mathbf{r}_\star$.

Because of the simplicity of the HPP microdynamics, (3.2) can be derived merely by intuition: the three terms on the right hand side simply express that the collision phase can either:

- leave the occupation state of channel i unchanged, or
- modify its state from 1 to 0 if channels i and $i+2$ are occupied and channels $i+1$ and $i+3$ are unoccupied, or
- modify its state from 0 to 1 if channels $i+1$ and $i+3$ are occupied and channels i and $i+2$ are unoccupied.

We note that (3.2) is compatible with the exclusion principle, since it guarantees that the n_is at time $t_\star + 1$ cannot be anything other than 0 or 1, provided the n_is at time $t_\star$ are 0 or 1.

For simplicity, we have chosen to write (3.2) for a microscopic evolution law where the collision phase is performed *before* the propagation phase. If the reverse order is chosen, the microdynamical equation takes almost the same form, except that each of the n_is and $\overline{n}_i$s on the right hand side must then be taken at a different node (namely at nodes $\mathbf{r}_\star - \mathbf{c}_i$), instead of being all taken at the same node $\mathbf{r}_\star$. This minor difference is generally unimportant, since the order of the two successive phases becomes macroscopically irrelevant when a large number of steps are executed sequentially.[2]

The microdynamical equation can also be written in logical form; it suffices to consider the n_is as logical variables with values 'TRUE' or 'FALSE' (instead of algebraic variables with values '1' or '0'), to obtain:

$$\begin{aligned} n_i(\mathbf{r}_\star + \mathbf{c}_i, t_\star + 1) = &\Big(n_i \textbf{ and } (n_{i+1} \textbf{ or } \overline{n}_{i+2} \textbf{ or } n_{i+3}) \Big) \\ \textbf{or } &\Big(\overline{n}_{i+1} \textbf{ and } (n_{i+2} \textbf{ and } \overline{n}_{i+2} \textbf{ and } n_{i+3}) \Big) \end{aligned} \tag{3.3}$$

Here, $\overline{n}_i$ now stands for the logical negation of n_i. Although equations (3.2) and (3.3) seem formally different, their content is the same. For many purposes, however, the algebraic form is more convenient.

2 Note that care must be taken as to when the iteration sequence is stopped, when performing correlation measurements on the lattice.

3.1.2 Microscopic properties

The properties listed below for the HPP model have been defined in a general framework in Section 2.3. Proofs are given for these properties only when non-trivial.

(i) The HPP model is deterministic, by construction.

(ii) It satisfies detailed balance and *a fortiori* semi-detailed balance (see Section 2.3.1).

(iii) It is self-dual since the only permitted collisions (1010) → (0101) and (0101) → (1010) are dual of each other, and have equal probabilities, namely one (see Section 2.3.2).

(iv) The only linearly independent collisional invariants are the mass (particle number) and the two components of the momentum:

$$m : \begin{pmatrix} 1 \\ 1 \\ 1 \\ 1 \end{pmatrix}, \quad p_x : \begin{pmatrix} 1 \\ 0 \\ -1 \\ 0 \end{pmatrix}, \quad p_y : \begin{pmatrix} 0 \\ 1 \\ 0 \\ -1 \end{pmatrix}. \tag{3.4}$$

In addition, the HPP model has several static and moving geometrical invariants (see Section 2.3.3), such as the x- (or y-) component of the momentum, summed over any line of lattice nodes with equal x- (or y-) coordinates (d'Humières, Qian and Lallemand, 1989, 1990).

(v) The model is clearly G-invariant (invariant under the crystallographic group of the lattice, see Section 2.3.4), and all four $\mathbf{c}_i$ vectors clearly belong to the same class, as defined in Section 2.3.4.

(vi) The HPP lattice gas model has third order crystallographic isotropy (see Section 2.3.5). As a justification, Table 3.1 gives the crystallographic tensors of order 1 to 4, contracted with a vector $\mathbf{x}$ with coordinates (x_1, x_2). Clearly, the contraction of the fourth order tensor is *not* proportional to

Table 3.1 Crystallographic isotropy properties of the HPP model. The second column gives the result of a full contraction of the nth order tensor with an arbitrary test vector $\mathbf{x}$. The third column summarizes the conclusion about isotropy.

Order n	$\sum T^{(n)}_{\alpha_1 \ldots \alpha_n} x_{\alpha_1} \ldots x_{\alpha_n}$		Status
1	0		isotropic
2	$2(x_1^2 + x_2^2)$	$(= 2\|x\|^2)$	isotropic
3	0		isotropic
4	$2(x_1^4 + x_2^4)$	$(\neq a\|x\|^4)$	**anisotropic**

$\|\mathbf{x}\|^4$; hence, by the criterion of Section 2.3.5, fourth order crystallographic isotropy definitely does not hold for the HPP lattice.

(vii) The HPP model is irreducible (see Section 2.3.6). Indeed, the only isometry in G that preserves $\mathbf{c}_1$, for example, is the mirror symmetry with respect to the x-direction, whose invariant sub-space only contains vectors proportional to $\mathbf{c}_1$. The same argument holds with $\mathbf{c}_2$, $\mathbf{c}_3$ and $\mathbf{c}_4$.

Since all $\mathbf{c}_i$s have the same modulus 1 and all the particles have the same mass, the kinetic energy observable is simply proportional to the mass observable. States with the same mass will also have the same kinetic energy, so, the energy conservation and the mass conservation are equivalent. As a consequence, the HPP model ignores thermal effects.

The degree of crystallographic isotropy of the HPP model (namely three) is not sufficient to produce isotropic large-scale dynamics as described by the Navier–Stokes equations for a Newtonian fluid.[3] This requires at least fourth order crystallographic isotropy (see Chapter 5). It is actually possible to construct models with fourth order isotropy on the square lattice. However, such models are not *minimal*, in the sense of the concept introduced in Section 1.3.1, and require a larger set of $\mathbf{c}_i$ vectors: in addition to the HPP $\mathbf{c}_i$s, which must now be endowed with multiplicity four (4 channels for each of the $\mathbf{c}_i$s), four vectors must be added, connecting each node to its next-nearest neighbors (1 channel per vector). Such 'modified' HPP models require $b = 20$ channels per node, and are not homokinetic. About ten years after the introduction of the HPP model, the 'anisotropy disease' has been cured in a less expensive way by models based on the triangular lattice, which are discussed in the next sections.

3.2 The FHP-1 model

The family of FHP models, based on an original idea of Frisch, Hasslacher, and Pomeau, provides a class of two-dimensional lattice gas models with the simplest structure compatible with the property of fourth order crystallographic isotropy producing proper (isotropic) large-scale dynamics (Frisch, Hasslacher and Pomeau, 1986). This family of models is based on the two-dimensional triangular lattice with unit lattice constant (see Figure 3.3). Several versions of the FHP model have been successively developed, with the same geometrical lattice structure, but with different collision rules. In this section, we present the earliest and simplest model of the family: the 'FHP-1' model.

The FHP-1 model is a 6-bit model. Each node has $b = 6$ channels, corre-

[3] Notice that the square lattice gas is well suited to model scalar transport such as diffusion, and has been used efficiently for a microscopic approach to reaction–diffusion phenomena (see Boon, Dab, Kapral, and Lawniczak, 1996).

sponding to the six directions of the triangular lattice, labeled by integers from 1 to 6. The $\mathbf{c}_i$s have unit modulus and their Cartesian components are:

$$\mathbf{c}_1 = \begin{pmatrix} 1 \\ 0 \end{pmatrix}, \quad \mathbf{c}_2 = \begin{pmatrix} \frac{1}{2} \\ \frac{\sqrt{3}}{2} \end{pmatrix}, \quad \mathbf{c}_3 = \begin{pmatrix} -\frac{1}{2} \\ \frac{\sqrt{3}}{2} \end{pmatrix},$$
$$\mathbf{c}_4 = \begin{pmatrix} -1 \\ 0 \end{pmatrix}, \quad \mathbf{c}_5 = \begin{pmatrix} -\frac{1}{2} \\ -\frac{\sqrt{3}}{2} \end{pmatrix}, \quad \mathbf{c}_6 = \begin{pmatrix} \frac{1}{2} \\ -\frac{\sqrt{3}}{2} \end{pmatrix}. \tag{3.5}$$

These vectors connect each node to its six nearest neighbors. Thus, the model's connectivity B and the number b of channels per node are both equal to the coordination number of the two-dimensional triangular lattice, namely six. Furthermore, the six connection vectors are identical to the six velocity vectors $\mathbf{c}_i$, the latter being all different and non-zero (see Section 2.1.1).

The FHP-1 model shares with the HPP model the following properties: The masses m_i of all the particles are equal, and by a suitable choice of the mass scale, the common value is taken as unity. The momentum $\mathbf{p}_i$ of each particle is equal to $\mathbf{c}_i$; this is physically consistent with unit mass and unit time steps. The kinetic energy of each particle is taken equal to $\frac{1}{2}$, again for physical consistency. The option that lattice constant, time step and particle mass be equal to unity completely defines a natural system of units for microscopic mechanical quantities.

The propagation phase in the FHP-1 model proceeds in exactly the same way as for the HPP model. An essential difference appears in the collision phase: while the collisions are deterministic for the HPP model, some stochasticity enters the FHP microdynamics. Indeed, two particles coming from opposite directions (for example 1 and 4) undergo a binary collision with an output state

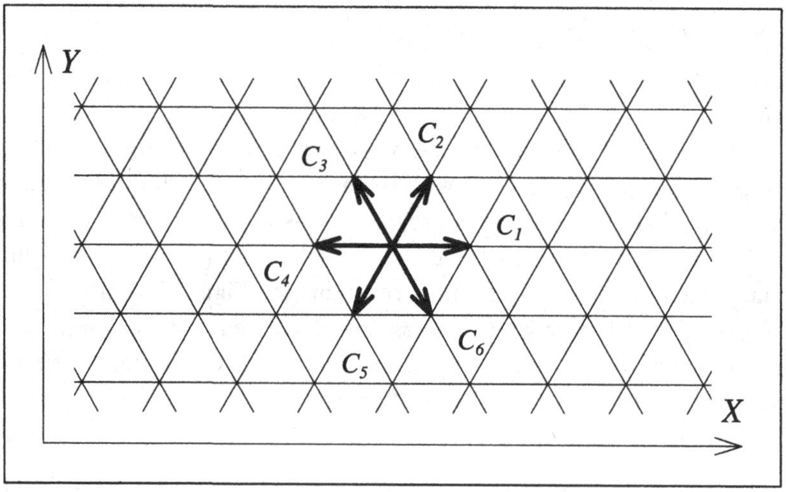

Figure 3.3 The FHP lattice with the corresponding $\mathbf{c}_i$s.

rotated by +60° or −60° (as shown in Figure 3.4(a)), with probabilities p and $1-p$ respectively. Another new feature is the introduction of deterministic triple collisions: three particles coming from three directions forming 120° angles between each other, will be deflected by 60° (as shown in Figure 3.4(b)). All other states remain unchanged. Note that, among the $2^6 = 64$ possible states, the five states shown in Figure 3.4 are the only ones that can undergo effective collisions. The collisional efficiency of FHP-1 is 7.81% (see Section 3.1 for definition).

3.2.1 The microdynamical equation

The microdynamical equation for the model FHP-1 exhibits the same structure as the microdynamical equation for the HPP model. It contains a few more terms because the number of possible collisions is larger. This equation can be obtained systematically from the general form (2.13), but it is a rather tedious operation, and the intuitive method described for the HPP model is preferable. One obtains:

$$\begin{aligned}
&\text{for } i = 1, \ldots, 6: \\
n_i(\mathbf{r}_\star + \mathbf{c}_i, t_\star + 1) \;=\;& n_i \\
&- \; n_i\overline{n}_{i+1}\overline{n}_{i+2}n_{i+3}\overline{n}_{i+4}\overline{n}_{i+5} \\
&+ \; \xi_{(+)} \; \overline{n}_i n_{i+1}\overline{n}_{i+2}\overline{n}_{i+3}n_{i+4}\overline{n}_{i+5} \\
&+ \; \xi_{(-)} \; \overline{n}_i\overline{n}_{i+1}n_{i+2}\overline{n}_{i+3}\overline{n}_{i+4}n_{i+5} \\
&- \; n_i\overline{n}_{i+1}n_{i+2}\overline{n}_{i+3}n_{i+4}\overline{n}_{i+5} \\
&+ \; \overline{n}_i n_{i+1}\overline{n}_{i+2}n_{i+3}\overline{n}_{i+4}n_{i+5}.
\end{aligned} \tag{3.6}$$

Here, $\overline{n}_i$ stands for $(1-n_i)$ and the index i is cyclic (modulo 6). All variables on the right hand side are taken implicitly at time $t_\star$ and node $\mathbf{r}_\star$. $\xi_{(+)}(\mathbf{r}_\star, t_\star)$

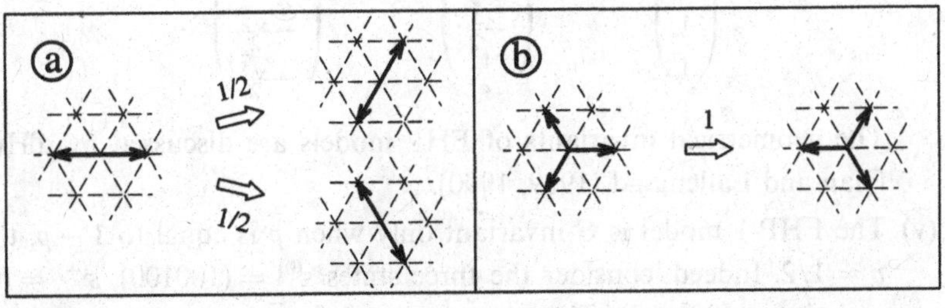

Figure 3.4 Collision rules for the FHP-1 model, reduced by symmetry. The full set of collisions can be reconstructed from the above reduced set by the application of all the isometries of $\mathcal{G}$. The numbers above the open arrows are the transition probabilities. The most commonly used value $p = 1/2$ has been chosen here, since it is the only value compatible with the G-invariance.

(respectively $\xi_{(-)}(\mathbf{r}_\star, t_\star)$) denotes a collection of Bernoulli random variables (with values 0 or 1) satisfying $\xi_{(+)}(\mathbf{r}_\star, t_\star) + \xi_{(-)}(\mathbf{r}_\star, t_\star) = 1$, at each time $t_\star$ and at each node $\mathbf{r}_\star$, and whose ensemble averages are all equal to p (respectively $1-p$).

3.2.2 Microscopic properties

The properties listed below for the FHP-1 model have been defined in a general framework in Section 2.3. Proofs are given for these properties only when non-trivial.

(i) The FHP-1 model is non-deterministic, except if $p = 1$ or $p = 0$.

(ii) The collision rule satisfies semi-detailed balance (2.17) for any value of the probability p between 0 and 1. Indeed, for any of the three Boolean states (100100), (010010) and (001001), there exist two possible pre-collision states, with respective probabilities p and $1-p$, and for all other Boolean states there exists one and only one possible pre-collision state with probability 1. The identity $\sum_s A(s \to s') = 1$ is thus satisfied. Note that when $p = 1/2$, the collision rule even satisfies detailed balance, since the relation $A(s \to s') = A(s' \to s)$ then holds for any pair of Boolean states s and s'.

(iii) The model is clearly not self-dual, as the dual of two-particle collisions are four-particle collisions which are ignored by the evolution rule of the FHP-1 model.

(iv) The only linearly independent collisional invariants are the mass (that is, the particle number) and the two components of the momentum:

$$m: \begin{pmatrix} 1 \\ 1 \\ 1 \\ 1 \\ 1 \\ 1 \end{pmatrix}, \quad p_x: \begin{pmatrix} 1 \\ \frac{1}{2} \\ -\frac{1}{2} \\ -1 \\ -\frac{1}{2} \\ \frac{1}{2} \end{pmatrix}, \quad p_y: \begin{pmatrix} 0 \\ \frac{\sqrt{3}}{2} \\ \frac{\sqrt{3}}{2} \\ 0 \\ -\frac{\sqrt{3}}{2} \\ -\frac{\sqrt{3}}{2} \end{pmatrix}. \tag{3.7}$$

The geometrical invariants of FHP models are discussed by d'Humières, Qian and Lallemand (1989, 1990).

(v) The FHP-1 model is G-invariant only when p is equal to $1-p$, that is, if $p = 1/2$. Indeed, consider the three states $s^{(0)} = (100100)$, $s^{(+)} = (010010)$ and $s^{(-)} = (001001)$. The mirror symmetry $\mathbf{g}_x$ with respect to the x-axis induces on the single-node phase-space γ a transformation $\mathcal{G}_x$ that leaves $s^{(0)}$ unchanged and permutes $s^{(+)}$ and $s^{(-)}$. Since $\mathcal{G}_x$ changes the collision $s^{(0)} \to s^{(+)}$ into the collision $s^{(0)} \to s^{(-)}$ and reciprocally, the probabilities of these two collisions must be equal in order that the model be G-invariant.

In addition, all six velocity vectors clearly belong to the same class, since any of these six vectors can be changed into any other by a rotation of some integer multiple of $\pi/3$.

(vi) The FHP-1 model has the fifth order crystallographic isotropy (see Section 2.3.5). As a justification, Table 3.2 gives the crystallographic tensors of order 1 to 6, contracted with a test vector $\mathbf{x}$ of coordinates (x_1, x_2). Clearly, the contraction of the fourth order tensor is proportional to $\|\mathbf{x}\|^4$, and the fifth order tensor is zero. Hence, by the criterion of Section 2.3.5, crystallographic isotropy holds up to the fifth order for the FHP lattice.

(vii) The FHP-1 model is irreducible (see Section 2.3.6). Indeed, the only isometry in G that preserves $\mathbf{c}_1$, for example, is the mirror symmetry with respect to the x-direction, whose invariant sub-space only contains vectors proportional to $\mathbf{c}_1$. The same argument holds for $\mathbf{c}_2$ to $\mathbf{c}_6$.

As for the HPP model, the energy observable is proportional to the mass observable and its conservation is a consequence of the mass conservation. So, the FHP-1 model has large-scale behavior ignoring thermal effects.

The basic reason for incorporating triple collisions in the FHP-1 model follows from the fact that without triple collisions the model would have an additional collisional invariant, namely $(1, -1, 1, -1, 1, -1)$, which is physically irrelevant. This spurious invariant would strongly affect the macroscopic dynamics of the model and would make it unphysical.

The degree of crystallographic isotropy of FHP-1 is larger than four, which is compatible with the possibility of having realistic large-scale dynamics governed in certain limits by the Navier–Stokes equation, provided the model be G-invariant, that is, for $p = 1/2$, which is the most commonly used value. Other values produce chirality.

Table 3.2 Crystallographic isotropy properties of the FHP models. The second column gives the result of a full contraction of the nth order tensor with an arbitrary test vector $\mathbf{x}$. The third column summarizes the conclusion about isotropy.

Order n	$\sum T^{(n)}_{\alpha_1 .. \alpha_n} x_{\alpha_1} ... x_{\alpha_n}$	Status
1	0	isotropic
2	$3(x_1^2 + x_2^2) \quad (= 3\|x\|^2)$	isotropic
3	0	isotropic
4	$\frac{9}{4}(x_1^4 + 2x_1^2x_2^2 + x_2^4) \quad (= \frac{9}{4}\|x\|^4)$	isotropic
5	0	isotropic
6	$\frac{3}{16}(11x_1^6 + 15x_1^4x_2^2 + 45x_1^2x_2^4 + 9x_2^6)$ $(\neq a\|x\|^6)$	**anisotropic**

3.3 The FHP-2 model

The FHP-2 model is a variant of the FHP-1 model that includes the possibility of one rest particle per node, in addition to the six moving particles of FHP-1. Each node then has $b = 7$ channels, corresponding to particles moving along the six directions of the triangular lattice and to the rest particle. The channels corresponding to moving particles are labeled by integers from 1 to 6, and the channel corresponding to the rest particle is labeled 0.

The six connection vectors are the same as for FHP-1. They do not include the vector $\mathbf{c}_0$ which is zero. As a consequence, the connectivity B, which is still equal to the coordination number of the triangular lattice (that is, 6), is now different from the number $b = 7$ of channels per node. The six vectors labeled by $i = 1$ to 6 have equal unit moduli, and the vector $\mathbf{c}_0$ corresponding to the rest particle obviously has zero modulus.

The masses m_i of all particles are equal to one and the momentum $\mathbf{p}_i$ of each particle is equal to $\mathbf{c}_i$. Thus, the momentum of the rest particle is zero. The kinetic energy is taken equal to $\frac{1}{2}$ for moving particles, and to 0 for the rest particle. As for the HPP and FHP-1 models, the above choices are physically consistent with unit particle mass and unit time step.

The propagation phase for moving particles is the same as for FHP-1, and does not affect the rest particles. The collision rules of the FHP-2 model are similar to the collision rules of FHP-1 with four additional collisions coupling moving and rest particles: (i) a moving particle arriving at a node occupied only by a rest particle produces a pair of moving particles with angles $+60°$ and $-60°$ measured from the direction of the incoming particle (Figure 3.5.e), (ii) the reverse of the former (Figure 3.5.f), (iii) the collisions of FHP-1 with a passive rest particle left unchanged by the collision (Figure 3.5.c and 3.5.d). *All* possible collisions of FHP-2 that cannot be deduced from one another by a lattice-preserving isometry are summarized in Figure 3.5. Only 22 among the $2^7 = 128$ possible states can undergo effective collisions. The collisional efficiency of FHP-2 is 17.2% (see Section 3.1 for definition).

3.3.1 The microdynamical equation

As for FHP-1, it is possible to derive the microdynamical equation of FHP-2 directly from the general form (2.13). The same result can be obtained more rapidly by the intuitive method which consists of writing, for each n_i, the balance for positive and negative contributions from each type of collision. The resulting microdynamical equation for the model FHP-2 has the same structure as for

FHP-1, except that there is an additional equation for n_0, and a few more terms on the right hand side. The result reads:

$$
\begin{array}{rcll}
\text{for } i = 1, \ldots, 6: & & & \\
n_i(\mathbf{r}_\star + \mathbf{c}_i, t_\star + 1) & = & & n_i \\
& - & & n_i\bar{n}_{i+1}\bar{n}_{i+2}n_{i+3}\bar{n}_{i+4}\bar{n}_{i+5} \\
& + & \xi_{(+)} & \bar{n}_i n_{i+1}\bar{n}_{i+2}\bar{n}_{i+3}n_{i+4}\bar{n}_{i+5} \\
& + & \xi_{(-)} & \bar{n}_i\bar{n}_{i+1}n_{i+2}\bar{n}_{i+3}\bar{n}_{i+4}n_{i+5} \\
& - & & n_i\bar{n}_{i+1}n_{i+2}\bar{n}_{i+3}n_{i+4}\bar{n}_{i+5} \\
& + & & \bar{n}_i n_{i+1}\bar{n}_{i+2}n_{i+3}\bar{n}_{i+4}n_{i+5} \\
& - & & n_i\bar{n}_{i+1}\bar{n}_{i+2}\bar{n}_{i+3}\bar{n}_{i+4}\bar{n}_{i+5}n_0 \\
& + & & \bar{n}_i n_{i+1}\bar{n}_{i+2}\bar{n}_{i+3}\bar{n}_{i+4}\bar{n}_{i+5}n_0 \\
& + & & \bar{n}_i\bar{n}_{i+1}\bar{n}_{i+2}\bar{n}_{i+3}\bar{n}_{i+4}n_{i+5}n_0 \\
& - & & n_i\bar{n}_{i+1}n_{i+2}\bar{n}_{i+3}\bar{n}_{i+4}\bar{n}_{i+5}\bar{n}_0 \\
& - & & n_i\bar{n}_{i+1}\bar{n}_{i+2}\bar{n}_{i+3}n_{i+4}\bar{n}_{i+5}\bar{n}_0 \\
& + & & \bar{n}_i n_{i+1}\bar{n}_{i+2}\bar{n}_{i+3}\bar{n}_{i+4}n_{i+5}\bar{n}_0
\end{array}
\tag{3.8}
$$

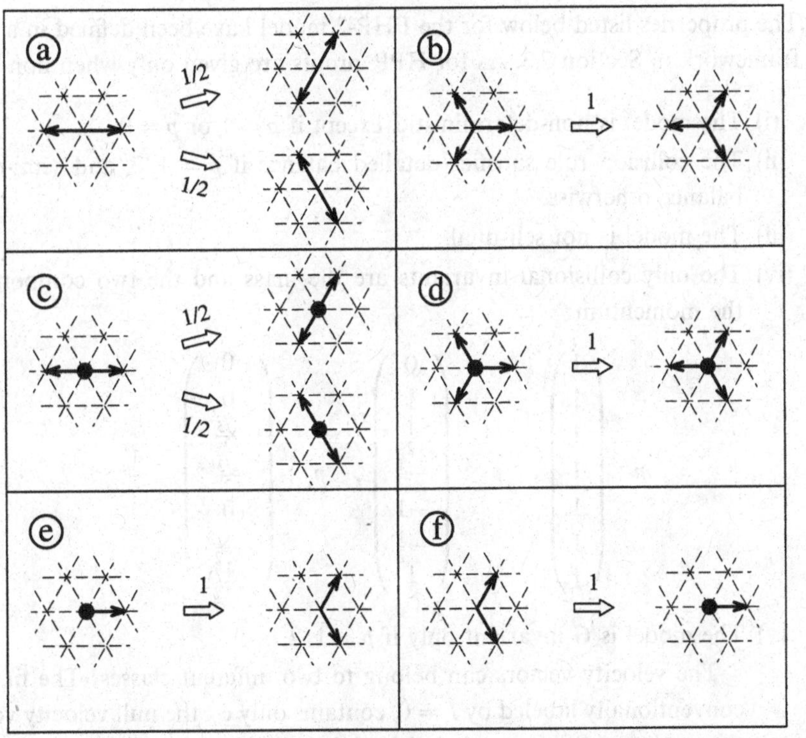

Figure 3.5 Collision rules for the FHP-2 model, reduced by symmetry. The full set of collisions can be reconstructed from the above reduced set by the application of all the isometries of $\mathcal{G}$. Black dots represent rest particles.

for $i = 0$:

$$\begin{array}{rcl} n_0(\mathbf{r}_\star, t_\star + 1) & = & n_0 \\ & - & \sum_{j=1}^{6} \; n_j \overline{n}_{j+1} \overline{n}_{j+2} \overline{n}_{j+3} \overline{n}_{j+4} \overline{n}_{j+5} n_0 \\ & + & \sum_{j=1}^{6} \; n_j \overline{n}_{j+1} n_{j+2} \overline{n}_{j+3} \overline{n}_{j+4} \overline{n}_{j+5} \overline{n}_0 \end{array} \qquad (3.9)$$

Here, the index i is cyclic (modulo 6), in such a way that index $i = 7$ is equivalent to index 1 (and not 0); other notations are as for FHP-1 (see Section 3.2.1).

Note that in (3.8) five terms are independent of n_0, which reflects that the binary and triple collisions of FHP-1 occur indifferently with or without a passive rest particle. The strict application of (2.13) would have required two pairs of random Bernoulli variables $(\xi^{(1)}_{(+)}, \xi^{(1)}_{(-)})$ and $(\xi^{(2)}_{(+)}, \xi^{(2)}_{(-)})$, for binary head-on collisions with and without passive rest particle respectively. In fact, since these two pairs of random variables have the same statistics, they are taken equal: this leads to the above form for the microdynamical equation of FHP-2.

3.3.2 Microscopic properties

The properties listed below for the FHP-2 model have been defined in a general framework in Section 2.3. As for HPP, proofs are given only when non-trivial.

(i) The model is non-deterministic, except if $p = 1$ or $p = 0$.

(ii) The collision rule satisfies detailed balance if $p = 1/2$, and semi-detailed balance otherwise.

(iii) The model is not self-dual.

(iv) The only collisional invariants are the mass and the two components of the momentum:

$$m : \begin{pmatrix} 1 \\ 1 \\ 1 \\ 1 \\ 1 \\ 1 \\ 1 \end{pmatrix}, \quad p_x : \begin{pmatrix} 0 \\ 1 \\ \frac{1}{2} \\ -\frac{1}{2} \\ -1 \\ -\frac{1}{2} \\ \frac{1}{2} \end{pmatrix}, \quad p_y : \begin{pmatrix} 0 \\ 0 \\ \frac{\sqrt{3}}{2} \\ \frac{\sqrt{3}}{2} \\ 0 \\ -\frac{\sqrt{3}}{2} \\ -\frac{\sqrt{3}}{2} \end{pmatrix}. \qquad (3.10)$$

(v) The model is G-invariant only if $p = 1/2$.

The velocity vectors can belong to two different classes: The first class, conventionally labeled by $I = 0$, contains only $\mathbf{c}_0$, the null velocity vector of the rest particle; the second class, labeled by $I = 1$, contains the non-zero velocity vectors $\mathbf{c}_1$ to $\mathbf{c}_6$.

(vi) The model has fifth order crystallographic isotropy, for the same reason as FHP-1, since the rest particles, which have zero velocity, do not modify the crystallographic tensors of any order.

(vii) The model is irreducible. Indeed, the only isometry in G that preserves $\mathbf{c}_1$, for example, is the mirror symmetry with respect to the x-direction, whose invariant subspace only contains vectors proportional to $\mathbf{c}_1$. The same argument holds for $\mathbf{c}_2$ to $\mathbf{c}_6$. For $\mathbf{c}_0$ which belongs to Class $I = 0$ however, the reasoning is as simple, albeit different: all the isometries in G leave $\mathbf{c}_0$ unchanged. The only vector left unchanged by *all* the isometries in the point group G of the triangular lattice is of course the null vector, which is $\mathbf{c}_0$.

As for FHP-1, the degree of crystallographic isotropy of FHP-2 is compatible with the possibility of large-scale behavior governed (within certain limits) by the Navier–Stokes equation, provided that the model be G-invariant (that is, for $p = 1/2$).

FHP-1 and FHP-2 have the same microscopic properties, and the proofs are identical. The essential difference lies in the number of possible effective collisions which is larger for FHP-2 (22 versus 5), and yields a higher collisional efficiency (17.2% versus 7.81%). This difference in collisional efficiency has a major consequence at the macroscopic level: The kinematic shear viscosity is smaller for FHP-2 than for FHP-1. This decrease in shear viscosity allows numerical simulations at higher Reynolds number (for equal numerical investment). Another difference between FHP-1 and FHP-2 is that the kinematic bulk viscosity is zero for FHP-1 and non-zero for FHP-2.

The kinetic energy observable in FHP-2 is *not* proportional to the mass observable because the velocity of rest particles is zero. The kinetic energy is not a collisional invariant for FHP-2, since binary collisions between a rest particle and a moving particle create some kinetic energy. Moreover, the only collisions that conserve the total mass, momentum, and kinetic energy are those which also conserve the number of rest particles, that is, those for which the rest particles are only passive spectators. Recovering the kinetic energy conservation by allowing only this type of collisions would cancel the benefit of rest particles: they would remain static for ever, being affected neither by the propagation, nor by the collision phase. To recover energy conservation without inhibiting the dynamics of rest particles, we need to assign to the particles some kind of 'internal degree of freedom' which is never excited (zero energy) for moving particles, and always excited for rest particles. The energy associated with the excitation of the internal degree of freedom of rest particles is chosen to be equal to the kinetic energy of a moving particle (1/2 in natural lattice units). So, the total energy observable (kinetic plus internal) is proportional to the mass observable and is thus conserved. It is precisely because the FHP-2 model allows for collisional transfer of internal energy to kinetic energy that bulk viscosity exists in the FHP-2 gas. However, the energy conservation does not lead to an additional relevant macroscopic equation, since mass and total energy are simply proportional, and there can be no effective collisional transfer of kinetic energy. Consequently there are no thermal effects in the FHP-2 lattice gas.

Triple collisions are not mandatory here, as they are for FHP-1 where they remove the spurious collisional invariant. Indeed, the two-particle collisions involving a rest particle are sufficient to eliminate the spurious invariance.[4]

3.4 The FHP-3 model

A further variant of the 7-bit FHP-2 model is FHP-3 where the collision rules are designed to include as many collisions as possible, under the constraint of having the same collisional invariants as with FHP-2. Figure 3.6 shows the collision rules of FHP-3. As for Figures 3.2, 3.4 and 3.5, the lattice-preserving isometries are used to reduce the number of collisions shown. In addition to the isometries, the self-duality property, that holds for FHP-3 but not for FHP-1 and FHP-2, has also been used to further reduce the number of configurations in the figure. A complete collision table could be reconstructed from Figure 3.6, by applying duality combined with the isometries of G. When this reconstruction is performed, it can be shown that 76 among the $2^7 = 128$ possible states can undergo effective collisions. The collisional efficiency of FHP-3 is 59.4 % (see Section 3.1 for definition).

3.4.1 The microdynamical equation

It is possible to write a microdynamical equation for FHP-3, but the operation is tedious and useless. One would obtain an equation with the same form as for FHP-2, but with many more terms to describe all possible collisions.

3.4.2 Microscopic properties

The properties listed below for the FHP-3 model have been defined in a general framework in Section 2.3. Proofs are omitted since they are either trivial or identical for all FHP models.

(i) The model is non-deterministic, except if $p = 1$ or $p = 0$.
(ii) The collision rule satisfies detailed balance if $p = 1/2$ and semi-detailed balance otherwise.
(iii) The model is self-dual, because the set of possible collisions is complete. This is the major difference with FHP-1 and FHP-2, at least at the microscopic level.

[4] For a detailed discussion of the collisional and geometrical invariants of the FHP models, see d'Humières, Qian and Lallemand (1989) and (1990).

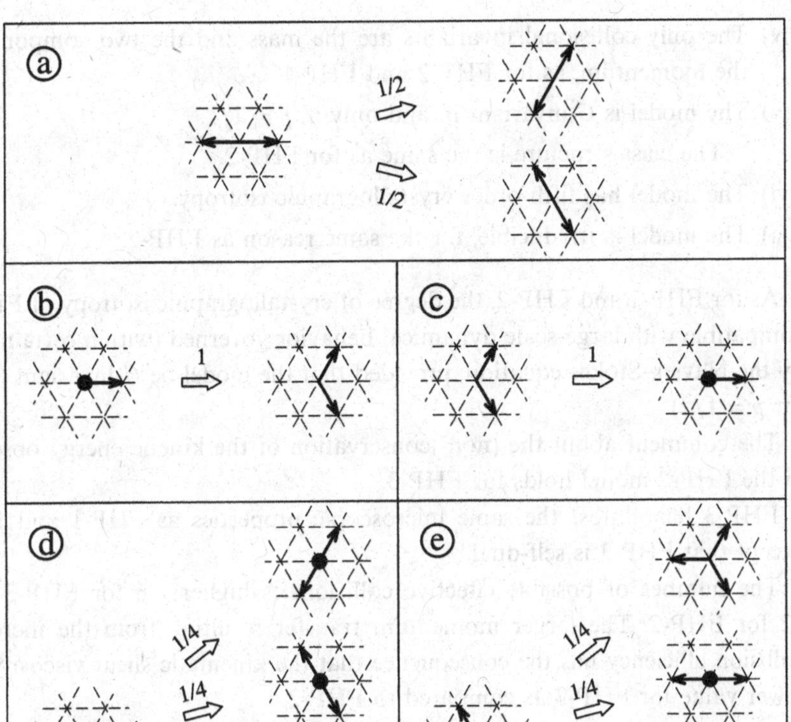

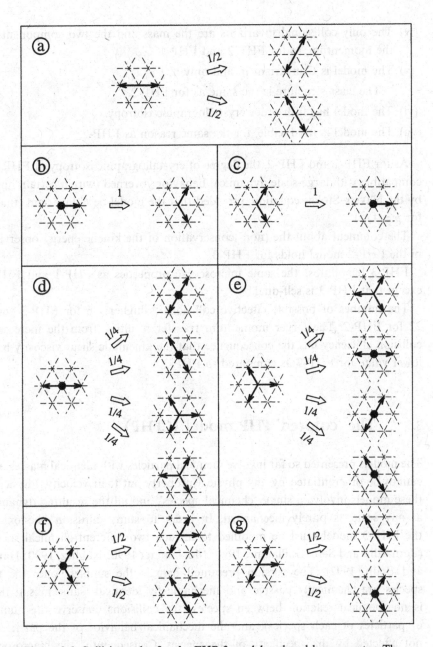

Figure 3.6 Collision rules for the FHP-3 model, reduced by symmetry. The full set of collisions can be reconstructed from the above reduced set by the application of all the isometries of $\mathcal{G}$, combined, when necessary, with the duality operation (exchanges between particles and holes). Black dots represent rest particles.

(iv) The only collisional invariants are the mass and the two components of the momentum, as for FHP-2 and FHP-1.

(v) The model is G-invariant if, and only if, $p = 1/2$.

The class structure is the same as for FHP-2.

(vi) The model has fifth order crystallographic isotropy.

(vii) The model is irreducible, for the same reason as FHP-2.

As for FHP-1 and FHP-2, the degree of crystallographic isotropy of FHP-3 is compatible with large-scale dynamical behavior governed (within certain limits) by the Navier–Stokes equation, provided that the model be G-invariant (that is, for $p = 1/2$).

The comment about the (non-)conservation of the kinetic energy observable in the FHP-2 model holds for FHP-3.

FHP-3 has almost the same microscopic properties as FHP-1 and FHP-2, except that FHP-3 is self-dual.

The number of possible effective collisions is higher: 76 for FHP-3 versus 22 for FHP-2. The larger momentum transfer resulting from the increase in collision efficiency has the consequence that the kinematic shear viscosity has a lower value for FHP-3 as compared to FHP-2.

3.5 The 'colored' FHP model (CFHP)

The models presented so far involve identical particles with identical masses, that cannot be differentiated by any physical property but their velocity. Physically, these models involve a single 'chemical species', and all the resulting dynamics, at any scale, is purely mechanical. In order to study diffusion phenomena, the FHP-3 model must be modified to include two different chemical species (Bernardin and Sero-Guillaume, 1990; McNamara, 1990; Noullez, 1990; Hanon and Boon, 1997). The version presented here is the simplest one: the two species are chemically passive and mechanically identical (same mass); there is no chemical reaction between species (the collisions conserve the number of particles of each species), and the mechanical behavior of the particles is not affected by the specificity of the species. Consequently, the macroscopic motions of the mixture are identical to those of a single component fluid such as FHP-3, and each species acts as a passively diffusing scalar with respect to the mixture. To get a better picture, we can associate colors (say red and blue) to the two species, and consider the CFHP ('colored' FHP) model as an FHP-3 model whose particles are painted red or blue, without modifying their motions and collisions. Indeed, the FHP-3 collision rules apply to the particles arriving at a node, independently of their color. Then, the color attribute is attached randomly to the outgoing particles, with the constraint to conserve the number of red and blue particles.

There exist several variations on the theme of colored lattice gases, e.g. incorporation of reactive processes (see Boon *et al.*, 1996) or surface tension effects (see Rothmann and Zaleski, 1994). They will not be considered here (see Chapter 11).

The encoding of the CFHP model raises a little problem. Indeed, each channel can be *a priori* in three different states: empty, occupied by a red particle, or occupied by a blue particle. This situation does not seem compatible with a binary encoding. One way to recover the Boolean formalism is to consider that each of the seven channels of the FHP-3 model is now a double channel: one 'sub-channel' for the red particles, and one 'sub-channel' for the blue ones. Fourteen Boolean quantities per lattice node are now required, as illustrated in Figure 3.7. The (sub-)channels 0 to 6 encode the presence of red particles, and the (sub-)channels 7 to 13, the presence of blue particles. The microdynamical equation of the CFHP model is designed to avoid the 'forbidden' situations where both red and blue sub-channels are occupied simultaneously. Of course, initially, the lattice configuration has to be preset so that each pair of sub-channels contains at most one particle (either red or blue). This rule expresses an exclusion principle per pair of sub-channels.

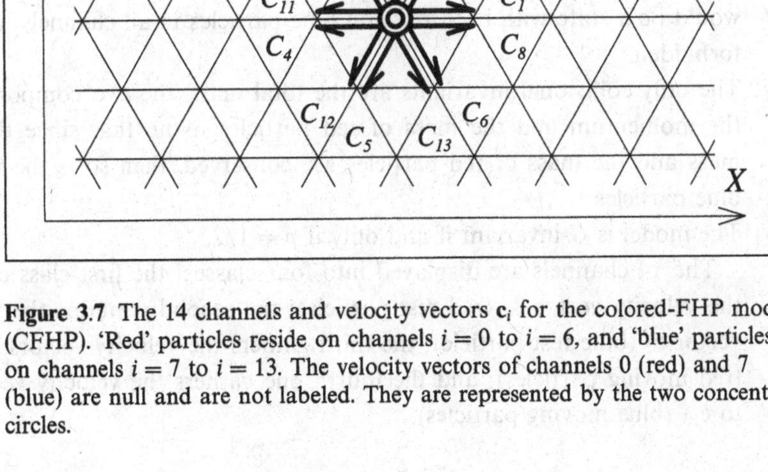

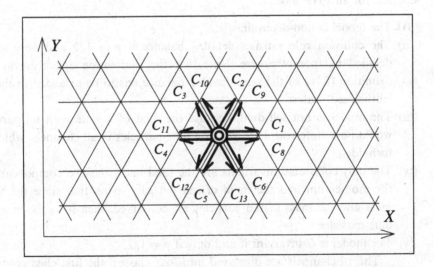

Figure 3.7 The 14 channels and velocity vectors $\mathbf{c}_i$ for the colored-FHP model (CFHP). 'Red' particles reside on channels $i = 0$ to $i = 6$, and 'blue' particles on channels $i = 7$ to $i = 13$. The velocity vectors of channels 0 (red) and 7 (blue) are null and are not labeled. They are represented by the two concentric circles.

3.5.1 The microdynamical equation

Rather than writing explicitly a voluminous microdynamical equation for CFHP, we describe its evolution as the following sequence of operations:

(i) At each node, the number of red particles is computed.
(ii) At each node and for each of the seven pairs of subchannels, the 'OR' operator is applied so as to obtain seven Boolean quantities.
(iii) The collision operator of the FHP-3 model is applied to these seven Boolean quantities.
(iv) The color attribute is then randomly redistributed, while conserving the number of red particles computed at the first step.
(v) The usual propagation operator is applied.

This evolution rule clearly guarantees that if the initial state does not have more that one particle per pair of subchannels (red or blue, but not both), then so will the final state. From now on, we will assume this is the case. Thus, the single-node phase space γ is restricted to the 3^7 'authorized' states, among the 2^{14} *a priori* possible states of a model with $b = 14$.

3.5.2 Microscopic properties

The properties listed below for the CFHP model have been defined in a general framework in Section 2.3. Proofs are omitted since they are either trivial or identical for all FHP models.

(i) The model is non-deterministic.
(ii) The collision rule satisfies detailed balance if $p = 1/2$ and only semi-detailed balance otherwise. Note that the summation appearing in the definition (2.16) of the semi-detailed balance has to be extended to the 3^7 'authorized' states.
(iii) The model is not self-dual, because the dual of a state with no particle would be a state with both red and blue particles in all channels, which is forbidden.
(iv) The only collisional invariants are the total mass, the two components of the momentum and the mass of red particles. Note that since the total mass and the mass of red particles are conserved, then so is the mass of blue particles.
(v) The model is G-invariant if and only if $p = 1/2$.

The 14 channels are displayed into four classes: the first class contains the velocity vector $\mathbf{c}_0$ (red rest particle); the second contains the velocity vector $\mathbf{c}_7$ (blue rest particle); the third gathers the velocity vectors $\mathbf{c}_1$ to $\mathbf{c}_6$ (red moving particles); and the fourth one gathers the velocity vectors $\mathbf{c}_8$ to $\mathbf{c}_{13}$ (blue moving particles).

(vi) The model has fifth order crystallographic isotropy, as FHP-3 and for the same reasons.

(vii) The model is irreducible, for the same reason as FHP-3.

3.6 The GBL model

The two-dimensional HPP and FHP models do not contain thermal effects, because there is no energy conservation independently of mass conservation (the energy observable is proportional to the mass observable). So these models are restricted to the description of non-thermal fluids.

We now introduce a lattice gas model with non-trivial thermodynamics: the '19-bit model', constructed by Grosfils, Boon and Lallemand (1992); for consistency in acronymic notation, we shall refer to this model as the 'GBL model'. The main feature of the model is that it is a multi-speed lattice gas residing on the two-dimensional triangular lattice. Figure 3.8 shows how the velocity vectors connect any lattice node to (i) itself (one rest particle per node; $\|\mathbf{c}_0\| = 0$), (ii) its six nearest neighbors ($\|\mathbf{c}_i\| = 1$, $i = 1,\ldots,6$), (iii) its six next-nearest neighbors ($\|\mathbf{c}_i\| = \sqrt{3}$, $i = 13,\ldots,18$), (iv) its six next-to-next-nearest neighbors ($\|\mathbf{c}_i\| = 2$, $i = 7,\ldots,12$).

The number of channels b is equal to 19, whereas the connection number B is 18, since there is one rest particle per node. Both numbers are independent of the coordination number of the triangular lattice (that is, 6).

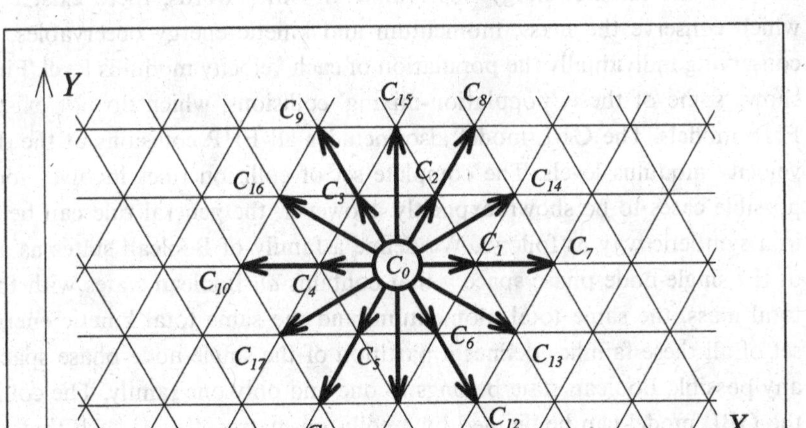

Figure 3.8 The GBL lattice with the corresponding $\mathbf{c}_i$s.

The $\mathbf{c}_i$s have moduli 0, 1, $\sqrt{3}$ and 2 and their Cartesian components are:

$$\mathbf{c}_0 = \begin{pmatrix} 0 \\ 0 \end{pmatrix},$$

$$\begin{aligned}
&\mathbf{c}_1 = \begin{pmatrix} 1 \\ 0 \end{pmatrix}, \quad \mathbf{c}_2 = \begin{pmatrix} \frac{1}{2} \\ \frac{\sqrt{3}}{2} \end{pmatrix}, \quad \mathbf{c}_3 = \begin{pmatrix} -\frac{1}{2} \\ \frac{\sqrt{3}}{2} \end{pmatrix}, \\
&\mathbf{c}_4 = \begin{pmatrix} -1 \\ 0 \end{pmatrix}, \quad \mathbf{c}_5 = \begin{pmatrix} -\frac{1}{2} \\ -\frac{\sqrt{3}}{2} \end{pmatrix}, \quad \mathbf{c}_6 = \begin{pmatrix} \frac{1}{2} \\ -\frac{\sqrt{3}}{2} \end{pmatrix}, \\
&\mathbf{c}_7 = \begin{pmatrix} 2 \\ 0 \end{pmatrix}, \quad \mathbf{c}_8 = \begin{pmatrix} 1 \\ \sqrt{3} \end{pmatrix}, \quad \mathbf{c}_9 = \begin{pmatrix} -1 \\ \sqrt{3} \end{pmatrix}, \\
&\mathbf{c}_{10} = \begin{pmatrix} -2 \\ 0 \end{pmatrix}, \quad \mathbf{c}_{11} = \begin{pmatrix} -1 \\ -\sqrt{3} \end{pmatrix}, \quad \mathbf{c}_{12} = \begin{pmatrix} 1 \\ -\sqrt{3} \end{pmatrix}, \\
&\mathbf{c}_{13} = \begin{pmatrix} \frac{3}{2} \\ -\frac{\sqrt{3}}{2} \end{pmatrix}, \quad \mathbf{c}_{14} = \begin{pmatrix} \frac{3}{2} \\ \frac{\sqrt{3}}{2} \end{pmatrix}, \quad \mathbf{c}_{15} = \begin{pmatrix} 0 \\ \sqrt{3} \end{pmatrix}, \\
&\mathbf{c}_{16} = \begin{pmatrix} -\frac{3}{2} \\ \frac{\sqrt{3}}{2} \end{pmatrix}, \quad \mathbf{c}_{17} = \begin{pmatrix} -\frac{3}{2} \\ -\frac{\sqrt{3}}{2} \end{pmatrix}, \quad \mathbf{c}_{18} = \begin{pmatrix} 0 \\ -\sqrt{3} \end{pmatrix}.
\end{aligned} \tag{3.11}$$

The masses m_i of the particles are equal to unity; their momenta are equal to the $\mathbf{c}_i$s, and their kinetic energy is equal to $\frac{1}{2}\|\mathbf{c}_i\|^2$. It is clear that the kinetic energy observable $(0, \frac{1}{2}, \ldots, \frac{1}{2}, 2, \ldots, 2, \frac{3}{2}, \ldots, \frac{3}{2})$ is *not* proportional to the mass observable $(1, \ldots, 1)$. In contrast to FHP-2 and FHP-3, rest particles carry no internal energy.[5] Collisions can couple the different velocity modulus levels, and conserve the kinetic energy observable. In other words, there exist collisions which conserve the mass, momentum and kinetic energy observables, *without* conserving individually the population of each velocity modulus level. Figure 3.9 shows some of these 'population-mixing' collisions, which do not exist in the FHP models. The GBL model also includes all FHP collisions at the different velocity modulus levels. The complete set of collision rules involves too many possible cases to be shown explicitly. However, the general rule can be defined in a synthetic way as follows. We define a family of Boolean states as a subset of the single-node phase space γ that contains *all* Boolean states with the same total mass, the same total momentum, and the same total kinetic energy. The set of all these families defines a partition of the single-node phase space, since any possible Boolean state belongs to one and only one family. The collision of the GBL model can be defined by a collision matrix $A(s \to s')$ such that:

- $A(s \to s') = 0$ if s and s' do not belong to the same family.
- $A(s \to s') = \left(card(\mathcal{F})\right)^{-1}$ if both s and s' belong to the same family $\mathcal{F}$ containing $card(\mathcal{F})$ elements.

[5] Similarly as the particles of the hard sphere gas of classical statistical mechanics.

An exhaustive count shows that 517 750 among the $2^{19} = 524\,288$ possible states can undergo effective collisions. The collisional efficiency of GBL is 94.4% (which is different from 517 750/524 288 because ineffective collisions can occur with probabilities different from 0 or 1). Note that an increase in collisional efficiency (raising its value up to 98.9%) can be gained by imposing that ineffective collisions are only admitted for Boolean states belonging to families with only one element. The probability for an effective collision $s \to s'$ is then $A(s{\to}s') = (card(\mathscr{F})-1)^{-1}$, if both s and s' belong to a family with more than one element.

3.6.1 The microdynamical equation

It would be a formidable task to write down explicitly the microdynamic equations for the GBL model, under its extensive form (as for the FHP models).

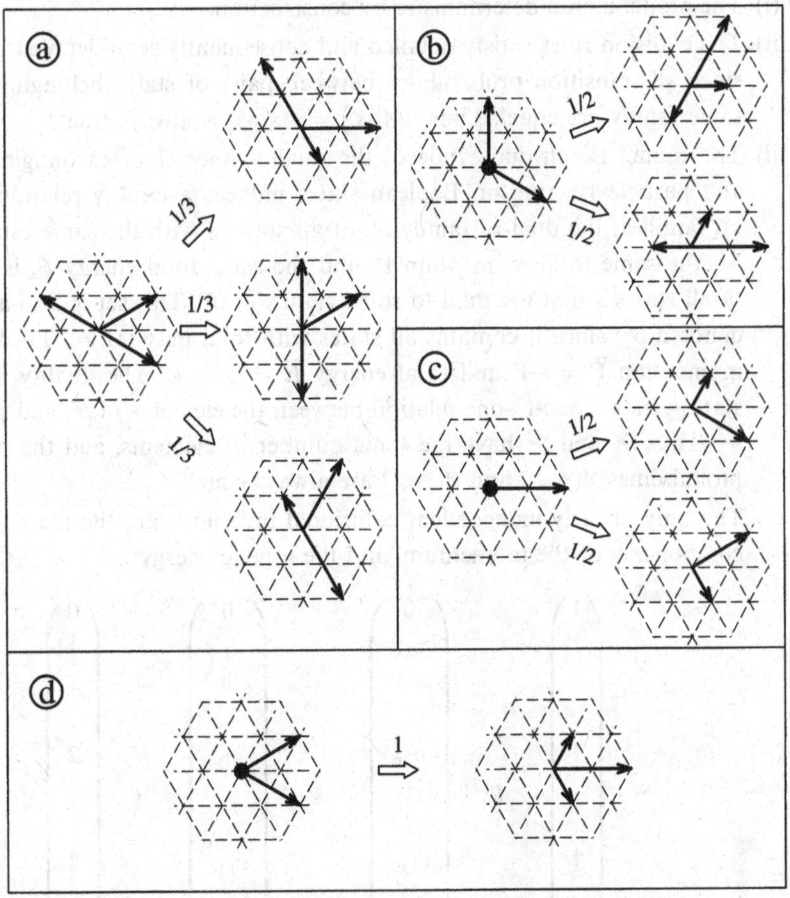

Figure 3.9 Examples of elementary 'population-mixing' collisions for the GBL model. Black dots represent rest particles.

Furthermore, the operation would be useless, since all macroscopically relevant quantities can be computed from expressions involving the collision matrix $A(s \to s')$, suitably summed. These sums can be computed by hand, or by a symbolic manipulator, even if the number of collisions is too large to allow the explicit writing of the microdynamical equations. The only useful microdynamical equation for the GBL model is the general form given in (2.12) or (2.13).

3.6.2 Microscopic properties

As for the FHP models, the properties listed below for the GBL model have been defined in a general framework in Section 2.3. Proofs are given only when non-trivial.

(i) The model is non-deterministic by construction.

(ii) The collision rules satisfy detailed and consequently semi-detailed balance, since all transition probabilities between pairs of states belonging to the same family are equal. Thus $A(s \to s') = A(s' \to s)$ is always true.

(iii) The model is self-dual. Indeed, the duality operation (exchanging holes and particles) acting on Boolean states induces a duality relation acting on families: the dual of family $\mathscr{F}$ of all states s with the same total mass M, the same total momentum $\mathbf{P}$, and the same total energy E, is the set of all states $\bar{s}$ that are dual to some state s of $\mathscr{F}$. This set is also a family denoted $\overline{\mathscr{F}}$, since it contains all states with total mass $\overline{M} = 19 - M$, total momentum $\overline{\mathbf{P}} = -\mathbf{P}$ and total energy $\overline{E} = 24 - E$. The duality between states yields a one-to-one relation between the elements of $\mathscr{F}$ and those of $\overline{\mathscr{F}}$. Thus $\mathscr{F}$ and $\overline{\mathscr{F}}$ have the same number of elements, and the collision probabilities $A(s \to s')$ and $A(\bar{s} \to \bar{s}')$ are always equal.

(iv) The only linearly independent collisional invariants are the mass, the two components of the momentum, and the kinetic energy:

$$m: \begin{pmatrix} 1 \\ 1 \\ \cdot \\ \cdot \\ 1 \\ \cdot \\ \cdot \\ 1 \\ \cdot \\ \cdot \\ 1 \end{pmatrix}, \quad p_x: \begin{pmatrix} 0 \\ c_{1x} \\ \cdot \\ \cdot \\ c_{7x} \\ \cdot \\ \cdot \\ c_{13x} \\ \cdot \\ \cdot \\ c_{18x} \end{pmatrix}, \quad p_y: \begin{pmatrix} 0 \\ c_{1y} \\ \cdot \\ \cdot \\ c_{7y} \\ \cdot \\ \cdot \\ c_{13y} \\ \cdot \\ \cdot \\ c_{18y} \end{pmatrix}, \quad e: \begin{pmatrix} 0 \\ \frac{1}{2} \\ \cdot \\ \cdot \\ 2 \\ \cdot \\ \cdot \\ \frac{3}{2} \\ \cdot \\ \cdot \\ \frac{3}{2} \end{pmatrix}. \tag{3.12}$$

(v) The model is G-invariant. The proof is quite similar to the proof of self-duality. Any isometry $\mathbf{g}$ of G induces a relation between families with the same number of elements. The G-invariance follows immediately.

The class structure within the $\mathbf{c}_i$ vectors is more complex than for FHP models. The class $I = 0$ only contains $\mathbf{c}_0$, the (null) velocity vector of the rest particle. The class $I = 1$ contains the velocity vectors $\mathbf{c}_1$ to $\mathbf{c}_6$, which have unit modulus. The class $I = 2$ contains vectors $\mathbf{c}_7$ to $\mathbf{c}_{12}$, with modulus equal to 2. The class $I = 3$ contains vectors $\mathbf{c}_{13}$ to $\mathbf{c}_{18}$, with modulus $\sqrt{3}$.

(vi) The model exhibits fifth order crystallographic isotropy, since its set of velocity vectors can be viewed as the superposition of a zero vector plus three groups of six vectors, each group being obtained from the six vectors of the FHP model by a similitude operation (rotation and rescaling). The degree of crystallographic isotropy is thus the same as for FHP models.

(vii) The GBL is irreducible. Indeed, the only isometry in G that preserves $\mathbf{c}_1$ is the mirror symmetry with respect to the x-direction, whose invariant subspace only contains vectors proportional to $\mathbf{c}_1$. The same argument holds for vectors $\mathbf{c}_2$ to $\mathbf{c}_{18}$. For $\mathbf{c}_0$, the same argument as for FHP-2 holds.

The degree of crystallographic isotropy and the non-trivial energy conservation property renders the GBL model suitable for large-scale thermo-hydrodynamic simulations of two-dimensional fluid dynamics.

3.7 Three-dimensional models

The models described so far are all constructed on two-dimensional lattices. Thus they are appropriate for the investigation of phenomena which are relevant in two dimensions. Obviously, many phenomena are not amenable to a two-dimensional description (e.g. the onset of three-dimensional turbulent flow), and require the introduction of lattice gas models with higher dimensionality. In two dimensions, models based on the triangular lattice with hexagonal symmetry satisfy the fourth order crystallographic isotropy, but we noticed that the HPP (square) model does not, and thus produces anisotropic large-scale dynamics. In three dimensions, the problem of isotropy is more complicated: none of the fourteen Bravais lattices can sustain a lattice gas model that satisfies the fourth order crystallographic isotropy, at least within the scope of minimal models, that is, with the smallest number of channels per node compatible with the lattice local symmetries (see Section 1.3.1). Indeed, minimal models based on the three cubic Bravais lattices (simple, body-centered, and face-centered) do not satisfy the crystallographic isotropy beyond the third order. Moreover, minimal models based on the two tetragonal lattices (simple and body-centered), the four orthorhombic lattices (simple, base-centered, body-centered, and face-centered), the two monoclinic lattices (simple and body-centered), the triclinic lattice and the trigonal lattice, do not even satisfy the second order crystallographic isotropy. The three-dimensional hexagonal Bravais lattice can serve as a basis for minimal models which satisfy the second order crystallographic isotropy only if its aspect ratio is equal to $\sqrt{3/2}$, but even in this case, the fourth order condition is not satisfied.

So, strategies must be developed to overcome this difficulty, and three-dimensional lattice gas models can indeed be constructed with fully isotropic large-scale dynamics. These strategies are discussed in the next three sections.

3.7.1 Models with multiple links

A first solution to the anisotropy problem is to broaden the scope of the search by including models that are not minimal, such as models where propagation connects nodes to nearest and next-nearest neighbors with different 'weights' on the different connecting links. These weights can be set by using *multiple-link* models, that is, models with $b > B$ where two or more distinguishable particles can have the same velocity vector,[6] so that some velocity vectors are counted twice or more. The price to pay for this recovered isotropy is an increase in the number of channels per node, leading to more complex collision rules. With this idea, one can construct several models based on the three-dimensional cubic lattices (simple, body-centered and face-centered), with propagation allowed to nearest and next-nearest neighbors via multiple links with multiplicities tuned to obtain fourth order crystallographic isotropy.

To clarify this type of solution, we describe in detail its application to the body-centered cubic lattice with unit lattice constant. We build the set of connection vectors with the vectors connecting one node to its nearest neighbors: $(\pm\frac{1}{2}, \pm\frac{1}{2}, \pm\frac{1}{2})$ *and* to its next-nearest neighbors: $(\pm 1, 0, 0)$, $(0, \pm 1, 0)$ and $(0, 0, \pm 1)$. The connection number B is thus equal to 14 (8 vectors with modulus $\frac{\sqrt{3}}{2}$, and 6 with modulus 1). We now consider that the links to nearest neighbors and to next-nearest neighbors have multiplicity ζ_1 and ζ_2 respectively (ζ_1 and ζ_2 are positive integers). In other words, each of the 8 connection vectors with modulus $\frac{\sqrt{3}}{2}$ corresponds to ζ_1 different channels, and each of the 6 connection vectors with modulus 1 corresponds to ζ_2 different channels. The second and fourth order crystallographic tensors, fully contracted with an arbitrary vector $\mathbf{x} = (x_1, x_2, x_3)$ yield respectively:

$$\sum_{i=1}^{b} (\mathbf{c}_i \cdot \mathbf{x})^2 = (2\zeta_1 + 2\zeta_2)(x_1^2 + x_2^2 + x_3^2), \tag{3.13}$$

[6] This is not in conflict with the exclusion principle, since particles with equal velocities are assumed to be distinguishable.

and

$$\sum_{i=1}^{b}(\mathbf{c}_i \cdot \mathbf{x})^4 = \left(\frac{\zeta_1}{2} + 2\zeta_2\right)(x_1^4 + x_2^4 + x_3^4) + \frac{3\zeta_1}{2}(2x_1^2x_2^2 + 2x_2^2x_3^2 + 2x_3^2x_1^2). \tag{3.14}$$

Clearly, the right hand side of (3.13) is proportional to $\|\mathbf{x}\|^2$ for any choice of ζ_1 and ζ_2, but the right hand side of (3.14) is proportional to $\|\mathbf{x}\|^4$ only if $\frac{1}{2}\zeta_1 + 2\zeta_2 = \frac{3}{2}\zeta_1$, that is, for $\zeta_1 = 2\zeta_2$. The easiest way to satisfy this condition is to choose $\zeta_1 = 2$ and $\zeta_2 = 1$, which leads to $b = 8\zeta_1 + 6\zeta_2 = 22$ channels per node.

The same idea can be applied to the simple cubic lattice, with propagation to nearest and next-nearest neighbors with suitable weights. It yields a model with $B = 18$ connections (6 to nearest neighbors and 12 to next-nearest ones) and $b = 24$ channels per node (each of the six links to nearest neighbors is a double channel).

3.7.2 Models with biased collisions

D. Levermore (personal communication, 1986) suggested that large scale isotropy can be obtained without multi-channel connections (and thereby less prohibitively since then $b = B$) by introducing a bias in the collision rules such that the statistical weight (mean population) of the channels corresponding to velocity vectors pointing to nearest neighbors be twice the statistical weight of the channels corresponding to velocity vectors pointing to next-nearest neighbors. Strictly, such non-homokinetic models do *not* satisfy the fourth order crystallographic isotropy, but only a 'statistically weighted fourth order crystallographic isotropy', which can be sufficient for fully isotropic large-scale dynamics.

3.7.3 The FCHC models

The most efficient and most used three-dimensional lattice gas models with correct isotropy follow from a detour into four-dimensional space. d'Humières, Lallemand and Frisch (1986) noticed that there exists at least one Bravais lattice in four dimensions that has the proper isotropy properties. This Bravais lattice known as the *F4* lattice in field theory, and is frequently called the 'face-centered-hyper-cubic' (FCHC) lattice. It can be defined as the set of all points of $\mathbb{R}^4$ with signed integer Cartesian coordinates with even sum. The name FCHC follows from the analogy with the three-dimensional face-centered cubic lattice which is defined as the subset of points of $\mathbb{R}^3$ with signed integer Cartesian coordinates with even sum.[7]

The FCHC lattice can be generated by the four vectors $(1,1,0,0)$, $(1,-1,0,0)$, $(0,0,1,1)$ and $(0,0,1,-1)$. The point symmetry group of the FCHC lattice contains 1152 isometries which can be generated by combinations of:

(i) The four symmetries S_1, S_2, S_3, and S_4 with respect to the four hyper-planes $x_1 = 0$, $x_2 = 0$, $x_3 = 0$, and $x_4 = 0$ (coordinate reversals).
(ii) The six symmetries P_{12}, P_{13}, P_{14}, P_{23}, P_{24}, and P_{34} with respect to the hyper-planes $x_1 - x_2 = 0$, $x_1 - x_3 = 0$, $x_1 - x_4 = 0$, $x_2 - x_3 = 0$, $x_2 - x_4 = 0$, and $x_3 - x_4 = 0$ (coordinate exchanges).
(iii) The symmetries Σ_1 and Σ_2 with respect to the hyper-planes $x_1+x_2-x_3-x_4 = 0$, and $x_1 - x_2 - x_3 - x_4 = 0$ respectively.

This set of symmetries generates the point group G of the FCHC lattice, but it is not a minimal set since only five of them, for example $(S_1, P_{12}, P_{13}, P_{14}, \Sigma_1)$ are sufficient to generate G. A complete discussion on the symmetry group of the FCHC lattice is given by Hénon (1987a,b).

The minimal set of velocity vectors that can be built on the four generating vectors is:

$$\mathbf{c}_{1,\dots,4} = \begin{pmatrix} \pm 1 \\ \pm 1 \\ 0 \\ 0 \end{pmatrix}, \quad \mathbf{c}_{5,\dots,8} = \begin{pmatrix} \pm 1 \\ 0 \\ \pm 1 \\ 0 \end{pmatrix}, \quad \mathbf{c}_{9,\dots,12} = \begin{pmatrix} \pm 1 \\ 0 \\ 0 \\ \pm 1 \end{pmatrix},$$
$$\mathbf{c}_{13,\dots,16} = \begin{pmatrix} 0 \\ \pm 1 \\ \pm 1 \\ 0 \end{pmatrix}, \quad \mathbf{c}_{17,\dots,20} = \begin{pmatrix} 0 \\ \pm 1 \\ 0 \\ \pm 1 \end{pmatrix}, \quad \mathbf{c}_{21,\dots,24} = \begin{pmatrix} 0 \\ 0 \\ \pm 1 \\ \pm 1 \end{pmatrix}. \tag{3.15}$$

Here, the numbering of the $\mathbf{c}_i$s is arbitrarily based on the lexicographic order of the alphabet $(+1,-1,0)$. Note that for operational convenience in numerical simulations, another order may be chosen in such a way that $\mathbf{c}_{i+12} = -\mathbf{c}_i$ for $i = 1,\dots,12$. The reason for this choice is related to technical details concerning the practical implementation of bounce-back reflections on solid obstacles (see Chapter 10).

A three-dimensional lattice gas model can be constructed from the four-dimensional FCHC lattice by the following dimension-reducing procedure:

(i) We consider the portion of the four-dimensional lattice limited by the two hyper-planes $x_4 = +1$ and $x_4 = -1$. This (hyper-)slice of lattice corresponds

[7] The acronym FCHC is slightly ambiguous; in the Appendix, Section A.2, we show that the lattice should be named 'two-dimensional face-centered body-centered hyper-cubic' lattice. For simplicity and convenience, we nevertheless use FCHC as commonly accepted.

to one lattice period in the fourth direction, and tiles exactly the FCHC lattice by successive translations of vector (0, 0, 0, 2).

(ii) We impose periodic conditions in the fourth direction; in other words, we cancel the fourth component of the $\mathbf{c}_i$s and project all the nodes in this portion onto the hyper-plane $x_4 = 0$.

Figure 3.10 illustrates this procedure for a simple case: the construction of a one-dimensional lattice from a two-dimensional square lattice.

Figure 3.11 shows how the three-dimensional (pseudo-four-dimensional) model so constructed is merely the three-dimensional cubic lattice, with double channels connecting nearest neighbors and single channels connecting next-nearest neighbors. There are still $b = 24$ channels per node in three dimensions, but six of them are double channels ($B = 18$).

The three-dimensional model constructed by projection of a four-dimensional lattice can also be viewed as a multiple-link model based on the simple cubic lattice with propagation to nearest and next-nearest neighbors with multiplicity 2 and 1 respectively. While it is easier to describe this model as a multiple-link

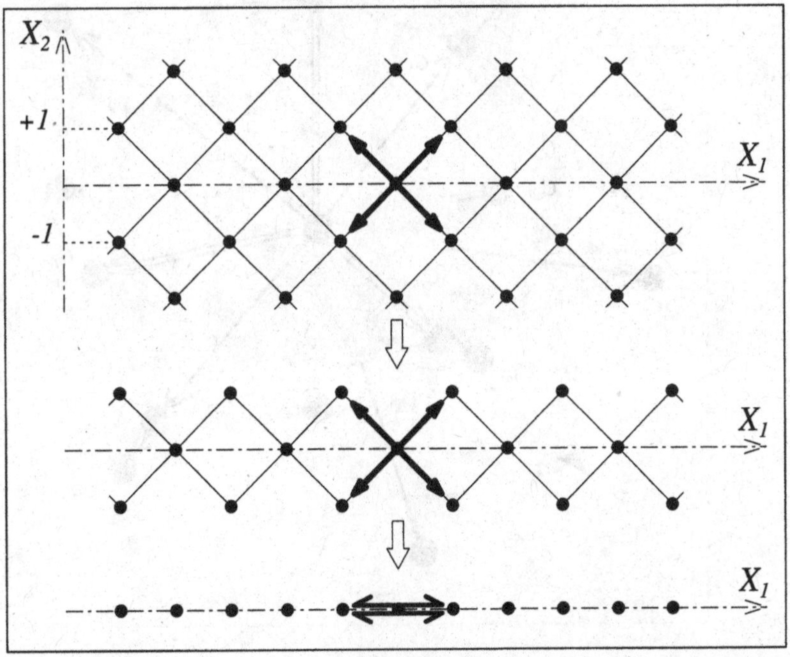

Figure 3.10 This picture illustrates in a simple case the section-projection procedure to construct a lattice gas model in $D-1$ dimensions from a model in D dimensions. We start with the HPP two-dimensional lattice gas model (see Section 3.1) and construct a one-dimensional model. The same procedure can be used to obtain a three-dimensional model from the FCHC four-dimensional Bravais lattice.

model (which requires no four-dimensional gymnastics), it is more convenient for theoretical analysis to work with a model originating from a four-dimensional homokinetic model (all $\mathbf{c}_i$s having equal moduli).

Note that the section-projection procedure can be applied once more to construct a two-dimensional model. The two-dimensional model so obtained has 20 moving particles and 4 distinguishable rest particles corresponding to the four-dimensional velocity vectors $\mathbf{c}_{21}$ to $\mathbf{c}_{24}$ whose projection on the two-dimensional space ($x_3 = 0$, $x_4 = 0$) is zero. This two-dimensional model is equivalent to the 'modified' HPP model described earlier (see Section 3.1) with four additional rest particles.

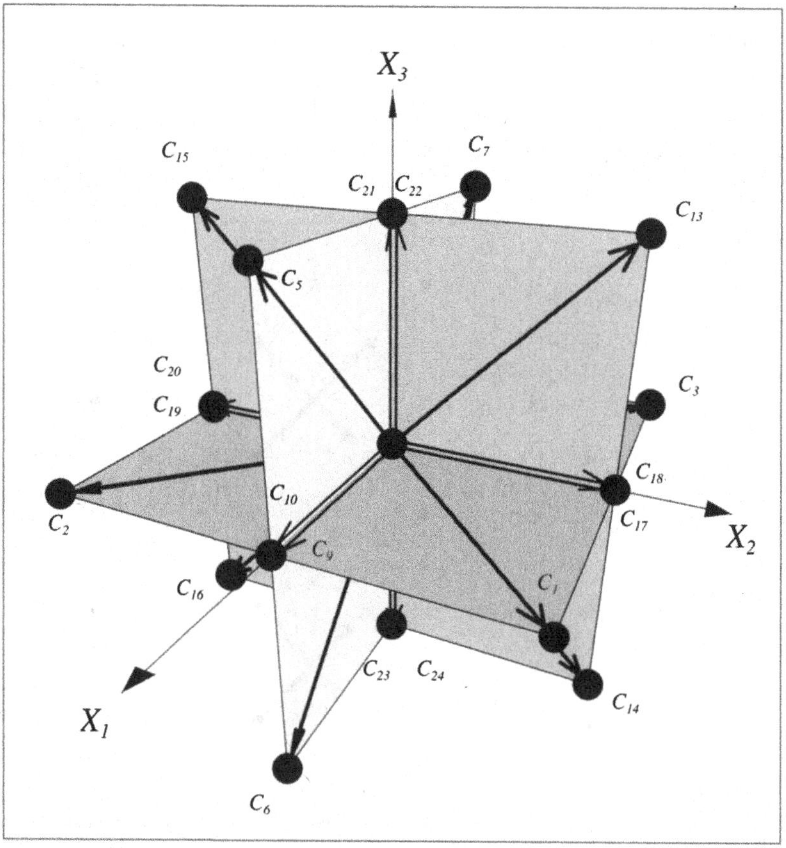

Figure 3.11 Three-dimensional lattice gas model obtained by projection of the four-dimensional FCHC lattice. Single arrows represent vectors with zero-component in the fourth dimension. Double arrows represent pairs of vectors such as $(1,0,0,1)$ and $(1,0,0,-1)$ which differ only by their component (± 1) in the fourth dimension and whose three-dimensional projections are thus identical. Hidden vectors are not labeled. Black dots represent the nodes of the three-dimensional lattice.

3.7.4 Collision rules

Designing and implementing collision rules is much more involved in three dimensions than for two-dimensional lattice gas models. The most obvious source of difficulties is the value of b which is generally larger in three dimensions. The number of possible states per node grows exponentially (2^b) with the number of channels. Even after reduction using the invariance properties of the lattice, the number of non-trivial cases to be treated remains quite large. For example, the FHP-3 model (see Section 3.4) with 7 channels per node has $2^7 = 128$ possible states. If the model is to be made G-invariant, one can restrict the analysis to 26 states which cannot be interchanged by isometries of the point group G. This number reduces to 13 if, in addition, the model must be self-dual. It is then easy to design, optimize, store and finally implement collision rules conserving the correct observables, by simply considering each case separately. For the three-dimensional FCHC model with 24 channels per node, the number of possible states rises to $2^{24} = 16\,777\,216$. When the model is assumed to be G-invariant, the number of cases to study is still as high as 18 736, which reduces to 10 805 if self-duality is imposed (Hénon, 1992). It is clear that 'hand-manufactured' collision rules are no longer possible. The design and optimization of collision rules for three-dimensional models require new strategies to handle the problem in a global and systematic way.

Even had the design problem been solved, the storage and implementation problem would persist. Indeed, since the collision operator $A(s \to s')$ is formally a $2^b \times 2^b$ matrix, the linear size of the complete information to be stored for run-time grows exponentially with b. The full collision matrix for a non-deterministic model with 24 channels per node would represent $2^{2\times 24} \simeq 2.8\times 10^{14}$ real numbers between 0.0 and 1.0, which by far exceeds the random access storage capacity of the most powerful computer systems presently known, even if each entry is stored on eight bits only. As for the design, new strategies are required to drastically reduce the amount of data to be stored and accessed at run-time.

Solutions exist which overcome the problem of designing and implementing collision rules for three-dimensional models. These solutions belong to three classes: *algorithmic strategies, look-up table strategies* and *compound strategies.*

3.7.4.1 Algorithmic strategies

These strategies amount to storing (almost) nothing, and computing the post-collision state for each node at each time step, according to a suitable algorithm. For the FCHC model, a collision rule involving an algorithmic strategy was introduced by M. Hénon (1987a). It is called the 'isometric algorithm' because it uses lattice-preserving isometries in a suitable way to perform collisions. This strategy has been implemented and used for validation tests on the FCHC lattice gas model (Rivet, 1987a), but because of its lack of numerical efficiency, the isometric algorithm was not further exploited.

3.7.4.2 Look-up table strategies

For deterministic models, a unique post-collision state corresponds to each pre-collision state. So the collision matrix contains only 0s and 1s, and each line (or column) contains only one non-zero element. Therefore, instead of storing $2^b \times 2^b$ real numbers between 0.0 and 1.0, it suffices to store only *the* unique post-collision state assigned to each of the 2^b pre-collision states. The storage of this look-up table requires $b \times 2^b$ bits, which is still a fairly large number for most three-dimensional models. For instance, the look-up table for the simplest FCHC model (with $b = 24$) requires about 50 megabytes storage capacity, which represents a non-negligible part of the random access memory of most supercomputers. These strategies involving a pre-computed look-up table are very 'memory-consuming', but potentially rather fast (see Chapter 10). The construction of the table can be quite time-consuming, but it is done once for all.

Strategies based on pre-computed look-up tables can be adapted to 'weakly non-deterministic' models by using more than one look-up table, with random call to each table at each time step and possibly at each node. This solution tends to become heavy and prohibitively bulky if the choice involves a large number of tables.

A considerable drawback of deterministic look-up table strategies is that it may be difficult to meet the important G-invariance property. A careful design of the look-up table can however yield some 'approximate G-invariance' by introducing some 'frozen randomness' in the design of the table.

3.7.4.3 Compound strategies (or reduced-table strategies)

Compound methods use features such as symmetries and/or duality to reduce the number of entries in the look-up table. These strategies combine the advantages and drawbacks of both look-up table and algorithmic strategies. They are less 'space-consuming' than straight look-up table methods, since only a reduced table has to be stored. On the other hand, they are more 'time-consuming' since for each node some computation is required to recast the pre-collision state into an entry of the reduced table.

3.7.5 The microdynamical equation

As a consequence of the quite large number of non-trivial collisions in the FCHC model, the microdynamical equation cannot be written, in practice, in an extensive form as for HPP or FHP models. Of course, the concept of an exact equation governing the FCHC microscopic dynamics still holds, but its explicit form would involve more lines than the present book contains; the only useful expression is the general form given by (2.12) or (2.13).

3.7.6 Microscopic properties

Several FCHC models can be designed with the same geometry and the same connection vectors, but with different collision rules, and various numbers of rest particles (i.e. with different numbers of channels per node). The simplest FCHC models have 24 channels per node (no rest particles). More complex models can have up to several tens of indistinguishable rest particles per node, in addition to the 24 moving particles (see comment on indistinguishable particles in Section 1.3.3). Only a few of these possible FCHC models have actually been implemented and used for simulations. The interested reader will find in Table 3.3 a summary of the microscopic properties of FCHC models. Most important is that, independently of these specific properties, the whole FCHC class exhibits the following properties:

(i) The class structure may depend on the number of rest particles. However, the 24 velocity vectors of the moving particles always belong to the same class. Indeed, let us take $\mathbf{c}_1$ (see Equation (3.15) for the definition of the $\mathbf{c}_i$ vectors). It is clear that $\mathbf{c}_2$ to $\mathbf{c}_4$ belong to the same class as $\mathbf{c}_1$, since the application of combinations of the symmetries S_1 and S_2 to $\mathbf{c}_1$ generates these three vectors. The application of P_{23} to vectors $\mathbf{c}_1$ to $\mathbf{c}_4$ generates vectors $\mathbf{c}_5$ to $\mathbf{c}_8$, and so on. Thus, any vector $\mathbf{c}_1$ to $\mathbf{c}_{24}$ can be changed into any other vector $\mathbf{c}_i$ by an isometry in G, which proves that all 24 vectors belong to the same class.

(ii) The mass and the *four* components of the momentum are the only linearly independent collisional invariants. The energy observable is proportional to the mass observable and so is trivially conserved for models without

Table 3.3 Microscopic properties of the most important FCHC models. The models with $b = 26$ admit up to *three* indistinguishable rest particles coded with *two* Boolean quantities (see comment in Section 1.3.3).

Label	b	Strategy	Deterministic	S.D.B.	Duality	G-invar.
FCHC-1	24	Algo.	No	Yes	Yes	Exact
FCHC-2	24	Table	Yes	Yes	Yes	Approx.
FCHC-3	24	Table	Yes	Yes	Yes	Approx.
FCHC-4	24	Table	Yes	Yes	Yes	Approx.
FCHC-5	26	Table	Yes	Yes	Yes	Approx.
FCHC-6	24	Table	Yes	No	Yes	Approx.
FCHC-7	26	Table	Yes	No	Yes	Approx.
FCHC-8	26	Table	Yes	No	No	Approx.

rest particles. Energy conservation in FCHC models with rest particles is subject to the discussion presented for FHP-2 and FHP-3 (see Sections 3.3.2 and 3.4.2).

(iii) The models have up to fifth order crystallographic isotropy. Table 3.4 gives the values of the crystallographic tensors of order 1 to 6, fully contracted with an arbitrary vector.

(iv) The FCHC models are irreducible. Indeed, let us consider $\mathbf{c}_1$. The sub-group of G which leaves $\mathbf{c}_1$ unchanged contains, among others, the isometries S_3, S_4 and P_{12} (see Section 3.7.3). The only vectors which are invariant under S_3, S_4 and P_{12} have components x_3 and x_4 equal to zero, and components x_1 and x_2 equal one to each other. They are thus proportional to $\mathbf{c}_1$. The same argument holds for vectors $\mathbf{c}_2$ to $\mathbf{c}_{24}$. For the velocity vectors of rest particles (if any), the argument developed for the FHP-2 model is applicable.

The reader may anticipate the following question: in order that *three-dimensional* lattice gas large-scale dynamics be governed by the incompressible fluid Navier–Stokes equations, the collisional invariants must be the mass, the energy, the *three* components of the momentum, *and no other quantity* ; but what about the *fourth* component of the momentum which is also conserved in FCHC models ? In fact, this additional invariant is not spurious, as it corresponds to an additional macroscopic variable with its own large-scale dynamics, but which does not perturb the large-scale dynamics of the three-dimensional relevant macroscopic quantities. Indeed, consider a fully four-dimensional FCHC model with large-scale collective dynamics governed by the *four-dimensional* Navier–Stokes equations. Now, when cutting a (hyper-)slice in the four-dimensional lattice and wrapping it onto itself in the fourth dimension (see Section 3.7.3), the macroscopic variables no longer depend on the fourth component x_4 of

Table 3.4 Crystallographic isotropy properties of the FCHC models. The second column gives the result of a full contraction of the nth order tensor with an arbitrary test vector $\mathbf{x}$. The third column summarizes the conclusion about isotropy.

Order n	$\sum T^{(n)}_{\alpha_1 \dots \alpha_n} x_{\alpha_1} \dots x_{\alpha_n}$	Status
1	0	isotropic
2	$12\left(\sum_\alpha x_\alpha^2\right) \quad (= 12\|x\|^2)$	isotropic
3	0	isotropic
4	$12\left(\sum_\alpha x_\alpha^4 + 2\sum_{\alpha<\beta} x_\alpha^2 x_\beta^2\right) \quad (= 12\|x\|^4)$	isotropic
5	0	isotropic
6	$12\left(\sum_\alpha x_\alpha^6 + 5\sum_{\alpha<\beta} x_\alpha^2 x_\beta^4 + 5\sum_{\alpha<\beta} x_\alpha^4 x_\beta^2\right)$ $(\neq a\|x\|^6)$	**anisotropic**

the position. As a result, the four-dimensional Navier–Stokes equations separate into (i) the three-dimensional Navier–Stokes equations for the three-dimensional variables, and (ii) an advection–diffusion equation for the fourth component of the momentum density, which behaves as a passive (non-perturbing) diffusive scalar advected by the three-dimensional flow. This 'charge-free' passive scalar can be considered as a bonus offered by pseudo-four-dimensional models.

Chapter 4

Equilibrium statistical mechanics

In Chapter 2, we established the equations governing the microscopic dynamics of the lattice gas. This microscopic dynamics provides the basics of the procedure for constructing automata simulating the behavior of fluid systems (see Chapter 10). However, in order to use LGAs as a method for the analysis of physical phenomena, we must go to a level of description which makes contact with macroscopic physics. Starting from a microscopic formalism, we adopt the statistical mechanical approach.

We begin with a Liouville (statistical) description to establish the lattice Liouville equation which describes the automaton evolution in terms of configuration probability in phase space. We then define ensemble-averaged quantities and, in particular, the average population per channel whose space and time evolution is governed by the lattice Boltzmann equation (LBE), when we neglect pre-collision correlations as well as post-collision re-correlations (Boltzmann *ansatz*). The LBE will be seen to play in LGAs the same crucial role as the Boltzmann equation in continuous fluid theory; an H-theorem follows from which the existence of a Gibbsian equilibrium distribution is established when the semi-detailed balance is satisfied.

4.1 The Liouville description

In Chapter 2 we introduced the notion of Boolean field $n_i(\mathbf{r}_\star, t_\star)$ to obtain a complete microscopic description of the time-evolution of lattice gases. We also introduced the phase space Γ which is the set of all possible Boolean configurations of the whole lattice. The Boolean configuration of the lattice is a collection of $\mathcal{N} \times b$ Boolean quantities (with values 0 or 1) which encodes the full microscopic state of the $\mathcal{N}$ nodes of the lattice, with b channels per node. The Boolean configuration of the lattice is called its 'microstate' because it describes the state of the whole lattice at the microscopic level.

Now when we are interested in macroscopic behavior, such a detailed description in terms of microstates is generally unnecessary and often even useless. To describe the macroscopic states we need averaged quantities. Following the same line of reasoning as in classical statistical mechanics, it is logical to introduce a statistical description of lattice gases in terms of its 'macrostates'.

4.1.1 Macrostates

Most common laboratory probes (thermometers, pressure sensors, velocimeters, etc.) yield *time- and space-averaged* measures, simply because they have a finite macroscopic size and a finite response time, that is, they operate on a length scale and a time scale which are large compared to the characteristic microscopic lengths and times. For theoretical developments however, it is logical and convenient, following Gibbs, to consider *ensemble-averages* (Huang, 1963). This procedure requires the ergodic hypothesis assuming that ensemble-averages correctly describe macroscopic observables which are in practice time- and space-averages.

Formally, we define a 'macrostate' of the lattice as a function acting on the phase space Γ, such that it associates to each microstate $S \in \Gamma$ a real number between 0 and 1:

$$\begin{array}{rccc} \mathcal{P}: & \Gamma & \rightarrow & [0,1] \\ & S & \mapsto & \mathcal{P}(S), \end{array} \tag{4.1}$$

with the normalization property:

$$\sum_{S\in\Gamma} \mathcal{P}(S) = 1. \tag{4.2}$$

Because of this normalization property, the function $\mathcal{P}$ defining a macrostate can be viewed as a *probability distribution function*. The concept of macrostate is the basic tool for the Liouville (statistical) description of lattice gases.

4.1.2 Ensemble-averages

Having defined a probability measure in the phase space, we can now define ensemble-averaged quantities. Consider an arbitrary generalized observable (see Section 2.1.4), that is, a physical quantity whose measure at node $\mathbf{r}_\star$ at time $t_\star$ is an arbitrary function $q(n, \mathbf{r}_\star)$ of the Boolean field n, and possibly of the discrete space-variable. We define the 'ensemble-averaged' value $\langle q \rangle_{(\mathbf{r}_\star)}$ of the generalized observable q measured at node $\mathbf{r}_\star$ as:

$$\langle q \rangle_{(\mathbf{r}_\star)} = \sum_{S \in \Gamma} q(S, \mathbf{r}_\star)\mathscr{P}(S). \tag{4.3}$$

Consider now an ordinary observable[1] $q = (q_i, i = 1, \ldots, b)$ (see Section 2.1.3). Its value measured at node $\mathbf{r}_\star$ is $\sum_i q_i n_i(\mathbf{r}_\star)$. The ensemble-averaged value reads:

$$\langle q \rangle_{(\mathbf{r}_\star)} = \sum_{S \in \Gamma} \sum_{i=1}^{b} q_i S_{i(\mathbf{r}_\star)} \mathscr{P}(S),$$

which reduces after summation inversion to:

$$\langle q \rangle_{(\mathbf{r}_\star)} = \sum_{i=1}^{b} q_i \left\langle n_{i(\mathbf{r}_\star)} \right\rangle, \tag{4.4}$$

where $\langle n_{i(\mathbf{r}_\star)} \rangle$ is the (ensemble-)averaged population of channel i at node $\mathbf{r}_\star$:

$$\langle n_{i(\mathbf{r}_\star)} \rangle = \sum_{S \in \Gamma} S_{i(\mathbf{r}_\star)} \mathscr{P}(S). \tag{4.5}$$

We have now at our disposal the concept of ensemble-averaged quantities, which are defined for an arbitrary probability measure (macrostate) $\mathscr{P}$ with no restricting hypothesis, except for the normalization condition (4.2). The next step in the statistical description of lattice gases is the derivation of the master equation that governs the time-evolution of the macrostate $\mathscr{P}$.

4.1.3 The lattice Liouville equation

When the lattice gas evolution rule is applied, the probability distribution function is *a priori* time-dependent. The time-evolution of $\mathscr{P}$ is governed by a relation that can be established from the following arguments. The collision phase acts independently at each lattice node, and changes the state $S_{(\mathbf{r}_\star)}$ of node $\mathbf{r}_\star$ into the state $S'_{(\mathbf{r}_\star)}$ according to the transition probability $A\big(S_{(\mathbf{r}_\star)} \to S'_{(\mathbf{r}_\star)}\big)$ (see Chapter 2). So, the conditional probability of obtaining the microstate S' after the collision phase, knowing that the microstate was S before collision, is $\prod A\big(S_{(\mathbf{r}_\star)} \to S'_{(\mathbf{r}_\star)}\big)$, where the product extends to all lattice nodes $\mathbf{r}_\star \in \mathscr{L}$. We now assume that before collision, the statistics of the lattice can be described by the macrostate $\mathscr{P}(S, t_\star)$. So the probability to find the microstate S' after collision is $\sum \mathscr{P}(S, t_\star) \prod A\big(S_{(\mathbf{r}_\star)} \to S'_{(\mathbf{r}_\star)}\big)$, where the sum is taken over all possible microstates $S \in \Gamma$. Since the propagation operator $\mathscr{S}$ (notation introduced in Chapter 2) is a fully deterministic operator, the probability of finding the lattice in the microstate $\mathscr{S}(S')$ after the propagation phase is equal to the probability

[1] An ordinary observable is a linear, local and homogeneous generalized observable (see Section 2.1.4).

of finding the lattice in the microstate S' before propagation. We then obtain a *Chapman–Kolmogorov* equation for the time-evolution of the macrostate:

$$\mathscr{P}\left(\mathscr{S}\left(S'\right), t_\star + 1\right) = \sum_{S\in\Gamma} \mathscr{P}(S, t_\star) \prod_{\mathbf{r}_\star \in \mathscr{L}} A\left(S(\mathbf{r}_\star) \to S'(\mathbf{r}_\star)\right), \quad \forall S' \in \Gamma. \tag{4.6}$$

Although formally different from the Liouville equation for continuous systems, Equation (4.6) plays an equivalent role for lattice gases: it governs the time-evolution of the probability measure in phase space Γ. We therefore call (4.6) the 'lattice Liouville equation'.

We now address the problem of finding the equation governing the time-evolution of the ensemble-averaged populations of each channel at each node.

4.2 The Boltzmann description

The probability distribution function $\mathscr{P}$ contains the *full* statistics of the whole lattice, and the lattice Liouville equation which describes the evolution of $\mathscr{P}$ is an *exact* equation. However, the ensemble-averaged populations of each channel of each node are more directly useful than the full statistics.

For consistency with the usual notation in statistical mechanics, we denote by $f(\mathbf{c}_i, \mathbf{r}_\star, t_\star)$, or equivalently $f_i(\mathbf{r}_\star, t_\star)$, the ensemble-averaged populations $\langle n_i \rangle_{(\mathbf{r}_\star, t_\star)}$. These ensemble-averaged populations play the same role as the single particle distribution function $f(\mathbf{p}, \mathbf{r}, t)$ in kinetic theory of continuous gases. Indeed, for continuous gases, $f(\mathbf{p}, \mathbf{r}, t)d^3r d^3p$ is the average number of particles present at time t in a volume element d^3r about $\mathbf{r}$ and within the volume element d^3p about $\mathbf{p}$ in momentum-space. For lattice gases, $f(\mathbf{c}_i, \mathbf{r}_\star, t_\star)$ is the average number of particles present at time $t_\star$ at node $\mathbf{r}_\star$ with velocity $\mathbf{c}_i$ (no volume element is needed for lattice gases since both position space and momentum space are discrete).

4.2.1 The Boltzmann approximation

In classical statistical mechanics, one cannot derive an *exact* and *closed* equation for the time-evolution of the single particle distribution function $f(\mathbf{p}, \mathbf{r}, t)$ (see Huang, 1963). The difficulty is bypassed by using the *Boltzmann approximation* (the molecular chaos hypothesis) to obtain the closed Boltzmann equation. The situation is very much the same in lattice gases.[2] We must assume that the

[2] There is at least one exception: the diffusive lattice gas model constructed as a collection of random walkers on the lattice (Boon, Dab, Kapral and Lawniczak, 1995) for which an exact lattice Boltzmann equation is obtained without the factorization assumption. However, one can argue that the assumption is contained in the model itself through the randomization of the channel occupations at each time step, a procedure used to model the random walk of the particles on the lattice.

propagation phase shuffles the particles in such a way that their correlations vanish at each node, so that the pre-collision (post-propagation) state of each node be totally factorized. Obviously, right after a collision, particles are in a strongly correlated (post-collision) state precisely because of the effect of the collisions, but we assume that during the subsequent propagation phase, these correlations are completely damped out.

The Boltzmann approximation, for lattice gases as for real gases, is a strong hypothesis whose validity is expected to hold at sufficiently low gas density. For lattice gases however, the approximation is rather well verified even at moderate densities, at least for models satisfying semi-detailed balance (like the GBL model and the FCHC-3 model; see Chapter 7).

4.2.2 The lattice Boltzmann equation

By ensemble-averaging both sides of the microdynamic equation (2.12), we obtain the evolution equation for $f_i(\mathbf{r}_\star, t_\star)$:

$$f_i(\mathbf{r}_\star + \mathbf{c}_i, t_\star + 1) = \sum_{s,s' \in \gamma} s'_i A(s \to s') \left\langle \prod_{j=1}^{b} n_j^{s_j} \overline{n}_j^{\overline{s}_j} \right\rangle, \tag{4.7}$$

where the n_j s on the r.h.s. are implicitly taken at time $t_\star$ and node $\mathbf{r}_\star$. For clarity, these arguments are not written explicitly, and will be omitted systematically when no ambiguity results. We also use the more compact notation $f_i(\mathbf{r}_\star, t_\star)$ rather than $f(\mathbf{c}_i, \mathbf{r}_\star, t_\star)$, which is formally closer to the classical notation $f(\mathbf{p}, \mathbf{r}, t)$, keeping in mind that $f_i(\mathbf{r}_\star, t_\star)$ and $f(\mathbf{c}_i, \mathbf{r}_\star, t_\star)$ denote the same object: $\langle n_{i(\mathbf{r}_\star, t_\star)} \rangle$. To arrive at (4.7), we have used the fact that the $\xi(s \to s')$s in (2.12) are random Bernoulli variables, statistically independent from the Boolean field, and whose averages are the $A(s \to s')$s.

Equation (4.7) is an exact equation but, as it stands, there is no way to express explicitly the r.h.s. in terms of the f_is. Therefore, we now use the Boltzmann approximation (see Section 4.2.1), that is, we make the assumption that the pre-collision configuration is totally uncorrelated. With this assumption, the ensemble-average of the product on the r.h.s. of (4.7) can be factorized into the product of the ensemble-averages, to obtain:

$$f_i(\mathbf{r}_\star + \mathbf{c}_i, t_\star + 1) = \sum_{s,s' \in \gamma} s'_i A(s \to s') \prod_{j=1}^{b} f_j^{s_j} \overline{f}_j^{\overline{s}_j}, \tag{4.8}$$

where $\overline{f}_j$ stands for $1 - f_j$. This equation governs the time-evolution of the averaged populations f_i. It plays the same role as the Boltzmann equation in kinetic theory of continuous gases, and is therefore referred to as the 'lattice Boltzmann equation'.[3]

In the same way as there are alternative versions of the microdynamic equation, (2.12) and (2.13), the lattice Boltzmann equation can be rewritten as:

$$f_i(\mathbf{r}_\star + \mathbf{c}_i, t_\star + 1) - f_i(\mathbf{r}_\star, t_\star) = \sum_{s,s' \in \gamma} (s'_i - s_i) A(s \to s') \prod_{j=1}^{b} f_j^{s_j} \bar{f}_j^{\bar{s}_j}. \quad (4.9)$$

The equivalence between (4.8) and (4.9) is an immediate consequence of the identity:

$$f_i = \sum_{s,s' \in \gamma} s_i A(s \to s') \prod_{j=1}^{b} f_j^{s_j} \bar{f}_j^{\bar{s}_j}, \quad (4.10)$$

which follows directly from the normalization condition (2.9).

The analogy with the classical Boltzmann equation is more visible from (4.9): the l.h.s. is the first order finite difference expression of the convective derivative $(\partial_t + \mathbf{v} \cdot \nabla) f$, and the r.h.s. is a collision term which plays a role similar to the collision integral (see Huang, 1963). The main difference is that for ordinary gases, only binary collisions are taken into account, so that the collision integral is quadratic in the distribution function f, whereas for lattice gases collisions involving up to b particles are admitted. Consequently, the collision term for lattice gases is a polynomial of degree b in the f_is.

Following Boltzmann's reasoning, we now prove an 'H-theorem' for lattice gases satisfying semi-detailed balance. Namely, we show that one can define a scalar function H of the average populations $f_i(\mathbf{r}_\star)$ that can only increase or remain constant. The important consequence of the H-theorem is the existence of equilibria (Section 4.4).

4.3 The H-theorem

Usually one presents the H-theorem – in the context of ordinary gases in continuous space-time – by introducing the functional H of the single particle probability distribution f, $H = -k_B \int f \log f d^3p\, d^3r$, and one shows that this quantity cannot decrease if the time-evolution of f is governed by the Boltzmann equation (Huang, 1963). The functional H is interpreted as the statistical entropy of the system and the equilibrium states follow from the maximization of H under constraints related to conservation laws.

Here, we shall steer a different – although equivalent – course, which makes a

[3] The lattice Boltzmann equation can also be used – quite independently of its underlying Boolean microscopic dynamics – as the starting point for the simulation of fluid dynamics (Succi, Benzi and Higuera, 1991; Succi, 2000); see also Section 11.5 in Chapter 11.

detour via the theory of communication as established in 1948 by C.E. Shannon and subsequently formalized in a more rigorous presentation by A.I. Kinchin (Kinchin, 1957). This presentation may seem rather abstract at first sight, but it leads to a nice intuitive interpretation.[4]

We first briefly review the basic ideas of Shannon's theory of communication and postpone the connection with lattice gases to Section 4.3.2.

4.3.1 Some basics about communication and information

The goal of Shannon was to provide a global mathematical description of communication processes. A communication process includes the generation of a message and its encoding, the (possibly noisy) transmission through a physical channel, and its reception and decoding. No quantitative theory could be built about communication without an accurate tool to measure the amount of information contained in a message. We shall restrict our short description to this crucial notion of information measurement, because it is the most relevant feature for lattice gas statistics. The interested reader will find the complete theory in Shannon and Weaver (1969), and in Kinchin (1957).

4.3.1.1 The deterministic case

We introduce the notions of 'quantity of information' and of 'entropy of information', by considering first the simple case of a variable x whose value can be either $x^{(1)}$ or $x^{(2)}$. Suppose for example that the value of this variable is *known* to be $x^{(1)}$, then the quantity of information we have about the variable x is one *bit*, by definition. In addition, since we fully know the value of x, no freedom of choice is left concerning x. Therefore, we say that the freedom of choice, which Shannon calls the 'entropy of information', is zero *bits*. The 'bit'[5] is the standard unit of measure for the quantity of information. Conversely, if we do not know anything about the value of x, then we say that we have zero bits of information about x, and that the freedom of choice (entropy of information) concerning x is one bit. The word *entropy* is used here by analogy with the concept of entropy in thermodynamics, since in some sense it is a measure of the 'disorder' of the system in both situations.

Consider now a discrete variable x which can take any of the n values $x^{(i)}$, $i = 1, \ldots, n$. (Superscripts between parentheses label the possible values taken by the variable x; subscripts will be used in further sections to label different variables.) Suppose that an incomplete measure of x revealed that the actual value of the variable x can be either $x^{(1)}$ or $x^{(2)}$ or ... or $x^{(m)}$ ($m \leq n$). We

[4] The material in this section is inspired by a work by M. Hénon, published in an appendix of the article by Frisch, d'Humières, Hasslacher, Lallemand, Pomeau and Rivet (1987). Hénon takes the traditional route to present the H-theorem, and uses information theory only as an interpretation.

[5] The word 'bit', introduced by J.W. Tuckey, is the contraction of 'binary digit'.

define the 'entropy of information' H associated with this incomplete measure of x, as the logarithm to the base 2 of m: $H = \log_2 m$. We also define the 'quantity of information' given by the incomplete measure as $I = \log_2 n - \log_2 m$. Note that if $m = 1$, the value of x is fully determined, and therefore the entropy is zero and the quantity of information takes its maximum value $\log_2 n$. On the contrary, if $m = n$, the value of x is totally unknown; then the entropy takes its maximum value $\log_2 n$ and the quantity of information is zero. We emphasize the analogy with Gibbs' entropy: the entropy associated with a state of a macroscopic system is proportional to the logarithm of the volume of phase space actually accessible to the system. In the context of information theory, the entropy is proportional to the logarithm of the number of accessible values of the discrete variable x.

We can also give a more operational interpretation of the notion of freedom of choice (entropy), by using the language of computer science: assume that the number n of possible values of x is an integer power of 2, say 2^b. Suppose that an incomplete measure reveals that x can actually take 2^c of the 2^b *a priori* possible values (c is an integer smaller than b). Then, the number of computer memory bits necessary to describe the possible choices remaining after the measure is $\log_2 2^c = c$. Thus, the entropy of information H associated to this incomplete measure of x is just the number of memory bits required to store the remaining free choices. Here, the word 'bit' is used in the context of computer memory rather than in the context of information theory, but the meaning is essentially the same. The only difference is that, in computer science, memory sizes must be an *integer* number of bits, whereas this constraint does not exist in information theory, since the number of possible values of a variable need not necessarily be an integer power of 2.

4.3.1.2 The non-deterministic case

Next comes the case of variables for which the only available information is statistical. Consider as before a discrete variable x which can take any of the n values $x^{(i)}$, $i = 1,\dots,n$, but now the only information we have at our disposal is the set of probabilities $p^{(i)}$, $i = 1,\dots,n$ for each one of the n possible values $x^{(i)}$, $i = 1,\dots,n$. Shannon shows that under some reasonable constraints, the only possible measures of the freedom of choice and of the quantity of information associated with this statistical knowledge about x are, respectively:

$$\begin{aligned} H &= -k\sum_{i=1}^{n} p^{(i)} \log_2 p^{(i)}, \\ I &= k\log_2 n - H, \end{aligned} \tag{4.11}$$

where k is an arbitrary positive constant that will be chosen equal to one if H is to be expressed in bits. We shall not reproduce Shannon's proof, but rather use a simple argument to convince ourselves that this measure is 'reasonable' and consistent with the intuitive and natural results of the deterministic case. Indeed, suppose that an incomplete measure reveals that x can take indifferently any of the values $x^{(1)},\dots,x^{(m)}$ $(m < n)$, and no other value. This partial information is equivalent to the statistical information that the probabilities $p^{(1)},\dots,p^{(m)}$ are equal to $1/m$, and $p^{(m+1)},\dots,p^{(n)}$ are zero. With this set of probabilities, (4.11) yields $H = \log_2 m$, which is consistent with the result presented above for the deterministic case.

4.3.1.3 Correlated random Bernoulli variables

With the context of lattice gases in mind, we now address the problem of the quantity of information and entropy related to a collection $X = (x_i, i = 1,\dots,q)$ of q Boolean variables (with values 0 or 1) for which only statistical knowledge is available. The composite variable X thus takes 2^q possible values. We denote by Γ the set of all these 2^q values. Suppose that we *know* the statistics of X, namely the probabilities $\mathcal{P}(Y)$ that X takes the value $Y = (y_i, i = 1,\dots,q) \in \Gamma$. Then, the entropy and the quantity of information associated with this knowledge are, respectively:

$$\begin{aligned} H &= -\sum_{Y\in\Gamma} \mathcal{P}(Y)\log_2 \mathcal{P}(Y), \\ I &= \log_2 2^q + \sum_{Y\in\Gamma} \mathcal{P}(Y)\log_2 \mathcal{P}(Y) \\ &= q - H. \end{aligned} \tag{4.12}$$

Suppose now that the only available knowledge is not the full statistics of X, but the individual probabilities p_i, $i = 1,\dots,q$ that x_i take the value 1. According to (4.11), the entropy and the quantity of information associated with the knowledge of the individual variable x_{i_0} are, respectively:

$$\begin{aligned} H_{i_0} &= -\left[p_{i_0}\log_2 p_{i_0} + (1-p_{i_0})\log_2(1-p_{i_0})\right], \\ I_{i_0} &= 1 - H_{i_0}, \end{aligned} \tag{4.13}$$

since x_{i_0} is a Boolean variable which can take two values: 1 or 0, with respective probabilities p_i and $(1-p_i)$. The entropy and the quantity of information associated with the knowledge of *all* the individual probabilities are thus, respectively:

$$\begin{aligned} H^\star &= -\sum_{i=1}^{q}\left[p_i\log_2 p_i + \overline{p}_i\log_2\overline{p}_i\right], \\ I^\star &= \sum_{i=1}^{q} 1 + \left[p_i\log_2 p_i + \overline{p}_i\log_2\overline{p}_i\right] \\ &= q - H^\star, \end{aligned} \tag{4.14}$$

where $\overline{p}_i$ stands for $1 - p_i$. Since the knowledge of the individual statistics does not account for possible correlations between the variables, it is logical to expect that it contains less information than the knowledge of the full statistics of X. The proof for this quite intuitive statement is given in the Appendix, Section A.3.

We now apply these concepts to lattice gases.

4.3.2 The H-theorem for lattice gases

Consider a lattice gas with b channels per node, residing on a finite portion $\mathscr{L}$ of a Bravais lattice, containing $\mathcal{N}$ connected nodes. At any discrete time $t_\star$, this lattice gas can be seen as a collection of $b \times \mathcal{N}$ Boolean variables. We focus attention on a particular node $\mathbf{r}_\star$ for which we adopt a Liouville description: the information available about the state of this node is the time-dependent local probability distribution function $p(\mathbf{r}_\star, t_\star, s)$. At any discrete time $t_\star$, we can associate to this local statistical knowledge the following 'local entropy':

$$-\sum_{s\in\gamma} p(\mathbf{r}_\star, t_\star, s) \ln p(\mathbf{r}_\star, t_\star, s). \tag{4.15}$$

Notice that we now use natural logarithms rather than logarithms to the base 2. (The difference is a non-significant multiplicative constant that we shall ignore.) To simplify the notation, we shall omit the arguments $\mathbf{r}_\star, t_\star$ in $p(\mathbf{r}_\star, t_\star, s)$.

The effect of the collision phase on the local probability distribution function p is a linear transformation involving the collisional transition probabilities $A(s{\to}s')$ (see Chapter 2), that is, the post-collision local probability distribution function p' can be expressed as:

$$p'(s') = \sum_{s\in\gamma} p(s)A(s{\to}s'), \quad \forall s' \in \gamma. \tag{4.16}$$

We now use a basic property of convex functions: for any convex function ϕ, the following inequality holds:

$$\phi\left(\frac{\sum A(s{\to}s')\,p(s)}{\sum A(s{\to}s')}\right) \geq \frac{\sum A(s{\to}s')\,\phi\,(p(s))}{\sum A(s{\to}s')}, \quad \forall s' \in \gamma, \tag{4.17}$$

where the sums implicitly extend over all s in γ. If we assume that the semi-detailed balance condition $\sum_{s\in\gamma} A(s{\to}s') = 1$ is satisfied, then, after summation over all s', (4.17) reads:

$$\sum_{s'\in\gamma} \phi\left(\sum_{s\in\gamma} A(s{\to}s')\,p(s)\right) \geq \sum_{s'\in\gamma}\sum_{s\in\gamma} A(s{\to}s')\,\phi\,(p(s)). \tag{4.18}$$

Using (4.16) and the normalization condition $\sum_{s'\in\gamma} A(s{\to}s') = 1$, we obtain:

$$\sum_{s'\in\gamma} \phi\,(p'(s')) \geq \sum_{s\in\gamma} \phi\,(p(s)). \tag{4.19}$$

Equation (4.19), applied to the convex function $\phi(x) = -x \ln x$, implies that:

$$-\sum_{s' \in \gamma} p'(s') \ln p'(s') \geq -\sum_{s \in \gamma} p(s) \ln p(s). \tag{4.20}$$

This result proves that under the assumption of semi-detailed balance the collision phase increases the local entropy of the lattice gas, and therefore decreases the quantity of statistical information we have about the state of the lattice node.

Note that without the semi-detailed balance assumption the collision phase *can decrease* the local entropy of the node. As a simple example, consider a lattice gas model with a collision rule such that the post-collision state of any collision is always the same state s_0, whatever the pre-collision state. This unphysical model does not satisfy the semi-detailed balance condition. Suppose that, initially, this lattice gas is in a macrostate such that all configurations are equally probable, namely: $p(s) = 2^{-b}$, for all s in γ. Then, the local entropy initially has its maximum value $b \times \ln 2$. After the collision phase, the lattice is in a completely determined configuration, where all the nodes are in state s_0. Thus, the local entropy after the collision phase has decreased to its minimum value: zero. We now make the Boltzmann approximation, i.e. we neglect the correlations between particles entering a collision. Then, the local entropy at the node before collision is equal to the sum of the entropies in each channel:

$$-\sum_{i=1}^{b} \left[f_i \ln f_i + \overline{f}_i \ln \overline{f}_i\right] = -\sum_{s \in \gamma} p(s) \ln p(s). \tag{4.21}$$

Now, there is no reason that the post-collision state be uncorrelated; so, from the results of Section 4.3.1.3, we have, *after* the collision:

$$-\sum_{i=1}^{b} \left[f'_i \ln f'_i + \overline{f'_i} \ln \overline{f'_i}\right] \geq -\sum_{s' \in \gamma} p'(s') \ln p'(s'), \tag{4.22}$$

Combination of (4.20), (4.21) and (4.22) leads to the local form of the H-theorem:

$$-\sum_{i=1}^{b} \left[f'_i \ln f'_i + \overline{f'_i} \ln \overline{f'_i}\right] \geq -\sum_{i=1}^{b} \left[f_i \ln f_i + \overline{f}_i \ln \overline{f}_i\right], \tag{4.23}$$

where the average populations are taken at node $\mathbf{r}_\star$.

Local H-theorem: *Under the Boltzmann assumption and with the semi-detailed balance condition, the local quantity*

$$h(\mathbf{r}_\star, t_\star) = -\sum_{i=1}^{b} \left[f_i(\mathbf{r}_\star, t_\star) \ln f_i(\mathbf{r}_\star, t_\star) + \overline{f}_i(\mathbf{r}_\star, t_\star) \ln \overline{f}_i(\mathbf{r}_\star, t_\star)\right]$$

cannot decrease under the effect of the collision phase. This quantity is called the local entropy.

We now introduce the 'global entropy' H which is the sum of the local entropies $h(\mathbf{r}_\star)$, extended over all lattice nodes $\mathbf{r}_\star$. If the lattice is either infinite or finite with periodic boundary conditions, the propagation phase only moves information from node to node without loss or gain. It thus has no effect on the global entropy. As a consequence, the local H-theorem implies that the global entropy cannot decrease under the action of the full evolution operator (collision and propagation).

Global H-theorem: *Under the Boltzmann approximation and with the semi-detailed balance condition, the global quantity*

$$H(t_\star) = -\sum_{\mathbf{r}_\star \in \mathscr{L}} \sum_{i=1}^{b} \left[f_i(\mathbf{r}_\star, t_\star) \ln f_i(\mathbf{r}_\star, t_\star) + \overline{f}_i(\mathbf{r}_\star, t_\star) \ln \overline{f}_i(\mathbf{r}_\star, t_\star) \right]$$

cannot decrease under the action of the lattice gas evolution rule, at least if the lattice is either infinite or finite with periodic boundary conditions. This quantity is called the global entropy.

This theorem, proved under the Boltzmann approximation, implies that the evolution of the statistics of a lattice gas is *irreversible*.

Let us illustrate the H-theorem with a simple example. We consider a lattice gas residing on a finite lattice with periodic boundary conditions and prepared in a homogeneous macrostate. In the homogeneous macrostate, the global entropy $H(t_\star)$ is merely $\mathscr{N}$ times the local entropy $h(\mathbf{r}_\star, t_\star)$ which no longer depends on the position variable $\mathbf{r}_\star$. So the statement that H cannot decrease when semi-detailed balance is verified, at least within the Boltzmann approximation, also applies to h. Suppose now that we prepare the lattice gas in a non-equilibrium macrostate, for instance, by biasing in a systematic way one of the collision rules; then, at some later time, we restore the usual rules (satisfying detailed or semi-detailed balance), and in the course of time, we perform successive measurements which give us an estimate of the entropy. We observe, as shown in Figure 4.1, an increase in the numerical estimate of $h(t_\star)$ (see Tribel and Boon, 1997) towards an asymptotic plateau corresponding to the equilibrium value as predicted by Boltzmann's theory. Now, in Figure 4.1, we also observe that on top of the global increase of $h(t_\star)$ (in accordance with Boltzmann's theory) there are fluctuations where h can decrease locally; the reason is that the Boltzmann decorrelation hypothesis is only an approximation.

Along the same lines of reasoning, but *without* the Boltzmann approximation, one can also prove a weaker form of the H-theorem:

Liouville H-theorem: *With the semi-detailed balance condition, the quantity*

$$-\sum_{S \in \Gamma} \mathscr{P}(S, t_\star) \ln \mathscr{P}(S, t_\star)$$

cannot decrease under the action of the lattice gas evolution rule, at least if the lattice is either infinite or finite with periodic boundary conditions.

Now that we have established the H-theorem, the next step in the statistical mechanical description of the lattice gas is to investigate the existence of equilibrium situations, that is, of macrostates for which all statistically relevant quantities are stationary.

4.4 Global equilibrium macrostates

Hardy, de Pazzis and Pomeau (1973) showed that the HPP lattice gas model has very simple statistical equilibrium solutions that they call 'invariant states'. These states have homogeneous (node-independent) properties, and they show no equal time (static) correlations between channels of any pair, whether on the same node or not. Such global (homogeneous) equilibrium macrostates are analogous to the Maxwellian equilibrium states of real gases. We shall prove that the equilibria, for a quite general class of lattice gases (including the HPP model), are not characterized by the Maxwell–Boltzmann distribution of

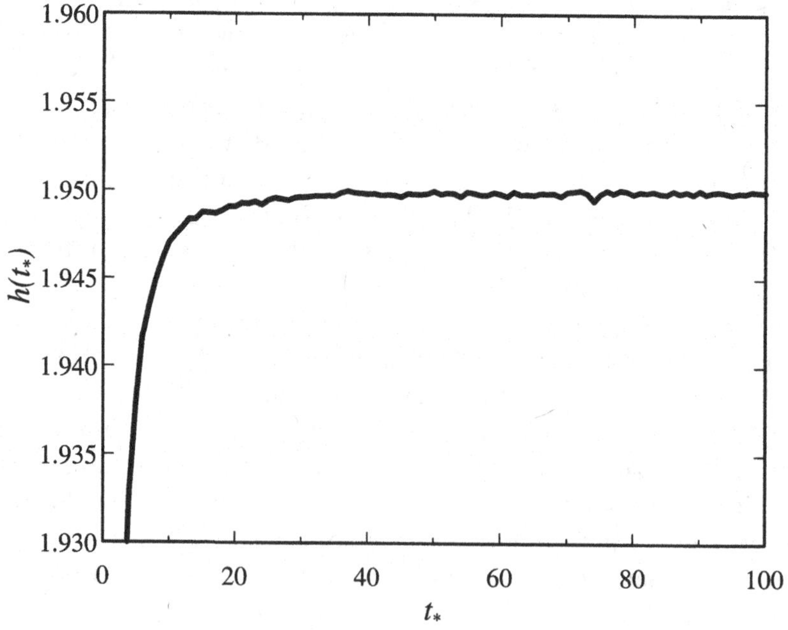

Figure 4.1 The local entropy $h(t_\star) = -\sum_{i=1}^{b}[f_i(t_\star)\ln f_i(t_\star) + \bar{f}_i(t_\star)\ln \bar{f}_i(t_\star)]$ as a function of time, for an FHP-1 lattice gas model on a lattice with 256 × 256 nodes. $t_\star$ is in the natural microscopic time unit: the automaton time step. (From Tribel and Boon, 1997.)

classical gases, but by the Fermi–Dirac distribution. As physical intuition would indicate, the equilibrium macrostates (also called 'invariant measures') play a crucial role in the theory of large scale collective motions.

In order to make the problem more easily tractable, we shall not attempt to find and characterize all possible equilibrium macrostates. Rather, we shall restrict the scope of our search to global (homogeneous) factorized macrostates, that is, macrostates whose statistical properties are node-independent, and for which no correlations exist between nodes and between channels. We shall adopt three different, although equivalent approaches:

- *The Liouville approach*: the equilibrium macrostates are characterized by their probability distribution functions $\mathscr{P}^{(eq)}$, which are required to be stationary solutions of the lattice Liouville equation.
- *The Boltzmann approach*: the equilibrium macrostates are characterized by their averaged populations $f_i^{(eq)}$, which are required to be stationary solutions of the Boltzmann equation.
- *The variational approach*: the equilibrium macrostates are characterized by their averaged populations $f_i^{(eq)}$, which are required to maximize the global entropy of the lattice, under conservation constraints.

4.4.1 The Liouville approach

We look for stationary solutions of the lattice Liouville equation among the restricted class of homogeneous and totally factorized distributions of the form:

$$\mathscr{P}(S) = \prod_{\mathbf{r}_\star \in \mathscr{L}} \prod_{i=1}^{b} f_i^{S_i(\mathbf{r}_\star)} \overline{f}_i^{\,\overline{S}_i(\mathbf{r}_\star)}, \tag{4.24}$$

where the f_is should be taken for the moment as b real numbers between 0 and 1, and where overlined quantities $\overline{f}_i$ and $\overline{S}_{i(\mathbf{r}_\star)}$ denote the complements: $1 - f_i$, and $1 - S_{i(\mathbf{r}_\star)}$, respectively. Note that the distribution (4.24) can be interpreted as the probability distribution of a collection of $b \times \mathscr{N}$ independent random Bernoulli variables with mean values f_i, $i = 1, \ldots, b$ which depend on the channel index i, but not on the node position $\mathbf{r}_\star$. The expression (4.24) is obviously the most general form for a fully factorized and spatially homogeneous equilibrium measure.

Notice that the identity:

$$f_i = \sum_{S \in \Gamma} S_{i(\mathbf{r}_\star)} \prod_{\mathbf{r}_\star' \in \mathscr{L}} \prod_{j=1}^{b} f_j^{S_j(\mathbf{r}_\star')} \overline{f}_j^{\,\overline{S}_j(\mathbf{r}_\star')}, \qquad \forall i = 1, \ldots, b, \qquad \forall \mathbf{r}_\star \in \mathscr{L}, \tag{4.25}$$

which follows straightforwardly from simple algebra, means that the probabilities f_i in (4.24) are the ensemble-averaged populations $\langle n_i \rangle$ of each channel.

A homogeneous factorized probability distribution such as (4.24) is completely determined by b real numbers: the averaged populations f_i. In order that this probability distribution be an equilibrium distribution, these b real numbers must verify the following set of $2^{(b\times\mathcal{N})}$ relations (stationary lattice Liouville equation):

$$\mathscr{P}\left(\mathscr{S}\left(S'\right)\right)=\sum_{S\in\Gamma}\mathscr{P}(S)\prod_{\mathbf{r}_\star\in\mathscr{L}}A\left(S(\mathbf{r}_\star)\rightarrow S'(\mathbf{r}_\star)\right),\qquad \forall S'\in\Gamma. \tag{4.26}$$

To proceed, we must assume that the lattice $\mathscr{L}$ is either infinite or finite with periodic boundary conditions (see Chapter 2). Under this assumption, all the nodes of the lattice are equivalent, and the free streaming operator only transfers information from node to node without any loss or gain. Since the equilibrium solutions are supposed to be spatially uniform (the f_is do not depend on the node position $\mathbf{r}_\star$), it becomes clear that $\mathscr{P}(\mathscr{S}(S'))=\mathscr{P}(S')$, for any microstate S' in Γ. If we now look for fully factorized solutions of the form given by (4.24), we must solve the following set of equations:

$$\prod_{\mathbf{r}_\star\in\Gamma}\prod_{i=1}^{b}f_i^{S_i'}\overline{f}_i^{\overline{S}_i'}=\sum_{S\in\Gamma}\prod_{\mathbf{r}_\star\in\mathscr{L}}A(S\rightarrow S')\prod_{j=1}^{b}f_j^{S_j}\overline{f}_j^{\overline{S}_j},\qquad \forall S'\in\Gamma, \tag{4.27}$$

where S and S' are implicitly taken at node $\mathbf{r}_\star$. As shown in the Appendix, Section A.4, the above set of $2^{b\times\mathcal{N}}$ equations is strictly equivalent to the set of 2^b equations:

$$\prod_{i=1}^{b}f_i^{s_i'}\overline{f}_i^{\overline{s}_i'}=\sum_{s\in\gamma}A(s\rightarrow s')\prod_{j=1}^{b}f_j^{s_j}\overline{f}_j^{\overline{s}_j},\qquad \forall s'\in\gamma. \tag{4.28}$$

The proof of equivalence does not involve any restricting hypothesis on the lattice gas model. Furthermore, the equivalence is rather intuitive and natural since (4.28) can be viewed as a local (single-node) version of (4.27). The physical content of (4.28) is that the single-node equilibrium probability distribution $\prod_{i=1}^{b}f_i^{s_i}\overline{f}_i^{\overline{s}_i}$ is left unchanged by the application of the collision operator. Indeed, the l.h.s. of (4.28) is the probability of finding the node in state s' *before* collision, and the r.h.s. can be understood as the probability of finding the node in state s' *after* collision.

At this point, it is not obvious whether the set (4.28) has a solution, since it imposes 2^b conditions to only b variables (the f_is). But we shall show that (4.28) reduces to at most (and in fact less than) b independent relations. This 'little miracle' requires the assumption of the semi-detailed balance property (Chapter 2). With this assumption, (4.28) is strictly equivalent to the following set of b equations:

$$\sum_{s,s'\in\gamma}(s_i'-s_i)A(s\rightarrow s')\prod_{j=1}^{b}f_j^{s_j}\overline{f}_j^{\overline{s}_j}=0,\qquad \forall i=1,\ldots,b, \tag{4.29}$$

which is itself equivalent to

$$\sum_{i=1}^{b} \log\left(\frac{f_i}{1-f_i}\right)(s'_i - s_i)A(s \to s') = 0, \qquad \forall s, s' \in \gamma. \tag{4.30}$$

Proving the equivalence between (4.28), (4.29) and (4.30) is not a trivial matter; the proof involves a trick introduced in the context of gas models with discrete velocities (Gatignol, 1975), and is given in the Appendix, Section A.5.

Before proceeding any further, it is useful to discuss the physical content of (4.29) and (4.30). Equation (4.29) expresses that for a global equilibrium, the ensemble-averaged population of any channel is left unchanged by collisions. This interpretation becomes clear when we use the identity (see proof in the Appendix, Section A.6):

$$f_i = \sum_{s,s' \in \gamma} s_i A(s \to s') \prod_{j=1}^{b} f_j^{s_j} \bar{f}_j^{\bar{s}_j}, \qquad \forall i = 1, \ldots, b, \tag{4.31}$$

to rewrite (4.29) under the strictly equivalent form:

$$\sum_{s,s' \in \gamma} s'_i A(s \to s') \prod_{j=1}^{b} f_j^{s_j} \bar{f}_j^{\bar{s}_j} = f_i, \qquad \forall i = 1, \ldots, b. \tag{4.32}$$

The meaning of (4.30) is that the observable $(\log\left(\frac{f_i}{1-f_i}\right), i = 1, \ldots, b)$ is a *collisional invariant*, that is, an observable whose value measured at any node is left unchanged by the collision phase (see Section 2.3.3). The important consequence is that the conservation laws in a lattice gas model have a crucial incidence on its ability to exhibit global statistical equilibria, and thus macroscopic collective motions.

We must now consider the set of all collisional invariants of the lattice gas (see Section 2.3.3). According to the definition (2.19) of a collisional invariant, it is obvious that the set of all collisional invariants of a lattice gas model is a vector subspace of $\mathbb{R}^b$. Its dimension δ lies *a priori* between 0 and b, but the limit cases $\delta = 0$ and $\delta = b$ correspond to rather pathological (unphysical) models. The case $\delta = 0$ corresponds to lattice gas models with no other collisional invariant than the null observable $(0, \ldots, 0)$, and so there is no factorized uniform statistical equilibrium other than the trivial one: $f_i = \frac{1}{2}$, $i = 1, \ldots, b$. The other limit, $\delta = b$, corresponds to lattice gas models with no efficient collisions (the post-collision state is always identical to the pre-collision state), for which any factorized homogeneous distribution is an equilibrium.

Let us now consider a basis $(q^{[\kappa]}, \kappa = 1, \ldots, \delta)$ of the vector subspace of collisional invariants. Any given collisional invariant is a unique linear combination of the form $\sum_{\kappa=1}^{\delta} \lambda^{[\kappa]} q^{[\kappa]}$. According to (4.30), $(\log\left(\frac{f_i}{1-f_i}\right), i = 1,\ldots,b)$ is a collisional invariant. So, there exist δ real numbers $\lambda^{[\kappa]}$ such that:

$$\log\left(\frac{f_i}{1-f_i}\right) = -\sum_{\kappa=1}^{\delta} \lambda^{[\kappa]} q_i^{[\kappa]}, \qquad \forall i = 1,\ldots,b, \tag{4.33}$$

which yields:

$$\begin{aligned} f_i = f_i^{(\mathrm{eq})} &\equiv \frac{1}{1+\exp\left(\sum_{\kappa=1}^{\delta} \lambda^{[\kappa]} q_i^{[\kappa]}\right)} \\ &\equiv \psi_{\mathrm{FD}}\left(\sum_{\kappa=1}^{\delta} \lambda^{[\kappa]} q_i^{[\kappa]}\right), \qquad \forall i = 1,\ldots,b. \end{aligned} \tag{4.34}$$

Here $\psi_{\mathrm{FD}}(x) = (1+e^x)^{-1}$ is the Fermi–Dirac distribution. The fact that we obtain a Fermi–Dirac distribution instead of the Maxwell–Boltzmann distribution of continuous gases, is a consequence of the exclusion principle. Equation (4.34) describes a family of homogeneous and fully factorized equilibrium solutions of the lattice Liouville equation (4.6), depending on δ arbitrary parameters $\lambda^{[\kappa]}$, where δ is the number of linearly independent collisional invariants of the model.

The parameterization by the $\lambda^{[\kappa]}$s comes out naturally in passing from (4.30) to (4.33). In Section 4.4.3, these arbitrary coefficients will be interpreted as Lagrange multipliers in the global entropy maximization procedure.

Two important features must be emphasized:

- We have proved, *without* recourse to the Boltzmann approximation, that lattice gas models satisfying the semi-detailed balance condition possess homogeneous, factorized, equilibrium macrostates with Fermi–Dirac distributions. However, the proof does not preclude the possibility that other kinds of equilibrium macrostates exist, neither does it give information about their stability.
- Provided the collision rules satisfy the semi-detailed balance condition, the homogeneous factorized equilibrium macrostates *do not depend on the details of the collision rules*, but only on the conserved quantities (collisional invariants). Consequently, lattice gas models with different collision rules sharing the same conservation properties and verifying the semi-detailed balance have the same global equilibria.

4.4.2 The lattice Boltzmann approach

A lattice gas in a factorized uniform equilibrium state can be seen as a collection of $b \times \mathcal{N}$ independent Bernoulli random variables. Since the probability distribution of a Bernoulli variable is fully determined by a single number (its averaged value), we can address the problem of equilibrium situations in terms of the ensemble-averaged populations f_i as well as in terms of the probability distribution $\mathscr{P}$. In other words, it is equivalent to searching for equilibrium probability distributions satisfying the stationary lattice Liouville equation, and to searching for equilibrium averaged populations satisfying the stationary lattice Boltzmann equation.

Remember that the lattice Liouville equation is an *exact* equation whereas the lattice Boltzmann equation is *not*. However, when we consider only factorized statistics, the lattice Boltzmann equation becomes *exact*.

From (4.8) it is clear that the time- and space-independent averaged populations f_i, $i = 1, \ldots, b$ must satisfy:

$$f_i = \sum_{s,s' \in \gamma} s'_i A(s \to s') \prod_{j=1}^{b} f_j^{(\mathrm{eq})s_j} \overline{f}_j^{(\mathrm{eq})\overline{s}_j}, \qquad \forall i = 1, \ldots, b, \tag{4.35}$$

or equivalently, with identity (4.31):

$$\sum_{s,s' \in \gamma} (s'_i - s_i) A(s \to s') \prod_{j=1}^{b} f_j^{(\mathrm{eq})s_j} \overline{f}_j^{(\mathrm{eq})\overline{s}_j} = 0, \qquad \forall i = 1, \ldots, b. \tag{4.36}$$

Since (4.36) is identical to (4.29), it follows that the lattice Boltzmann approach is equivalent to the lattice Liouville approach if semi-detailed balance is satisfied; consequently, the equilibrium solutions of the lattice Boltzmann equation are Fermi–Dirac distributions.

4.4.3 The variational approach

As shown in Section 4.3.2, the global entropy H of the lattice gas cannot decrease under the action of the evolution rule, which result is valid only under the Boltzmann approximation, and if the semi-detailed balance condition is satisfied. We search for equilibrium distributions within the class of factorized homogeneous macrostates for which the expression for the global entropy reduces to (see Section 4.3.2):

$$H = -\mathscr{N} \sum_{i=1}^{b} \left[f_i \ln f_i + \overline{f}_i \ln \overline{f}_i \right], \tag{4.37}$$

where f_i does not depend on the position variable $\mathbf{r}_\star$ because of the homogeneity of the equilibrium state.

In order that the macrostate be an equilibrium solution, the global entropy must be maximum, under constraints imposed by conservation. Indeed, consider a basis $(q^{[\kappa]}, k = 1, \ldots, \delta)$ of the vector space of collisional invariants. The quantities $\langle q^{[\kappa]} \rangle = \sum q_i^{[\kappa]} f_i$ are constant, because the $q^{[\kappa]}$ s are collisional invariants.

So we seek b average populations f_i which maximize H under the constraints that the averaged collisional invariants $\sum q_i^{[\kappa]} f_i$ have given values. We then introduce δ Lagrange multipliers $\lambda^{[\kappa]}$ for these δ constraints, and search for average populations f_i that maximize the quantity:

$$-\sum_{i=1}^{b} [f_i \ln f_i + \overline{f}_i \ln \overline{f}_i] - \sum_{\kappa=1}^{\delta} \lambda^{[\kappa]} \sum_{i=1}^{b} q_i^{[\kappa]} f_i.$$

It is a simple matter to show that the solution is the Fermi–Dirac distribution (4.34).

In conclusion, we have shown that the lattice Liouville approach, the lattice Boltzmann approach and the variational approach yield strictly equivalent results when semi-detailed balance is satisfied.

The expression (4.34) for the family of global factorized equilibrium distributions gives the equilibrium mean populations f_i in terms of the Lagrange multipliers $\lambda^{[\kappa]}$. This is natural from a theoretical point of view, but practically it would be more convenient to have these expressions in terms of physically relevant and accessible quantities such as the *macroscopic variables* of the lattice gas, that is, in terms of the macroscopic densities of the collisional invariants (see Section 2.1.3). For example, consider a lattice gas model for standard fluid dynamics, for which the collisional invariants are the mass, the momentum and the energy. The physically relevant variables are the mass density, momentum density and energy density, and it is useful (for example to set up the initial lattice state in a simulation) to have the mean population of each channel in terms of these mass, momentum and energy densities. The next section is dedicated to this question.[6]

4.5 Natural parameterization of equilibria

We consider a lattice gas prepared in a homogeneous, fully factorized equilibrium macrostate $\mathscr{P}$ such that:

$$\mathscr{P}(S) = \mathscr{P}^{(\mathrm{eq})}(S) \equiv \prod_{\mathbf{r}_\star \in \mathscr{L}} \prod_{i=1}^{b} f_i^{(\mathrm{eq}) S_i(\mathbf{r}_\star)} \overline{f}_i^{(\mathrm{eq}) \overline{S}_i(\mathbf{r}_\star)}, \tag{4.38}$$

where the equilibrium mean populations $f_i^{(\mathrm{eq})}$ are given by the Fermi–Dirac distribution (4.34).

We define the macroscopic variables $\rho^{[\kappa]}$ as the *macroscopic densities per node of the collisional invariants* $q^{[\kappa]}$, that is, the microscopic densities per node $\rho^{\star}_{q^{[\kappa]}}$,

6 Section 4.5 contains material which is important for lattice gas practice, but it is rather technical and may be omitted at first reading.

ensemble-averaged over the probability distribution $\mathscr{P}^{(\mathrm{eq})}$ (see Section 2.1.3 for the notion of density of an observable):

$$\rho^{[\kappa]} \equiv \left\langle \rho^{\star}_{q^{[\kappa]}} \right\rangle = \sum_{i=1}^{b} q_i^{[\kappa]} f_i^{(\mathrm{eq})}.$$

This definition of the macroscopic variables leads to the following set of δ non-algebraic relations between the macroscopic variables $\rho^{[\kappa]}$ and the Lagrange multipliers $\lambda^{[\kappa]}$:

$$\rho^{[\kappa]} = \sum_{i=1}^{b} q_i^{[\kappa]} \psi_{\mathrm{FD}} \left(\sum_{\kappa'=1}^{\delta} \lambda^{[\kappa']} q_i^{[\kappa']} \right), \qquad \forall \kappa = 1, \ldots, \delta. \tag{4.39}$$

If we are able to solve this set of equations, i.e. if we can write the $\lambda^{[\kappa]}$s as functions of the $\rho^{[\kappa]}$s, then we can express the equilibrium distribution (4.34) in terms of the macroscopic variables $\rho^{[\kappa]}$.

Unfortunately, the general solution to (4.39) cannot be cast in an explicit form, and we have no exact expression for the equilibrium distribution in terms of macroscopic variables. However, approximations can be obtained by perturbation methods. Indeed, suppose that we have an exact solution ($\lambda^{[\kappa]}_{(0)}$, $\kappa = 1, \ldots, \delta$) for (4.39), corresponding to some particular values ($\rho^{[\kappa]}_{(0)}$, $\kappa = 1, \ldots, \delta$) of the macroscopic variables. Then it is usually possible to find an approximate solution ($\lambda^{[\kappa]}$, $\kappa = 1, \ldots, \delta$), corresponding to macroscopic variables ($\rho^{[\kappa]}$, $\kappa = 1, \ldots, \delta$) close to ($\rho^{[\kappa]}_{(0)}$, $\kappa = 1, \ldots, \delta$).

We now sketch the asymptotic expansion leading to this approximate solution. Let η ($\ll 1$) be an expansion parameter such that the perturbed macroscopic variables read:

$$\rho^{[\kappa]} = \rho^{[\kappa]}_{(0)} + \eta \rho^{[\kappa]}_{(1)}, \qquad \kappa = 1, \ldots, \delta. \tag{4.40}$$

(The subscript between parentheses refers to the order in the expansion scheme, whereas the superscript between square brackets refers to the index of the associated collisional invariant.)

We look for solutions of (4.39) with the following form:

$$\lambda^{[\kappa]} = \lambda^{[\kappa]}_{(0)} + \eta \lambda^{[\kappa]}_{(1)} + \eta^2 \lambda^{[\kappa]}_{(2)} + \mathcal{O}(\eta^3), \qquad \kappa = 1, \ldots, \delta. \tag{4.41}$$

The expansion will be described in detail up to second order in η, which is sufficient for hydrodynamics.[7]

For simplicity, we introduce compact notations for the Taylor expansion of the Fermi–Dirac distribution $\psi_{\mathrm{FD}}(x)$, namely f_{0i}, f_{1i} and f_{2i} for the values of

[7] The method can be applied to any order, but with rapidly increasing algebraic complexity.

the Fermi–Dirac distribution and of its first and second derivatives respectively, computed for the unperturbed solution ($\lambda_{(0)}^{[\kappa]}$, $\kappa = 1, \ldots, \delta$):

$$
\begin{aligned}
f_{0i} &\equiv \psi_{\mathrm{FD}}\left(\sum_{\kappa=1}^{\delta} \lambda_{(0)}^{[\kappa]} q_i^{[\kappa]}\right) \equiv \psi_{\mathrm{FD}}(x), \\
f_{1i} &\equiv \frac{d}{dx}\psi_{\mathrm{FD}}\left(\sum_{\kappa=1}^{\delta} \lambda_{(0)}^{[\kappa]} q_i^{[\kappa]}\right), \\
f_{2i} &\equiv \frac{d^2}{dx^2}\psi_{\mathrm{FD}}\left(\sum_{\kappa=1}^{\delta} \lambda_{(0)}^{[\kappa]} q_i^{[\kappa]}\right).
\end{aligned}
\tag{4.42}
$$

Using the explicit form $1/(1+\exp x)$ for ψ_{FD}, it is easy to show that:

$$
f_{1i} = -f_{0i}(1-f_{0i}) \quad \text{and} \quad f_{2i} = f_{0i}(1-f_{0i})(1-2f_{0i}). \tag{4.43}
$$

The Taylor expansion of the equilibrium populations is straightforward:

$$
\begin{aligned}
f_i &\equiv \psi_{\mathrm{FD}}\left(\sum_{\kappa=1}^{\delta} \lambda^{[\kappa]} q_i^{[\kappa]}\right) \\
&= f_{0i} \\
&+ \eta\, f_{1i} \sum_{\kappa=1}^{\delta} \lambda_{(1)}^{[\kappa]} q_i^{[\kappa]} \\
&+ \eta^2 \left(f_{1i} \sum_{\kappa=1}^{\delta} \lambda_{(2)}^{[\kappa]} q_i^{[\kappa]} + \frac{1}{2} f_{2i} \left(\sum_{\kappa=1}^{\delta} \lambda_{(1)}^{[\kappa]} q_i^{[\kappa]} \right)^2 \right) \\
&+ O(\eta^3), \qquad i = 1, \ldots, b.
\end{aligned}
\tag{4.44}
$$

It then follows that (4.39), to first order in η, leads to the following linear set for ($\lambda_{(1)}^{[\kappa]}$, $\kappa = 1, \ldots, \delta$):

$$
\sum_{\kappa'=1}^{\delta} M^{[\kappa,\kappa']} \lambda_{(1)}^{[\kappa']} = \rho_{(1)}^{[\kappa]}, \tag{4.45}
$$

where M is a real δ by δ square matrix with elements:

$$
M^{[\kappa,\kappa']} = \sum_{i=1}^{b} q_i^{[\kappa]} q_i^{[\kappa']} f_{1i}, \qquad \kappa = 1, \ldots, \delta, \qquad \kappa' = 1, \ldots, \delta. \tag{4.46}
$$

This matrix is invertible (see proof in the Appendix, Section A.7); so the first order linear set (4.45) has a unique solution ($\lambda_{(1)}^{[\kappa]}$, $\kappa = 1, \ldots, \delta$), which is explicitly given in terms of the macroscopic variables.

The second order of the expansion in powers of η leads to a similar linear set for the $\lambda_{(2)}^{[\kappa]}$s, given in terms of the $\lambda_{(1)}^{[\kappa]}$s:

$$
\sum_{\kappa'=1}^{\delta} M^{[\kappa,\kappa']} \lambda_{(2)}^{[\kappa']} = -\frac{1}{2} \sum_{i=1}^{b} q_i^{[\kappa]} f_{2i} \left(\sum_{\kappa''=1}^{\delta} \lambda_{(1)}^{[\kappa'']} q_i^{[\kappa'']} \right)^2. \tag{4.47}
$$

This set also has a unique solution. The structure is similar for higher orders, but the r.h.s. grows more and more cumbersome.

The next step is to re-introduce the expressions of the perturbations $\lambda_{(1)}^{[\kappa]}$ and $\lambda_{(2)}^{[\kappa]}$ into (4.44) to obtain the desired approximate expression to the average populations f_i.

This expansion is quite general, but rather formal. We now turn to examples so that these notions become more practical. We first apply the general expansion scheme to a class of simple lattice gases which contains most models with fluid-like non-thermal macroscopic properties (Section 4.5.1). Then we deal with the more complex case of thermal models (Section 4.5.2).

4.5.1 Low-speed equilibria for single-species non-thermal models

We illustrate the formalism for a specific but wide class of lattice gases that display fluid-like purely kinetic (non-thermal) macroscopic behavior. These models, which are of practical interest for fluid dynamics simulations, are defined by the following set of properties:

(i) The model is G-invariant (see Section 2.3.4).
(ii) The model has crystallographic isotropy, at least up to second order (see Section 2.3.5).
(iii) The model verifies the irreducibility condition (see Section 2.3.6).
(iv) The model verifies the semi-detailed balance condition (see Section 2.3.1).[8]
(v) All particles have unit mass.[9]
(vi) The model has $\delta = D + 1$ linearly independent collisional invariants (see Section 2.3.3), namely the mass and the D components of the momentum:

$$q^{[0]} = \begin{pmatrix} 1 \\ \cdot \\ \cdot \\ \cdot \\ 1 \end{pmatrix}, \qquad q^{[\alpha]} = \begin{pmatrix} c_{1\alpha} \\ \cdot \\ \cdot \\ \cdot \\ c_{b\alpha} \end{pmatrix}, \tag{4.48}$$

where the Greek index α refers to Cartesian coordinates and ranges from 1 to D.[10]

Property (v) – all particles have same mass – leads to considerably lighter algebra. The formalism, however, extends very simply to the case where only

[8] Property (iv) excludes models FCHC-6, -7, and -8 (see Table 3.3).
[9] Property (v) excludes models FCHC-5, -7, and -8 (see Table 3.3), because they have rest particles with mass 1 and 2.
[10] Property (vi) excludes the thermal GBL model which has an additional independent collisional invariant (the kinetic energy, see Section 3.6.2), and will be treated in Section 4.5.2.

moving particles have the same mass, while the mass of rest particles can take different values (for instance, for models FCHC-5, -7, and -8).

We now use property (ii) (crystallographic isotropy up to order 2) to define a characteristic number, ξ_2, which will be useful in the subsequent calculations. Property (ii) implies that the crystallographic tensor $\sum_i c_{i\alpha}c_{i\beta}$ is isotropic (see Section 2.3.5 for a definition of the notion of isotropic tensors); so there exists a number ξ_2 such that:

$$\sum_{i=1}^{b} c_{i\alpha}c_{i\beta} \quad = \quad \xi_2\delta_{\alpha\beta}. \tag{4.49}$$

The characteristic number ξ_2 depends on the geometrical structure of the lattice gas automaton model, and not on the collision rules. Table 4.1 in Section 4.5.2 gives the numerical values of ξ_2 for several models.

The macroscopic variables associated with the collisional invariants are ρ, the *mass density per node* and $\mathbf{j}$, the *momentum density per node*:

$$\begin{aligned} \rho &\equiv \rho^{[0]} = \sum_{i=1}^{b} f_i, \\ j_\alpha &\equiv \rho^{[\alpha]} = \sum_{i=1}^{b} c_{i\alpha} f_i \qquad (\alpha = 1, \ldots, D). \end{aligned} \tag{4.50}$$

We choose to study nearly equally distributed equilibria, that is, global equilibria for which the average population of all channels is close to a common value. The basic unperturbed equilibrium is a macrostate in which all channels have equal probability to be occupied: $f_{0i} = d, \ \forall i = 1, \ldots, b$. The parameter d $(0 \leq d \leq 1)$ can be interpreted as the 'particle density per channel'.

These equilibria are also called 'low-speed equilibria', because the mean velocity of particles at any node is small, as soon as the populations of all the channels on that node are almost equal. For instance, if there are approximately as many upward moving particles as downward moving particles, the mean velocity is obviously small.

The values of the macroscopic variables corresponding to the unperturbed (fully equally distributed) equilibrium are:[11]

$$\begin{aligned} \rho^{[0]}_{(0)} &= \rho = d\,b, \\ \rho^{[\alpha]}_{(0)} &= 0, \quad \alpha = 1, \ldots, D. \end{aligned} \tag{4.51}$$

[11] Remember that the subscript between parentheses refers to the order in the expansion scheme, whereas the superscript between square brackets refers to the index of the associated collisional invariant.

For these values of the macroscopic variables, it is easy to verify that the following Lagrange multipliers satisfy (4.39):

$$\lambda_{(0)}^{[0]} = \log\left(\frac{1-d}{d}\right),$$
$$\lambda_{(0)}^{[\alpha]} = 0, \qquad \alpha = 1,\ldots,D. \tag{4.52}$$

We look for nearly equally distributed equilibrium states for which the mass density is still ρ, but the momentum density is now $\mathbf{j} = \mathbf{0} + \eta \mathbf{j}_{(1)}$, where η is an expansion (small) parameter. The perturbations ($\rho_{(1)}^{[\kappa]}$, $\kappa = 1,\ldots,\delta$) are:

$$\rho_{(1)}^{[0]} = 0,$$
$$\rho_{(1)}^{[\alpha]} = j_{(1)\alpha}, \quad \alpha = 1,\ldots,D. \tag{4.53}$$

From the unperturbed values (4.52) of the Lagrange multipliers, we can compute the values of the expansion coefficients f_{0i} (using (4.42) and (4.48)), and f_{1i} and f_{2i} (using (4.43)):

$$f_{0i} = d,$$
$$f_{1i} = -d(1-d), \tag{4.54}$$
$$f_{2i} = d(1-d)(1-2d).$$

As a consequence of the fact that all particles have equal mass, these quantities are independent of the channel index i.

It is now straightforward to compute the $\delta \times \delta$ real matrix M defined in (4.46). We find:

$$M^{[\kappa,\kappa']} = -d(1-d)\begin{pmatrix} b & 0 & . & . & 0 \\ 0 & \xi_2 & . & . & 0 \\ . & . & . & . & . \\ . & . & . & . & . \\ 0 & 0 & . & . & \xi_2 \end{pmatrix}, \tag{4.55}$$

Since $M^{[\kappa,\kappa']}$ is diagonal, the solution of the first order problem (4.45) is simply:

$$\lambda_{(1)}^{[0]} = 0,$$
$$\lambda_{(1)}^{[\alpha]} = -\frac{1}{d(1-d)}\frac{j_{(1)\alpha}}{\xi_2}, \qquad \alpha = 1,\ldots,D. \tag{4.56}$$

To compute the second order perturbations $\lambda_{(2)}^{[\kappa]}$, we need to evaluate the r.h.s. of the second order equation (4.47). This operation does not involve any conceptual difficulty, but requires some care with the algebraic manipulations. Section A.8 in the Appendix sketches this operation for the more complicated case of thermal models. For non-thermal models, the algebra is similar, but simpler.

The resulting values for the second order corrections $\lambda_{(2)}^{[\kappa]}$ to the Lagrange multipliers are:

$$\lambda_{(2)}^{[0]} = \frac{1-2d}{2d^2(1-d)^2} \frac{\mathbf{j}_{(1)}^2}{b\,\xi_2}, \qquad \lambda_{(2)}^{[\alpha]} = 0, \qquad \alpha = 1,\ldots,D. \tag{4.57}$$

The final expression for the perturbed mean populations f_i can be obtained from (4.44), using the expressions (4.56) and (4.57) for the first and second order perturbations $\lambda_{(1)}^{[\kappa]}$ and $\lambda_{(2)}^{[\kappa]}$ to the Lagrange multipliers. The resulting expression reads, in terms of the macroscopic variables $\rho = bd$ and $\mathbf{j}$:[12]

$$\begin{aligned} f_i &= \frac{\rho}{b} \\ &+ \frac{1}{\xi_2} j_\alpha c_{i\alpha} \\ &+ \frac{(b-2\rho)}{2\rho(b-\rho)} j_\alpha j_\beta \left(c_{i\alpha}c_{i\beta} - \frac{\xi_2}{b}\delta_{\alpha\beta}\right) \\ &+ \mathcal{O}(j^3). \end{aligned} \tag{4.58}$$

We can rewrite (4.58) in terms of the 'fluid velocity' vector $\mathbf{u} = \mathbf{j}/\rho$, to obtain:

$$\begin{aligned} f_i &= \frac{\rho}{b} \\ &+ \frac{\rho}{\xi_2} u_\alpha c_{i\alpha} \\ &+ \rho G(\rho) u_\alpha u_\beta \left(c_{i\alpha}c_{i\beta} - \frac{\xi_2}{b}\delta_{\alpha\beta}\right) \\ &+ \mathcal{O}(u^3), \end{aligned} \tag{4.59}$$

where the density-dependent factor $G(\rho)$ is given by:[13]

$$G(\rho) = \frac{(b-2\rho)}{2(b-\rho)} \frac{b}{\xi_2^2}. \tag{4.60}$$

Notice the similarity between (4.59) and the low-speed expansion of the Maxwell–Boltzmann distribution in classical statistical mechanics:

$$f(\mathbf{c}) = \left(\frac{m}{2\pi k_B T}\right)^{3/2} \exp\left(-\frac{(\mathbf{c}-\mathbf{u})^2}{2k_B T}\right), \tag{4.61}$$

[12] Remember that Einstein's convention is used for implicit summation over repeated Greek indices.

[13] This factor and the 'non-Galilean factor' $g(\rho)$ in Chapter 5 differ by a multiplicative constant.

which gives:

$$f(\mathbf{c}) \simeq \text{const.}\left[1+\frac{m}{k_B T}u_\alpha c_\alpha + \frac{1}{2}\left(\frac{m}{k_B T}\right)^2 u_\alpha u_\beta \left(c_\alpha c_\beta - \frac{k_B T}{m}\delta_{\alpha\beta}\right) + \mathcal{O}(u^3)\right]. \tag{4.62}$$

4.5.2 Nearly equally distributed equilibria for thermal models

We now apply the formalism to a class of single-species lattice gases with thermal effects; this class is defined by the following properties:

(i) The model is G-invariant (see Section 2.3.4).
(ii) The model has crystallographic isotropy at least up to *fourth* order (see Section 2.3.5).
(iii) The model verifies the irreducibility condition (see Section 2.3.6).
(iv) The model verifies the semi-detailed balance condition (see Section 2.3.1).
(v) All particles have unit mass.
(vi) The model has $\delta = D + 2$ linearly independent collisional invariants (see Section 2.3.3), namely the mass, the D components of the momentum, and the kinetic energy:[14]

$$\begin{pmatrix} 1 \\ \cdot \\ \cdot \\ \cdot \\ 1 \end{pmatrix}, \quad \begin{pmatrix} c_{1\alpha} \\ \cdot \\ \cdot \\ \cdot \\ c_{b\alpha} \end{pmatrix}, \quad \begin{pmatrix} \frac{1}{2}\mathbf{c}_1^2 \\ \cdot \\ \cdot \\ \cdot \\ \frac{1}{2}\mathbf{c}_b^2 \end{pmatrix}. \tag{4.63}$$

This set of properties is almost the same as in Section 4.5.1, except that fourth order crystallographic isotropy is now required, and that energy conservation is taken into account. The GBL model for thermo-hydrodynamics is a typical example of this class of LGAs.

We use property (ii) (crystallographic isotropy up to order 4) to define two additional characteristic numbers, ξ_4 and ξ_6 (ξ_2 was defined in Section 4.5.1). Indeed, property (ii) implies that the crystallographic tensors $\sum_i c_{i\alpha}c_{i\beta}$ and $\sum_i c_{i\alpha}c_{i\beta}c_{i\gamma}c_{i\delta}$ are isotropic (see Section 2.3.5 for a definition of isotropic tensors). Moreover, two velocity vectors corresponding to two channels belonging to the same class (see Section 2.3.4) are connected by an isometry of the crystallographic point group G of the underlying lattice. So all velocity vectors in a given class have the same modulus, and the quantity $\frac{\mathbf{c}_i^2}{2}$ depends only on the class index I of $\mathbf{c}_i$. As a consequence, the fourth order tensor $\sum_i(\mathbf{c}_i^2/2)c_{i\alpha}c_{i\beta}c_{i\gamma}c_{i\delta}$ is a crystallographic tensor (as defined in Section 2.3.5), and is therefore isotropic

[14] The Greek index α refers to Cartesian coordinates and ranges from 1 to D.

according to property (ii). Since the three crystallographic tensors defined above are isotropic, there exist three numbers, ξ_2, ξ_4 and ξ_6, such that:

$$\begin{aligned}
\sum_{i=1}^{b} c_{i\alpha}c_{i\beta} &= \xi_2\delta_{\alpha\beta}, \\
\sum_{i=1}^{b} c_{i\alpha}c_{i\beta}c_{i\gamma}c_{i\delta} &= \frac{1}{D(D+2)}\left(4\xi_4 + \frac{D^2\xi_2^2}{b}\right) \\
&\quad \times\left(\delta_{\alpha\beta}\delta_{\gamma\delta} + \delta_{\alpha\gamma}\delta_{\beta\delta} + \delta_{\alpha\delta}\delta_{\beta\gamma}\right), \\
\sum_{i=1}^{b} \frac{\mathbf{c}_i^2}{2} c_{i\alpha}c_{i\beta}c_{i\gamma}c_{i\delta} &= \frac{1}{D(D+2)}\left(4\xi_6 + \frac{6D\xi_2\xi_4}{b} + \frac{D^3\xi_2^3}{2b^2}\right) \\
&\quad \times\left(\delta_{\alpha\beta}\delta_{\gamma\delta} + \delta_{\alpha\gamma}\delta_{\beta\delta} + \delta_{\alpha\delta}\delta_{\beta\gamma}\right).
\end{aligned} \tag{4.64}$$

The three characteristic numbers ξ_2, ξ_4 and ξ_6 depend on the geometrical structure of the lattice gas automaton model, and not on the collision rules. The apparently strange choice for the definition of ξ_4 and ξ_6 will become clear below. Table 4.1 gives the values of ξ_2, ξ_4 and ξ_6 for various LGA models.

It is technically more convenient to choose an orthogonal basis for the space of collisional invariants, that is, a set of δ collisional invariants $q^{[\kappa]}$ such that

Table 4.1 The geometrical characteristic coefficients ξ_2, ξ_4 and ξ_6 for various lattice gas models. (See Equation (4.64) for definitions, and Chapter 3 for a description of the models).

Model name	D	b	ξ_2	ξ_4	ξ_6
HPP	2	4	2	—	—
FHP-1	2	6	3	0	0
FHP-2 and -3	2	7	3	$\frac{3}{14}$	$-\frac{15}{196}$
GBL	2	19	24	$\frac{165}{19}$	$-\frac{795}{361}$
FCHC-1 to -4	4	24	12	0	0

$\sum_{i=1}^{b} q_i^{[\kappa]} q_i^{[\kappa']} = 0$ when $\kappa \neq \kappa'$. The simplest and most physical choice is:

$$q^{[0]} = \begin{pmatrix} 1 \\ \cdot \\ \cdot \\ \cdot \\ 1 \end{pmatrix}, \qquad q^{[\alpha]} = \begin{pmatrix} c_{1\alpha} \\ \cdot \\ \cdot \\ \cdot \\ c_{b\alpha} \end{pmatrix} \qquad (\alpha = 1, \ldots, D),$$

$$q^{[D+1]} = \begin{pmatrix} e_1 \equiv \frac{1}{2}\mathbf{c}_1^2 - \frac{D\xi_2}{2b} \\ \cdot \\ \cdot \\ \cdot \\ e_b \equiv \frac{1}{2}\mathbf{c}_b^2 - \frac{D\xi_2}{2b} \end{pmatrix}. \tag{4.65}$$

The collisional invariants $q^{[0]}$ and $q^{[\alpha]}$ are the mass and the momentum of the particles, respectively; however, $q^{[D+1]}$ is not the kinetic energy, but a combination of the kinetic energy and of the mass, and is also a collisional invariant which we call the 'effective energy'. It can be verified that the definition of the microscopic effective energies e_i guarantees the desired orthogonality property.

The macroscopic variables associated with these collisional invariants are respectively the 'mass density per node' ρ, the 'momentum density per node' $\mathbf{j}$, and the 'effective energy density per node' w, (a linear combination of the kinetic energy and mass densities, which we shall call 'energy' for short):

$$\begin{aligned} \rho \equiv \rho^{[0]} &= \sum_{i=1}^{b} f_i \\ j_\alpha \equiv \rho^{[\alpha]} &= \sum_{i=1}^{b} c_{i\alpha} f_i, \qquad (\alpha = 1, \ldots, D), \\ w \equiv \rho^{[D+1]} &= \sum_{i=1}^{b} e_i f_i = \sum_{i=1}^{b} \frac{1}{2}\mathbf{c}_i^2 f_i - \frac{D\xi_2}{2b} \sum_{i=1}^{b} f_i. \end{aligned} \tag{4.66}$$

We study nearly equally distributed equilibria, and we proceed along the same lines as in the previous section. Here the values of the macroscopic variables for the unperturbed equilibrium are:

$$\begin{aligned} \rho_{(0)}^{[0]} &= \rho = d\,b, \\ \rho_{(0)}^{[\alpha]} &= 0, \quad \alpha = 1, \ldots, D, \\ \rho_{(0)}^{[D+1]} &= 0, \end{aligned} \tag{4.67}$$

and the Lagrange multipliers satisfying (4.39) are:

$$\begin{aligned} \lambda^{[0]}_{(0)} &= \log\left(\frac{1-d}{d}\right), \\ \lambda^{[\alpha]}_{(0)} &= 0, \quad \alpha = 1,\ldots,D, \\ \lambda^{[D+1]}_{(0)} &= 0. \end{aligned} \tag{4.68}$$

We now look for nearly equally distributed equilibrium states with mass density ρ, momentum density $\mathbf{j} = \mathbf{0} + \eta \mathbf{j}_{(1)}$, and energy density $w = 0 + \eta w_{(1)}$, where η is an expansion (small) parameter. The perturbations $(\rho^{[\kappa]}_{(1)}, \kappa = 1,\ldots,\delta)$ are now:

$$\begin{aligned} \rho^{[0]}_{(1)} &= 0, \\ \rho^{[\alpha]}_{(1)} &= j_{(1)\alpha}, \quad \alpha = 1,\ldots,D, \\ \rho^{[D+1]}_{(1)} &= w_{(1)}. \end{aligned} \tag{4.69}$$

The values of the expansion coefficients f_{0i}, f_{1i} and f_{2i} are the same as given in (4.54), and the $\delta \times \delta$ real matrix M defined in (4.46) reads:

$$M^{[\kappa,\kappa']} = -d(1-d)\begin{pmatrix} b & 0 & . & . & 0 & 0 \\ 0 & \xi_2 & . & . & 0 & 0 \\ . & . & . & . & . & . \\ . & . & . & . & . & . \\ 0 & 0 & . & . & \xi_2 & 0 \\ 0 & 0 & . & . & 0 & \xi_4 \end{pmatrix}. \tag{4.70}$$

The choice of an orthogonal basis for the space of collisional invariants has the beneficial consequence to render M diagonal. In addition, all diagonal elements are non-zero, since the collisional invariants $q^{[\kappa]}$ are linearly independent. The particular choice for the definition of ξ_4 in (4.64) is now seen to be justified as we obtain simple expressions for the coefficients of M. The solution to the first order problem (4.45) is simply:

$$\begin{aligned} \lambda^{[0]}_{(1)} &= 0, \\ \lambda^{[\alpha]}_{(1)} &= -\frac{1}{d(1-d)}\frac{j_{(1)\alpha}}{\xi_2}, \quad \alpha = 1,\ldots,D, \\ \lambda^{[D+1]}_{(1)} &= -\frac{1}{d(1-d)}\frac{w_{(1)}}{\xi_4}. \end{aligned} \tag{4.71}$$

The computation of the second order perturbations $\lambda^{[\kappa]}_{(2)}$ requires the evaluation of the r.h.s. of the second order equation (4.47); the algebra is given in the

Appendix, Section A.8. The results read:

$$\begin{aligned}
\lambda^{[0]}_{(2)} &= \frac{1-2d}{2d^2(1-d)^2}\left(\frac{1}{b\xi_2}\mathbf{j}^2_{(1)} + \frac{1}{b\xi_4}w^2_{(1)}\right),\\
\lambda^{[\alpha]}_{(2)} &= \frac{1-2d}{2d^2(1-d)^2}\left(\frac{4}{D\xi_2^2}w_{(1)}j_{(1)\alpha}\right), \quad \alpha = 1,\ldots,D,\\
\lambda^{[D+1]}_{(2)} &= \frac{1-2d}{2d^2(1-d)^2}\left(\frac{2}{D\xi_2^2}\mathbf{j}^2_{(1)} + \frac{\xi_6}{\xi_4^3}w^2_{(1)}\right).
\end{aligned} \tag{4.72}$$

With (4.71) and (4.72), the final expression for the perturbed mean populations f_i is obtained in terms of the macroscopic variables $\rho = bd$, $\mathbf{j}$ and w:

$$\begin{aligned}
f_i &= \frac{\rho}{b}\\
&+ \frac{1}{\xi_2}j_\alpha c_{i\alpha} + \frac{1}{\xi_4}we_i\\
&+ \frac{b(b-2\rho)}{2\rho(b-\rho)}\Bigg[\frac{1}{\xi_2^2}j_\alpha j_\beta\left(c_{i\alpha}c_{i\beta} - \frac{\mathbf{c}_i^2}{D}\delta_{\alpha\beta}\right)\\
&+ \frac{2}{\xi_2\xi_4}wj_\alpha\left(e_i - \frac{2\xi_4}{D\xi_2}\right)c_{i\alpha}\\
&+ \frac{1}{\xi_4^2}w^2\left(e_i^2 - \frac{\xi_6}{\xi_4}e_i - \frac{\xi_4}{b}\right)\Bigg]\\
&+ \mathcal{O}(j^3) + \mathcal{O}(j^2w) + \mathcal{O}(jw^2) + \mathcal{O}(w^3).
\end{aligned} \tag{4.73}$$

or alternatively, in terms of the velocity $\mathbf{u} = \mathbf{j}/\rho$:

$$\begin{aligned}
f_i &= \frac{\rho}{b}\\
&+ \frac{\rho}{\xi_2}u_\alpha c_{i\alpha} + \frac{1}{\xi_2}we_i\\
&+ \rho G(\rho)\Bigg[u_\alpha u_\beta\left(c_{i\alpha}c_{i\beta} - \frac{\mathbf{c}_i^2}{D}\delta_{\alpha\beta}\right)\\
&+ \frac{2\xi_2}{\xi_4}\frac{w}{\rho}u_\alpha\left(e_i - \frac{2\xi_4}{D\xi_2}\right)c_{i\alpha}\\
&+ \frac{\xi_2^2}{\xi_4^2}\frac{w^2}{\rho^2}\left(e_i^2 - \frac{\xi_6}{\xi_4}e_i - \frac{\xi_4}{b}\right)\Bigg]\\
&+ \mathcal{O}(u^3) + \mathcal{O}(u^2w) + \mathcal{O}(uw^2) + \mathcal{O}(w^3),
\end{aligned} \tag{4.74}$$

where the factor $G(\rho)$ is defined in Equation (4.60).

This is the type of expression for the mean populations given in terms of the macroscopic variables (mass, momentum and energy densities) that we shall use to compute the fluxes of the same conserved quantities, in order to obtain the macrodynamic equations of the lattice gas. This result is also of practical interest for fluid dynamics simulations. Indeed (as in computational fluid dynamics (CFD) methods) the system must be initialized with values of the numerical variables which represent the desired *physical* initial situation, e.g. describing a field of given mass density, velocity and temperature. To initialize the Boolean field in a lattice gas simulation, we must decide the random occupation level (0 or 1) of each channel at each node, with the constraint that the resulting macroscopic fields match the desired initial physical situation. An efficient way is to use the approximate equilibrium mean populations in (4.74) as the probabilities for the random distribution of 0s and 1s; this procedure offers the advantage that the initial state of the Boolean field is close to a (local) equilibrium state, so that spurious transient regimes are avoided (or at least considerably shortened).

4.6 Statistical thermodynamics

Lattice gases with single (unitary) velocity modulus (such as the FHP and FCHC models) are well suited for the investigation of a large class of hydrodynamic phenomena. They are by nature *non-thermal* models: mass and momentum are the conserved physical quantities, and energy conservation is simply a consequence of mass conservation. So the notion of temperature is irrelevant in lattice gases with single non-zero velocity modulus. For a *thermal* lattice gas, obviously we must have a multi-speed model. While this requirement is a necessary condition (as fulfilled e.g. by the GBL model), the definition of a temperature in the lattice gas does not follow straightforwardly, and in fact it is a rather subtle matter to establish a *lattice gas thermodynamics*.[15]

In classical statistical thermodynamics, temperature T is introduced via the canonical distribution function where $\exp(-\beta\mathscr{H})$ (with $\mathscr{H}$ the total energy) contains the quantity $\beta = (k_B T)^{-1}$ (with k_B the Boltzmann constant). Can we legitimately expect a similar relation in the thermal lattice gas? One of the complications arises from the exclusion principle (as in the Fermi gas) which, in the lattice gas, imposes an upper bound to the energy. Consider for instance the equipartition:

$$\frac{3}{2}\beta^{-1} = \int d\mathbf{p}\, p^2\, e^{-\beta p^2} \Big/ \int d\mathbf{p}\, e^{-\beta p^2}, \tag{4.75}$$

[15] This problem was considered by Ernst (1991), Cercignani (1994), and Grosfils (1994).

whose analogue for the general class of models with a finite number of discrete velocities (including the class of lattice gases) would read:

$$\frac{D}{2}\beta^{-1} = \sum_i \mathbf{c}_i^2 e^{-\beta \mathbf{c}_i^2} \Big/ \sum_i e^{-\beta \mathbf{c}_i^2}, \tag{4.76}$$

where the sum is taken over all velocity channels. To yield a standard definition of the temperature, the above expression should be satisfied for any value of β, which is clearly not the case when i is discrete and finite.

Let us consider a volume element in the lattice gas at thermodynamic equilibrium; the size of the volume element – the region $\mathscr{L}$ of the lattice – is large compared to v_0, the elementary volume associated to a node. So we shall use the *grand canonical ensemble.* We start from the probability density in phase space, (4.38):

$$\mathscr{P}^{(\mathrm{eq})}(S) = \prod_{\mathbf{r}_\star \in \mathscr{L}} \prod_{i=1}^{b} f_i^{(\mathrm{eq})S_i(\mathbf{r}_\star)} \bar{f}_i^{(\mathrm{eq})\bar{S}_i(\mathbf{r}_\star)}, \tag{4.77}$$

where $f_i^{(\mathrm{eq})}$ is the Fermi–Dirac distribution (4.34):

$$f_i^{(\mathrm{eq})} = \psi_{\mathrm{FD}}\left(\sum_{\kappa=1}^{\delta} \lambda^{[\kappa]} q_i^{[\kappa]}\right) = \left[1 + \exp\left(\sum_{\kappa=1}^{\delta} \lambda^{[\kappa]} q_i^{[\kappa]}\right)\right]^{-1}, \tag{4.78}$$

and with the normalization: $\sum_{S\in\Gamma} \mathscr{P}^{(\mathrm{eq})}(S) = 1$, which is easily verified with (4.77) since the S_is take the value 0 or 1. We rewrite (4.77) as:

$$\mathscr{P}^{(\mathrm{eq})}(S) = \prod_{\mathbf{r}_\star \in \mathscr{L}} \prod_{i=1}^{b} \left(\frac{f_i^{(\mathrm{eq})}}{1 - f_i^{(\mathrm{eq})}}\right)^{S_i(\mathbf{r}_\star)} (1 - f_i^{(\mathrm{eq})}), \tag{4.79}$$

where

$$\left(\frac{f_i^{(\mathrm{eq})}}{1 - f_i^{(\mathrm{eq})}}\right) = \exp - \sum_{\kappa=1}^{\delta} \lambda^{[\kappa]} q_i^{[\kappa]}. \tag{4.80}$$

Since we consider the system at global equilibrium to be described with the grand canonical ensemble, we set:

$$-\sum_{\kappa=1}^{\delta} \lambda^{[\kappa]} q_i^{[\kappa]} = \alpha - \beta e_i, \tag{4.81}$$

where e_i is the energy of channel i, and α and β are the conjugate thermodynamic variables to be determined.[16] Then we can rewrite $\mathscr{P}^{(\mathrm{eq})}(S)$ as:

$$\mathscr{P}^{(\mathrm{eq})}(S) = \Xi^{-1} \prod_{\mathbf{r}_\star \in \mathscr{L}} \prod_{i=1}^{b} e^{(\alpha - \beta e_i) S_i(\mathbf{r}_\star)}, \tag{4.82}$$

16 We assume that there are no spurious invariants, and the variable conjugate to the momentum does not appear since the system is at global equilibrium.

where Ξ is the grand canonical partition function. From the normalization of $\mathscr{P}^{(eq)}(S)$, and with the definition: $z = \exp \alpha$, we obtain :

$$\Xi = \sum_{S \in \Gamma} \prod_{\mathbf{r}_\star \in \mathscr{L}} \prod_{i=1}^{b} (z\, e^{-\beta e_i})^{S_i(\mathbf{r}_\star)}. \tag{4.83}$$

We can commute the sum and the products provided we take exactly the S_is that are compatible with the state $S \in \Gamma$; since here $S_i = 0$ or 1, Ξ is the *Fermi–Dirac grand partition function* given by:

$$\Xi_{\mathrm{FD}} = \prod_{\mathbf{r}_\star \in \mathscr{L}} \prod_{i=1}^{b} (1 + z\, e^{-\beta e_i}). \tag{4.84}$$

It will be convenient to define:

$$\begin{aligned} Q &= \log \Xi_{\mathrm{FD}} \\ &= \sum_{\mathbf{r}_\star, i} \log (1 + z\, e^{-\beta e_i}) \\ &= \mathscr{N} \sum_i \log (1 + z\, e^{-\beta e_i}), \end{aligned} \tag{4.85}$$

where we have taken into account that the $\mathscr{N}$ nodes in the volume element V (i.e. in $\mathscr{L}$) have identical energy levels e_i. From the knowledge of the grand partition function, we can obtain the basic thermodynamic quantities of the system. Let us first compute the average number of particles in V:[17]

$$\begin{aligned} \langle N \rangle &= \left(\frac{\partial \log \Xi}{\partial \alpha} \right)_{V,\beta} = \left(\frac{\partial Q}{\partial \alpha} \right)_{V,\beta} \\ &= \mathscr{N} \sum_i \frac{e^{\alpha - \beta e_i}}{1 + e^{\alpha - \beta e_i}}. \end{aligned} \tag{4.86}$$

Since $\langle N \rangle = \mathscr{N} \sum_i \langle n_i(\mathbf{r}_\star) \rangle$, the average density per channel is given by:

$$\langle n_i(\mathbf{r}_\star) \rangle \equiv f_i^{(eq)} = \frac{1}{1 + e^{-\alpha + \beta e_i}}, \tag{4.87}$$

which is indeed the Fermi–Dirac distribution (per channel) for global equilibrium.[18]

By analogy with classical statistical thermodynamics (see e.g. McQuarrie, 1976), we can now compute the thermodynamic variables. The average density per node is obtained straightforwardly:

$$\rho = \frac{\langle N \rangle}{V} = \frac{\langle N \rangle}{\mathscr{N} v_0} = \sum_i f_i, \tag{4.88}$$

[17] Note that it is not essential to indicate explicitly that the partial derivatives in the thermodynamic expressions are to be taken at constant V since, in the lattice gas, expansion or compression cannot be performed as in classical thermodynamics; here $\mathscr{L}$ is fixed and V is necessarily constant.

[18] For simplicity, we shall omit the superscript in $f_i^{(eq)}$.

where we have used the elementary volume v_0, associated to a node.[19] To compute the average energy $\langle E\rangle$, the thermodynamic pressure p, and the entropy S, it is convenient to use the following algebraic relations:

$$1-f_i = \frac{e^{-\alpha+\beta e_i}}{1+e^{-\alpha+\beta e_i}},$$

$$\frac{f_i}{1-f_i} = e^{\alpha-\beta e_i},$$

$$\log\left(\frac{f_i}{1-f_i}\right) = \alpha-\beta e_i.$$

We find:

$$\langle E\rangle = \beta^{-2}\left(\frac{\partial Q}{\partial \beta^{-1}}\right)_{V,\alpha} = -\left(\frac{\partial Q}{\partial \beta}\right)_{V,\alpha} = \mathcal{N}\sum_i e_i f_i, \tag{4.89}$$

and

$$pV = \beta^{-1}\log\Xi = \beta^{-1}Q = -\mathcal{N}\beta^{-1}\sum_i \log(1-f_i), \tag{4.90}$$

which suggest that β can be interpreted as the reciprocal temperature.

Now starting from the expression for the global entropy (see Section 4.3.2):

$$S = -\mathcal{N}\sum_i \left[f_i\log f_i + (1-f_i)\log(1-f_i)\right], \tag{4.91}$$

and using (4.85), (4.86), and (4.89), it is easy to show that (4.91) can be written as:

$$S = Q + \beta\langle E\rangle - \alpha\langle N\rangle, \tag{4.92}$$

which is consistent with the classical thermodynamic expression for the entropy provided $\alpha = \beta\mu$; the consistency criterion determines μ as the chemical potential (and consequently $z = \exp(\beta\mu)$ as the fugacity), and we have:

$$\begin{aligned} S &= Q + \beta^{-1}\left(\frac{\partial Q}{\partial \beta^{-1}}\right)_{V,\mu} \\ &= Q + \beta^{-1}\left[\left(\frac{\partial Q}{\partial \beta}\right)_{V,\mu}\frac{\partial \beta}{\partial \beta^{-1}} + \left(\frac{\partial Q}{\partial \alpha}\right)_{V,\mu}\frac{\partial \alpha}{\partial \beta^{-1}}\right] \\ &= Q + \beta\langle E\rangle - \beta\mu\langle N\rangle. \end{aligned} \tag{4.93}$$

The basic relation in classical thermodynamics is the *equation of state*, $p = p(\rho, T)$, where T is defined through $\beta = (k_B T)^{-1}$ (here k_B is an arbitrary constant which can be set equal to one). However, it is not obvious how an

[19] Often we can omit writing v_0 explicitly as it can be taken equal to one; then $\mathcal{N}$ can be interpreted as the volume of the lattice gas.

explicit relation can be obtained for $p(\rho, T)$ in the lattice gas. Indeed, using (4.87), let us rewrite Equation (4.90) as:

$$p\,v_0 = \beta^{-1} \sum_i \log(1 + z\,e^{-\beta e_i}), \tag{4.94}$$

where z (or alternatively α) should be expressed in terms of ρ and T, i.e. $z = z(\langle N \rangle / V, \beta)$; this amounts to inverting Equation (4.86), but no explicit analytical solution can be obtained.

Nevertheless, there is an interesting limiting case that we can consider: the case of low densities and high temperatures. Then the number of accessible states is considerably larger than the number of particles and the mean occupation of an energy channel e_i is very small, i.e. $f_i \to 0$. From Equation (4.87), we see that this condition is realized if $z \to 0$, i.e.

$$f_i \Big|_{z\to 0} \equiv \langle n_i(\mathbf{r}_\star) \rangle \Big|_{z\to 0} = z\,e^{-\beta\,e_i}\,, \tag{4.95}$$

and the corresponding thermodynamic state is characterized by (see (4.88)) $\rho \to 0$ and positive T. Summation of (4.95) over i and using (4.88) allows us to eliminate z by writing:

$$\frac{f_i}{\rho}\Big|_{z\to 0} = \frac{e^{-\beta\,e_i}}{\sum_i e^{-\beta\,e_i}}\,. \tag{4.96}$$

We have thus retrieved Boltzmann statistics in the low density limit. Notice however that the status of the partition function $\sum_i e^{-\beta\,e_i}$ is different in lattice gases where – despite the fact that there are no non-local interactions – particles subject to the exclusion principle are not independent, and (4.85)) does not lead to the classical result:

$$Q_{\text{class}} = \frac{1}{N!} \left(\sum_i e^{-\beta\,e_i} \right)^N. \tag{4.97}$$

The reason is that in classical statistical mechanics one sums over indistinguishable particles (hence the factor $1/N!$ in (4.97)), whereas in lattice gases one sums over the lattice sites.

Another difference with classical statistical thermodynamics is the equation of state of the *ideal* gas which we are now in a position to compute from the above results. Using (4.90):

$$p\,V = \beta^{-1} Q = \mathcal{N} \beta^{-1} \sum_i \log(1 + e^{\alpha - \beta\,e_i}), \tag{4.98}$$

and expanding the log on the r.h.s. as:

$$\log(1 + x) = \sum_{m=1}^{\infty} \frac{1}{m} \left(\frac{x}{1 + x} \right)^m, \tag{4.99}$$

we obtain:

$$p v_0 = \beta^{-1} \sum_{m=1}^{\infty} \frac{1}{m} \sum_i (f_i)^m . \tag{4.100}$$

For the reason discussed above, the pressure of the 'ideal lattice gas' is larger than the intuitively expected value which would read:

$$p_0 = \beta^{-1} \frac{\langle N \rangle}{V} = \beta^{-1} \rho = \frac{\beta^{-1}}{v_0} \sum_i f_i . \tag{4.101}$$

Indeed, from (4.100), we have:

$$p v_0 = \beta^{-1} \left(\sum_i f_i + \frac{1}{2} \sum_i f_i^2 + \cdots \right) , \tag{4.102}$$

which we rewrite formally as:

$$p = \beta^{-1} \rho (1 + B_2 \rho + \cdots) , \tag{4.103}$$

with:

$$B_2 = \frac{1}{2} \rho^{-2} \sum_i f_i^2 . \tag{4.104}$$

Equation (4.103) is analogous to the virial expansion of classical statistical mechanics. However, the lattice gas is not a Newtonian system:[20] the notion of *interaction potential* is precluded in the classical mechanical sense,[21] and there is no virial theorem; therefore (4.103) is not *stricto sensu* a virial expansion. Nevertheless, we shall see in the next section that the coefficient B_2 can be expressed in terms of a correlation function and takes a form equivalent to the expression of the classical second virial coefficient. In fact, at this point, we can notice an interesting analogy. For lattice gases with single non-zero velocity modulus, $f_i^{(eq)} = d$, and (4.104) can also be written as $\rho B_2 = \frac{1}{2} d$; for the hard disk gas, one has $\rho B_2 = \frac{1}{2} N \pi \sigma^2 / V$, where σ is the hard disk diameter (see e.g. McQuarrie, 1976). So in both cases, the result is one half the relative average 'volume' where there is *one* particle of the gas.

The *kinetic pressure* p_K, defined as the density of momentum transfer:

$$p_K v_0 = \sum_i c_{i\alpha}^2 f_i = \frac{2}{D} \sum_i e_i f_i , \tag{4.105}$$

with D the space dimension, is different from the *thermodynamic pressure* p, (4.94). To illustrate the point, let us define:

$$c_s^2 = \frac{1}{bD} \sum_i c_i^2 , \tag{4.106}$$

[20] The lattice gas dynamics is not governed by Newton's equations of motion.

[21] Models have been developed where non-local interactions are incorporated in the dynamics; one can then define a quantity which is identified as the *potential of mean force* of the lattice gas (Tribel and Boon, 1995).

where the summation is taken over all velocity channels; for non-thermal models (with $|\mathbf{c}| = 1$), (4.106) reduces to $c_s^2 = 1/D$, and (4.105) becomes:

$$p_K v_0 = b\, c_\alpha^2\, d = \frac{b}{D}\, d, \quad \text{or} \quad p_K = \rho c_s^2. \tag{4.107}$$

Now setting $\beta^{-1} = c_s^2$ in (4.90), we find for the thermodynamic pressure:

$$p v_0 = -b c_s^2 \log(1-d). \tag{4.108}$$

In the low density limit ($\rho \to 0$), the thermodynamic pressure ($p \simeq \beta^{-1}\rho$, see (4.103)) is equal to the kinetic pressure (see (4.107)), and

$$\chi \equiv \frac{1}{\rho}\frac{\partial \rho}{\partial p} = \frac{1}{\rho c_s^2} \tag{4.109}$$

is the compressibility coefficient.[22] In this limit, and with $\beta^{-1} = c_s^2$, one retrieves the law of ideal gases $p = \rho\beta^{-1}$.

The difference between thermodynamic and kinetic pressures in thermal lattice gases is illustrated for the GBL model (see Section 3.6) in Figure 4.2; the kinetic pressure for the same model is shown as a function of $\theta = \exp(-\beta/2)$ in Figure 4.3.

Some authors (Chen, Zhao and Doolen, 1989, and Molvig *et al.*, 1989) have used the expression $p_K = \rho k_B T_K$, which defines a *kinetic temperature* T_K. As suggested by Figure 4.3 the kinetic temperature is a complicated function of ρ and of β, and must be distinguished from the *thermodynamic temperature* $T = (k_B\beta)^{-1}$. To discuss this point we consider the low density limit where the mean occupation of the energy channels e_i is small and is given by (4.95); the condition $z \to 0$ is realized for the thermodynamic state $\rho \to 0$ at β or

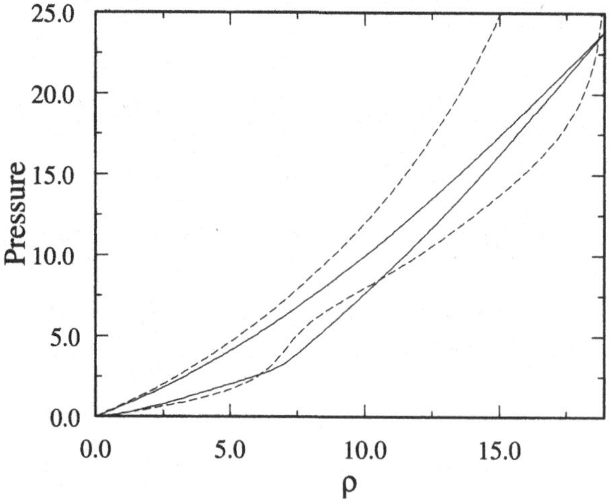

Figure 4.2 Kinetic pressure (full lines) and thermodynamic pressure (dashed lines) as a function of density (GBL model); isotherms for $\theta = \exp(-\beta/2) = 0.05$ (upper curves), 0.8 (lower curves).

[22] The notation c_s used here will be justified in Chapter 9 where we shall see that c_s is the speed of sound propagation, which is consistent with (4.109).

T fixed. Now $z \to 0$ also corresponds to the thermodynamic state where ρ is fixed and $T \to \infty$ ($\beta \to 0$); indeed, considering the expression $z = \rho \Big/ \sum_i e^{-\beta e_i}$ (obtained by summing (4.95) over i), we see that for $\beta \to 0$ or $T \to \infty$, i.e. $e^{-\beta e_i} = 1$, the condition is realized provided the sum in the denominator is large, that is, when there is a large number of energy levels. From Figure 4.3, we observe that $p_K/\rho = k_B T_K$ goes to a constant value as $\theta \to 1$, that is, when $T \to \infty$, the kinetic temperature goes to a constant value: $k_B T_K \to c_s^2$, whereas when $\theta \to 0$, i.e. $T \to 0$, T_K is a function of the density, $T_K(\rho)$. That $p_K/\rho = k_B T_K \to c_s^2$ when $\theta \to 1$ can be explained in physical terms: in lattice gases, energy is transferred through collisions between particles residing on channels with different energy levels; but when the temperature becomes infinite the channel occupation probability no longer depends on the energy level, so that the energy transfer is then governed by momentum transfer which produces sound propagation. We can now appreciate why the definition of a thermodynamic temperature in lattice gases (and more generally in discrete systems) in accordance with classical statistical thermodynamics is a delicate matter.

4.7 Static correlation functions

In statistical mechanics the static correlations of density fluctuations are described by the function:

$$\langle \delta\rho(\mathbf{r}_\star)\, \delta\rho(\mathbf{r}_\star{}')\rangle = \sum_{\mathbf{c},\mathbf{c}'} \langle \delta n(\mathbf{r}_\star, \mathbf{c})\, \delta n(\mathbf{r}_\star{}', \mathbf{c}')\rangle \equiv \sum_{i,j} \langle \delta n_i(\mathbf{r}_\star)\, \delta n_j(\mathbf{r}_\star{}')\rangle \, , \quad (4.110)$$

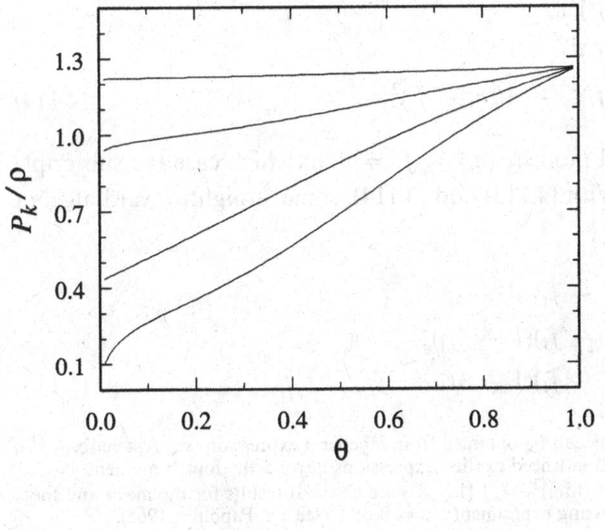

Figure 4.3 The kinetic pressure as a function of $\theta = \exp(-\beta/2)$ for the GBL model; lower to upper curves are for $\rho = 1, 7, 13$, and 18.

where $\delta n_i(\mathbf{r}_\star) = n_i(\mathbf{r}_\star) - f_i$, or equivalently by the *static structure factor* $S(\mathbf{k})$, the space-Fourier transform of (4.110):

$$\begin{aligned} \rho\, S(\mathbf{k}) &= \sum_{\mathbf{r}_\star,\mathbf{r}_\star'} e^{\imath\mathbf{k}\cdot\mathbf{r}_\star} e^{-\imath\mathbf{k}\cdot\mathbf{r}_\star'} \langle \delta\rho(\mathbf{r}_\star)\, \delta\rho(\mathbf{r}_\star') \rangle \\ &= \sum_{\mathbf{r}_\star,\mathbf{r}_\star'} \sum_{i,j} e^{\imath\mathbf{k}\cdot\mathbf{r}_\star} e^{-\imath\mathbf{k}\cdot\mathbf{r}_\star'} \langle \delta n_i(\mathbf{r}_\star)\, \delta n_j(\mathbf{r}_\star') \rangle . \end{aligned} \tag{4.111}$$

Higher order correlation functions may also be of interest. Now, at equilibrium, lattice gases exhibit no correlations between different states $\{\mathbf{r}_\star, \mathbf{c}\}$. So we are interested in the quantities $\langle (\delta n(\mathbf{c}))^n \rangle \equiv \langle (\delta n_i)^n \rangle$. Therefore we consider the single channel probability distribution, and use the cumulant expansion method, because through the connection between the cumulants and the central moments, we can readily obtain explicit expressions for the density fluctuation correlations. The generating function for the cumulants $\kappa_i^{(n)}$, i.e. the *log* of the characteristic function of the random variable s_i, reads:

$$\mathscr{G}(x) \equiv \log \langle \exp\{x n_i\} \rangle = \sum_{n=1}^{\infty} \kappa_i^{(n)} \frac{x^n}{n!}, \tag{4.112}$$

and the nth cumulant is given by:

$$\kappa_i^{(n)} = \left(\frac{\partial^n Q_i}{\partial \alpha^n} \right)_\beta , \tag{4.113}$$

with $Q_i = \log(1 + e^{\alpha - \beta e_i}) \equiv \log(1 + e^{\alpha - \beta e(\mathbf{c})})$. The relations between the first cumulants and moments of interest here are (Abramowitz and Stegun, 1965):

$$\begin{aligned} \kappa_i^{(1)} &= \langle n_i \rangle = f_i, \\ \kappa_i^{(2)} &= \langle (n_i - f_i)^2 \rangle , \\ \kappa_i^{(3)} &= \langle (n_i - f_i)^3 \rangle , \\ \kappa_i^{(4)} &= \langle (n_i - f_i)^4 \rangle - 3 \langle (n_i - f_i)^2 \rangle^2 . \end{aligned} \tag{4.114}$$

Note that, for single speed models, $\langle n_i \rangle = f_i = d$, in which case the subscripts of the κs are redundant. With (4.113) and (4.114), some straightforward algebra yields the results:[23]

$$\begin{aligned} \langle (\delta n_i)^2 \rangle &= f_i(1 - f_i), \\ \langle (\delta n_i)^3 \rangle &= f_i(1 - f_i)(1 - 2f_i), \\ \langle (\delta n_i)^4 \rangle &= f_i(1 - f_i)(1 - 3f_i + 3f_i^2). \end{aligned} \tag{4.115}$$

[23] Higher order cumulants can be obtained from a general expression (see Appendix A.11), but in practice we shall not need explicit expressions beyond the fourth moment. Notice also that $\langle n_i \rangle = f_i$ and $\langle (\delta n_i)^2 \rangle = f_i(1 - f_i)$ are classical results for the mean and the variance of the coin-tossing experiment: $n_i = 0$ or 1 (see e.g. Papoulis, 1965).

(See also (4.42) and (4.43).) It then follows from (4.111) and (4.114) that the static structure factor is given by:

$$\rho\, S(\mathbf{k}) = \sum_{\mathbf{r}_\star,\mathbf{r}_\star'} e^{i\mathbf{k}\cdot\mathbf{r}_\star} e^{-i\mathbf{k}\cdot\mathbf{r}_\star'} \sum_i \kappa_i^{(2)} \delta_{\mathbf{r}_\star,\mathbf{r}_\star'}, \tag{4.116}$$

or, with (4.115):[24]

$$\rho\, S(k) = \sum_i \kappa_i^{(2)} = \sum_i f_i(1 - f_i). \tag{4.117}$$

For LGAs with single velocity modulus ($f_i = d$), (4.117) reduces to:

$$S(k) = 1 - d. \tag{4.118}$$

We notice that the absence of structure in $S(\mathbf{k})$ (= const.) stems from the strictly local nature of the interactions.[25]

The 'virial' coefficient obtained in the previous section (see (4.103) and (4.104)) can now be expressed in terms of the static structure factor; indeed we can rewrite (4.117) as:

$$\sum_c f^2(c) \equiv \sum_i f_i^2 = \sum_i f_i - \rho\, S(k) = \rho\,(1 - S(k)), \tag{4.119}$$

and it follows from (4.104) that:

$$B_2 = \frac{1}{2}\rho^{-1}(1 - S(k)), \tag{4.120}$$

a result which is analogous to the classical expression obtained in continuous theory.[26]

The static structure factor is also related to the thermodynamic quantities β and χ discussed in the previous section. From (4.98), we have:

$$\left(\frac{\partial p}{\partial \rho}\right)_\beta = \beta^{-1} \sum_i \left(\frac{\partial}{\partial \alpha} \log(1 + e^{\alpha - \beta\, e_i}) \frac{\partial \alpha}{\partial \rho}\right)_\beta, \tag{4.121}$$

where:

$$\left(\frac{\partial}{\partial \alpha} \log(1 + e^{\alpha - \beta\, e_i})\right)_\beta = f_i\,. \tag{4.122}$$

Combining this result with (4.113) to obtain:

$$\left(\frac{\partial \rho}{\partial \alpha}\right)_\beta = \sum_i \left(\frac{\partial f_i}{\partial \alpha}\right)_\beta = \sum_i \kappa_i^{(2)}, \tag{4.123}$$

24 Rigorously $\rho\, S(\mathbf{k}) = \sum_i \kappa_i^{(2)}(1 - \delta(\mathbf{k}))$, which shows that exactly at $|\mathbf{k}| = 0$ (i.e. at infinite wavelength) $S(\mathbf{k}) = 0$ (Forster, 1975). But for all practical purposes the factor $(1 - \delta(\mathbf{k}))$ can be ignored.

25 LGAs with non-local interactions exhibit a **k**-dependent static structure factor reflecting the effect of the interaction range (Tribel and Boon, 1995).

26 In classical statistical mechanics, B_2 is expressed in terms of the radial distribution function $g(r)$, which is related to the static structure factor (see Huang, 1963, and Boon and Yip, 1980): $B_2 = -\frac{1}{2}\int_0^\infty d^3r\,(g(r) - 1) = \frac{1}{2}\rho^{-1}(1 - \lim_{k\to 0} S(k))$.

we find:

$$\left(\frac{\partial p}{\partial \rho}\right)_\beta = \beta^{-1} \sum_i f_i \Big/ \sum_i \kappa_i^{(2)} . \tag{4.124}$$

So the isothermal compressibility is given by:

$$\chi \equiv \frac{1}{\rho}\left(\frac{\partial \rho}{\partial p}\right)_\beta = \beta \rho^{-2} \sum_i \kappa_i^{(2)} , \tag{4.125}$$

and, using (4.117):

$$S(k) = \rho \beta^{-1} \chi . \tag{4.126}$$

This expression is an interesting result because it relates the mean squared density fluctuations to the thermodynamic quantities of the system. In continuous statistical mechanics, one has essentially the same relation (see e.g. Boon and Yip, 1980), the major difference being that the l.h.s. of (4.126) must be taken in the thermodynamic limit ($k \to 0$). In lattice gases, (4.126) is valid at all ks, which is a consequence of the lack of 'molecular structure' in LGAs.[27]

[27] For lattice gases whose microscopic dynamics includes non-local interactions (Tribel and Boon, 1995), the limit $k \to 0$ must also be taken in (4.126).

Chapter 5

Macrodynamics: Chapman–Enskog method

In Chapter 4, we gave a statistical mechanical analysis of lattice gases, and discussed the equilibrium properties; we described uniform uncorrelated statistical equilibria, and we showed that they have a Fermi–Dirac probability distribution. The existence of these uniform equilibrium solutions can be established *without* recourse to the Boltzmann approximation; the only required conditions are the semi-detailed balance and the existence of local invariants. These equilibrium solutions – the analogue of *global equilibria* in usual statistical mechanics – are uniform by construction: the macroscopic variables, i.e. the ensemble-averages of the local invariants, are space- and time-independent.

Now, we address the problem of space- and time-varying macrostates in the 'hydrodynamic limit', that is, macrostates for which macroscopic variables vary over space and time scales *much larger* than the characteristic microscopic scales (lattice spacing and time-step duration). This scale separation between microscopic and macroscopic variables is a crucial physical ingredient in the theory of lattice gas hydrodynamic equations, and it is at the core of the forthcoming derivation, which rests upon a discrete version of the well-known 'Chapman–Enskog method' (see Chapman and Cowling, 1970).

Although the principles of the method described hereafter apply to a wider class of models, we are led, at a certain point of the forthcoming algebraic procedure, to particularize our study to the (still wide) class of 'single-species thermal models' defined in Chapter 4, Section 4.5.2. The simpler case of non-thermal models is also treated, but in a less detailed manner (see Section 5.7).[1]

[1] We consider here single-species thermal or non-thermal lattice gases but the method can be extended to systems such as, multi-component lattice gases (see e.g. Section 7.7).

5.1 Local equilibria and the hydrodynamic limit

Consider a lattice gas with δ collisional invariants $q^{[\kappa]}$, $\kappa = 1, \ldots, \delta$, and satisfying the semi-detailed balance condition so as to be in accordance with the requirements for the existence of factorized global equilibrium states (see Chapter 4).

Its macrostate at time $t_\star = 0$ is described by a probability distribution function $\mathscr{P}$ (see Section 4.1.1) with the following properties:

- The macroscopic variables $\rho^{[\kappa]} \equiv \langle q^{[\kappa]} \rangle$, computed with the distribution $\mathscr{P}$ are δ given functions $\rho^{[\kappa]}(\mathbf{r}_\star)$ of the space variable $\mathbf{r}_\star$.
- The macroscopic fields $\rho^{[\kappa]}(\mathbf{r}_\star)$ vary slowly in space, which means that the typical distance over which the macroscopic fields vary significantly is much larger than the microscopic unit length, namely the mesh size.
- The distribution $\mathscr{P}$ depends on the macroscopic fields $\rho^{[\kappa]}(\mathbf{r}_\star)$ exactly in the same way as $\mathscr{P}^{(\mathrm{eq})}$ depends on the macroscopic variables $\rho^{[\kappa]}$ in the homogeneous case (see Section 4.5).

This kind of macrostate is clearly *not* an equilibrium state: it cannot be a stationary solution of the Liouville equation since the macroscopic variables vary with space. However, these macrostates are called *local equilibria* (in the same sense as in continuous statistical mechanics) because the fluid can be considered to be in equilibrium *locally*, that is, in a region which is small compared to the smallest macroscopic scale (yet large with respect to the microscopic scale). They are denoted by $\mathscr{P}^{(\mathrm{loc})}$. This terminology is justified by the fact that local equilibria have the same analytical form as global (homogeneous) equilibria, but with slowly varying parameters. So locally the fluid seems at equilibrium, whereas globally it evolves slowly towards a true uniform (global) equilibrium state.[2]

During this evolution, macroscopic phenomena can occur on different time scales. When the lattice gas shows fluid-like behavior, there may be elastic effects, such as sound propagation, with short (macroscopic) time scales, and viscous effects such as damping that occur on longer time scales. Both phenomena are of interest here. Since they have well-separated time scales, we can analyze them with *multi-scale* asymptotic methods.

[2] Of course, the situation is quite different in the presence of an external force (e.g. pressure gradient, gravity) which maintains the system out of equilibrium.

5.2 The multi-scale expansion for macrodynamics

5.2.1 The scale separation parameter

The theory developed in this chapter is valid in the *hydrodynamic limit*, the limit where the smallest macroscopic scale is large compared to the characteristic microscopic length (here typically the distance between two nearest nodes). Be it for lattice gases or for real fluids, the scale separation hypothesis is mandatory for a hydrodynamical description of the collective motions of the microscopic particles. So we introduce a scale separation parameter ϵ, defined as the ratio of the microscopic scale to the smallest macroscopic scale. The hydrodynamic limit corresponds to $\epsilon \ll 1$. This parameter will serve as an expansion parameter in the multi-scale expansion leading to the *macrodynamic* (hydrodynamic) description of lattice gases. This expansion is inspired by the Chapman–Enskog expansion of continuous gases.

5.2.2 Perturbed local equilibrium

In a weakly inhomogeneous situation, it is reasonable to consider that the actual time-dependent solution $\mathscr{P}(S,t)$ of the Liouville equation (4.6) deviates very little from the local equilibrium $\mathscr{P}^{(\mathrm{loc})}$ corresponding to the *same* space and time dependent macroscopic fields $\rho^{[\kappa]}(\mathbf{r}_\star, t_\star)$. This deviation must vanish when the macroscopic fields become homogeneous, that is, when the scale separation parameter vanishes. So it is natural to introduce the following expansion for $\mathscr{P}$ in powers of the scale separation ϵ:

$$\mathscr{P} = \mathscr{P}^{(0)} + \epsilon\mathscr{P}^{(1)} + \epsilon^2\mathscr{P}^{(2)} + \cdots, \tag{5.1}$$

where $\mathscr{P}^{(0)}$ is just $\mathscr{P}^{(\mathrm{loc})}$, and $\mathscr{P}^{(1)}$ and $\mathscr{P}^{(2)}$ are the first and second order deviations from local equilibrium.

The average populations f_i can be expanded in the same way:

$$f_i = f_i^{(0)} + \epsilon f_i^{(1)} + \epsilon^2 f_i^{(2)} + \cdots, \tag{5.2}$$

where $f_i^{(0)}$, $f_i^{(1)}$ and $f_i^{(2)}$ are computed respectively with the local equilibrium probability distribution $\mathscr{P}^{(0)} \equiv \mathscr{P}^{(\mathrm{loc})}$, and its first and second order deviations $\mathscr{P}^{(1)}$ and $\mathscr{P}^{(2)}$:

$$\begin{aligned} f_i^{(0)} &= \sum_{S\in\Gamma} S_i(\mathbf{r}_\star)\mathscr{P}^{(0)}(s), \\ f_i^{(1)} &= \sum_{S\in\Gamma} S_i(\mathbf{r}_\star)\mathscr{P}^{(1)}(s), \\ f_i^{(2)} &= \sum_{S\in\Gamma} S_i(\mathbf{r}_\star)\mathscr{P}^{(2)}(s). \end{aligned} \tag{5.3}$$

That the actual solution f_i and its lowest order approximation $f_i^{(0)}$ must lead to the same macroscopic fields implies that the first and second order corrections must not contribute to the macroscopic variables. Thus, the corrections $f_i^{(1)}$ and $f_i^{(2)}$ must verify:

$$\sum_{i=1}^{b} q_i^{[\kappa]} f_i^{(1)} = 0, \qquad \kappa = 1, \ldots, \delta$$

and

$$\sum_{i=1}^{b} q_i^{[\kappa]} f_i^{(2)} = 0, \qquad \kappa = 1, \ldots, \delta. \tag{5.4}$$

In terms of the probability distributions, this gives the conditions:

$$\sum_{S\in\Gamma} \sum_{i=1}^{b} q_i^{[\kappa]} S_i(\mathbf{r}_\star)\mathscr{P}^{(1)}(S) = 0, \qquad \kappa = 1, \ldots, \delta$$

and

$$\sum_{S\in\Gamma} \sum_{i=1}^{b} q_i^{[\kappa]} S_i(\mathbf{r}_\star)\mathscr{P}^{(2)}(S) = 0, \qquad \kappa = 1, \ldots, \delta. \tag{5.5}$$

5.2.3 Macroscopic space and time scales

Since, in the hydrodynamic limit, noticeable space variations of the macroscopic variables take place typically over distances of order ϵ^{-1}, it is natural to introduce a macroscopic space variable $\mathbf{r}_1$ defined by:

$$\mathbf{r}_1 = \epsilon\, \mathbf{r}, \tag{5.6}$$

where $\mathbf{r}$ is the continuous version of the discrete microscopic space variable $\mathbf{r}_\star$. Rescaling $\mathbf{r}$ into $\mathbf{r}_1$ amounts to choosing ϵ^{-1} as the new length unit, instead of the lattice constant which is the natural length unit for microscopic phenomena.

When the macroscopic variables are inhomogeneous, with gradients of order ϵ^{-1}, then, by physical intuition (based on our experience with real fluids), we anticipate two generic classes of macroscopic phenomena, superimposed on the underlying microscopic evolution: (i) non-linear and pressure effects such as acoustic propagation, and (ii) linear diffusive effects like viscous damping.

For a real Newtonian fluid, with given sound speed and viscosity (see e.g. Landau and Lifschitz, 1987), non-linear and pressure effects involve first order space derivatives. So, for inhomogeneous macroscopic fields with length scales of order ϵ^{-1}, the typical time scales over which non-linear and pressure effects occur should be of order ϵ^{-1}. This is particularly clear for sound waves: a macroscopic inhomogeneity with wavelength of order ϵ^{-1} will propagate with a given speed such that, at some given point, noticeable variations of the macroscopic variables (pressure, velocity) will occur over time scales of order ϵ^{-1}.

On the other hand, diffusive effects involve second order space derivatives. Thus, the inhomogeneities with typical length scales of order ϵ^{-1} undergo slow damping over time scales of order ϵ^{-2}. This is the case for sound attenuation in viscous fluids (viscous damping of sound waves).

From these physical arguments, we anticipate similar macroscopic dynamical processes in lattice gases and we introduce *two* auxiliary time variables t_1 and t_2:

$$t_1 = \epsilon t \qquad \text{and} \qquad t_2 = \epsilon^2 t. \tag{5.7}$$

At this point t_1 and t_2 are not two different time variables, but simply two expressions of the same time variable t with different scalings. To illustrate the argument, imagine a lattice gas whose (microscopic) time step would be one second. Then the second is the natural time unit for microscopic phenomena. Suppose now that the lattice gas is in an inhomogeneous macrostate with a space scale separation parameter ϵ of 1/60. Then, t_1 is just the time expressed in minutes and t_2 the time expressed in hours. For this lattice gas, the acoustic phenomena would evolve on time scales of a few minutes, and thus are better described in terms of a time variable expressed in minutes; diffusive effects would only be noticeable after hours, and therefore should be analyzed with a time variable expressed in hours.

Note that t_1 and t_2 are *not* independent variables: they are connected through their relations with t. But it is usual in multi-scale methods (see Bender and Orszag, 1978) to consider any function $F(\mathbf{r}, t)$ of space and time as a function $F(\mathbf{r}_1, t_1, t_2)$ of $\mathbf{r}_1$, t_1 and t_2, as if t_1 and t_2 were independent variables.[3] The time and space derivatives of an arbitrary function $F(\mathbf{r}, t)$ can thus be expressed as follows in terms of the derivatives of $F(\mathbf{r}_1, t_1, t_2)$:

$$\frac{\partial F}{\partial \mathbf{r}} = \epsilon \frac{\partial F}{\partial \mathbf{r}_1} \qquad \text{and} \qquad \frac{\partial F}{\partial t} = \epsilon \frac{\partial F}{\partial t_1} + \epsilon^2 \frac{\partial F}{\partial t_2}. \tag{5.8}$$

All the tools necessary for the multi-scale expansion are now ready; it remains to introduce the master equation to which the expansion will be applied.

[3] Strictly (from the mathematical point of view) the two functions should be labeled differently, but, as no ambiguity results, we keep the same notation.

5.2.4 The averaged microdynamic equation

A convenient way to derive the macrodynamic equations for the macroscopic variables $\rho^{[\kappa]}$ is to start from the microdynamic equation (2.13) rewritten as:

$$\begin{cases} n_i(\mathbf{r}_\star + \mathbf{c}_i\,,\, t_\star + 1) = n_i'(\mathbf{r}_\star, t_\star), \\ n_i'(\mathbf{r}_\star, t_\star) = n_i + \sum_{(s,s')\in\gamma^2} (s_i' - s_i)\xi(s\to s') \prod_{j=1}^{b} n_j^{s_j} \overline{n}_j^{\overline{s}_j}. \end{cases} \tag{5.9}$$

Here, n_i' denotes the Boolean population *after* collision. Note that the Boolean populations n_i' and n_i in the second equation are all taken at node $\mathbf{r}_\star$ and time $t_\star$. This formulation which separates the collision phase and the propagation phase, is strictly equivalent to (2.13), but conceptually easier to handle for the forthcoming averaging operations.

Ensemble-averaging the first equation in (5.9) with the probability distribution $\mathscr{P}$ of the corresponding macrostate gives:

$$f_i(\mathbf{r}_\star + \mathbf{c}_i\,,\, t_\star + 1) = f_i'(\mathbf{r}_\star, t_\star), \tag{5.10}$$

which merely expresses that the propagation (or free-streaming) phase just moves information, without loss or gain, from one node to another node. The second equation in (5.9), after ensemble-averaging over $\mathscr{P}$ and over all possible choices for non-deterministic collisions, yields:

$$f_i'(\mathbf{r}_\star, t_\star) = f_i(\mathbf{r}_\star, t_\star) + \Delta_i(\mathscr{P}, \mathbf{r}_\star, t_\star), \tag{5.11}$$

where the averaged collision term $\Delta_i(\mathscr{P}, \mathbf{r}_\star, t_\star)$ is:

$$\Delta_i(\mathscr{P}, \mathbf{r}_\star, t_\star) \equiv \left\langle \sum_{(s,s')\in\gamma^2} (s_i' - s_i) A(s\to s') \prod_{j=1}^{b} n_j^{s_j} \overline{n}_j^{\overline{s}_j} \right\rangle. \tag{5.12}$$

Indeed, the random variables $\xi(s\to s')$ involved in the collision phase (see Section 2.2.3) are statistically independent from the Boolean field. So, the averaging over all possible choices for non-deterministic collisions simply replaces the random variables $\xi(s\to s')$ by their mean values $A(s\to s')$ (the collisional transition probabilities). The brackets $\langle\cdots\rangle$ denote ensemble-averaging over the probability distribution $\mathscr{P}$, that is, summing over all possible realizations S of the Boolean field n, with each term weighted by its probability $\mathscr{P}(S, t_\star)$:

$$\Delta_i(\mathscr{P}, \mathbf{r}_\star, t_\star) = \sum_{S\in\Gamma} \mathscr{P}(S) \sum_{(s,s')\in\gamma} (s_i' - s_i) A(s\to s') \prod_{j=1}^{b} S_{j(\mathbf{r}_\star)}^{s_j}\, \overline{S}_{j(\mathbf{r}_\star)}^{\overline{s}_j}.$$

Since the product $\prod S_{j(\mathbf{r}_\star)}^{s_j} \overline{S}_{j(\mathbf{r}_\star)}^{\overline{s}_j}$ is zero unless $S_{(\mathbf{r}_\star)} = s$, the averaged collision term $\Delta_i(\mathscr{P}, \mathbf{r}_\star, t_\star)$ takes the simpler form:

$$\Delta_i(\mathscr{P}, \mathbf{r}_\star, t_\star) = \sum_{S\in\Gamma} \mathscr{P}(S) \sum_{s'\in\gamma} (s_i' - S_{i(\mathbf{r}_\star)}) A(S_{(\mathbf{r}_\star)}\to s'). \tag{5.13}$$

Without any further hypothesis, the averaged collision term Δ_i depends *a priori* on the full probability distribution $\mathscr{P}$, and not just on the mean populations $f_i(\mathbf{r}_\star, t_\star)$. Indeed, the average of the product $\prod_{j=1}^{b} n_j^{s_j} \overline{n}_j^{\overline{s}_j}$ in (5.9) cannot be factorized into the product of the averages, since the populations of different channels at the same node may be correlated. Consequently, a closed form for the averaged microdynamic equation cannot be obtained without the Boltzmann approximation (see Chapter 4), which assumes that particles entering a collision are uncorrelated; then the average of the product $\prod_{j=1}^{b} n_j^{s_j} \overline{n}_j^{\overline{s}_j}$ can be approximated by the product of the averages. With this *ansatz*, we obtain the lattice Boltzmann equation as the averaged microdynamic equation (see Chapter 4). Here we postpone the introduction of the Boltzmann approximation and keep the full expression (5.13) for the averaged collision term.

$\Delta_i(\mathscr{P}, \mathbf{r}_\star, t_\star)$ is clearly *linear* in $\mathscr{P}$: it is thus a linear function of $2^{b \cdot \mathscr{N}}$ real variables, the probabilities $\mathscr{P}(S, t_\star)$ of each of the $2^{b \cdot \mathscr{N}}$ possible configurations S of the full lattice gas at time $t_\star$. We rewrite (5.13) as:

$$\Delta_i(\mathscr{P}, \mathbf{r}_\star, t_\star) = \sum_{S \in \Gamma} L_{iS}(\mathbf{r}_\star)\, \mathscr{P}(S, t_\star), \tag{5.14}$$

where $L_{iS}(\mathbf{r}_\star)$ is a node-dependent $b \times 2^{b \cdot \mathscr{N}}$ real matrix with coefficients:

$$L_{iS}(\mathbf{r}_\star) = \sum_{s' \in \gamma} (s'_i - S_{i(\mathbf{r}_\star)}) A(S_{(\mathbf{r}_\star)} \to s'), \qquad \forall i = 1, \ldots, b, \qquad \forall S \in \Gamma. \tag{5.15}$$

To derive the macrodynamics of the lattice gas we thus start from the following equation for the averaged microdynamics:

$$f_i(\mathbf{r}_\star + \mathbf{c}_i, t_\star + 1) - f_i(\mathbf{r}_\star, t_\star) = \sum_{S \in \Gamma} L_{iS}(\mathbf{r}_\star)\, \mathscr{P}(S, t_\star). \tag{5.16}$$

Obviously the average collision term Δ_i vanishes for any global equilibrium $\mathscr{P}^{(\mathrm{eq})}$. It is less obvious, but nonetheless true, that it also vanishes for any *local* equilibrium $\mathscr{P}^{(\mathrm{loc})}$. Indeed, the collision process in a lattice gas, as defined in Section 1.2.3, is purely local. Therefore, the value $\Delta_i(\mathscr{P}, \mathbf{r}_\star)$ of the averaged collision term at node $\mathbf{r}_\star$ and time $t_\star$ depends only on the *single node* probability distribution for node $\mathbf{r}_\star$ at time $t_\star$, and not on the full information contained in $\mathscr{P}$. On the other hand, the definition of a local equilibrium (see Section 5.1) implies that for each node $\mathbf{r}_\star$ in $\mathscr{L}$, there exists a global equilibrium $\mathscr{P}^{(\mathrm{eq})}$ such that the single node distribution extracted from $\mathscr{P}^{(\mathrm{eq})}$ is identical to the single node distribution extracted from $\mathscr{P}^{(\mathrm{loc})}$ at $\mathbf{r}_\star$. Since all global equilibria make Δ_i vanish, it is clear that Δ_i also vanishes for local equilibria:

$$\Delta_i(\mathscr{P}^{(\mathrm{loc})}, \mathbf{r}_\star, t_\star) = 0, \quad \forall i = 1, \ldots, b, \quad \forall \mathbf{r}_\star \in \mathscr{L}, \quad \forall t_\star. \tag{5.17}$$

It is important to note that the conservation laws imply some constraints for Δ_i. Indeed (as in Section 4.4.1) consider a basis $(q^{[\kappa]}, \kappa = 1, \ldots, \delta)$ of the vector subspace of collisional invariants. Using (5.13) for $\Delta_i(\mathscr{P}, \mathbf{r}_\star)$, and Equation (2.19) as a definition of a collisional invariant, we can straightforwardly establish the following relation:

$$\sum_{i=1}^{b} q_i^{[\kappa]} \Delta_i(\mathscr{P}, \mathbf{r}_\star, t_\star) = 0, \quad \forall \kappa = 1, \ldots, \delta, \quad \forall \mathscr{P}, \tag{5.18}$$

which must be verified by the averaged collision term $\Delta_i(\mathscr{P}, \mathbf{r}_\star, t_\star)$. In terms of the matrix L_{iS} introduced in (5.14) and (5.16), (5.18) implies:

$$\sum_{i=1}^{b} q_i^{[\kappa]} L_{iS}(\mathbf{r}_\star) = 0, \quad \forall \kappa = 1, \ldots, \delta, \quad \forall S \in \Gamma, \quad \forall \mathbf{r}_\star \in \mathscr{L}, \tag{5.19}$$

which means that the rank of the $b \times 2^{b \cdot \mathscr{N}}$ matrix L_{iS} is smaller than b, and is at best equal to $b - \delta$, δ being the number of independent collisional invariants. In other words, the subspace of collisional invariants is included in the kernel of the adjoint of L_{iS}. This property will play an important role in the derivation of the macrodynamics of the lattice gas.

5.2.5 The expansion in powers of ϵ

NOTE: *We recall that Latin indices are used for channel labels, and that Greek indices are used to label space coordinates. In order to simplify the algebraic manipulations, we follow Einstein's convention for implicit summation over repeated Greek indices. For better readability, we use the compact notation $\partial_{1\alpha}$ for the partial derivative with respect to the coordinate α of rescaled space variable $\mathbf{r}_1$ as defined in (5.6), and ∂_{t_1} and ∂_{t_2} for the partial derivatives with respect to the rescaled time variables t_1 and t_2, respectively, as defined in (5.7).*

When the scale separation parameter ϵ is small (i.e. the averaged populations vary little from one node to the next), we can validly approximate the averaged population $f_i(\mathbf{r}_\star + \mathbf{c}_i, t_\star + 1)$ in Equation (5.16), by its Taylor expansion around $f_i(\mathbf{r}_\star, t_\star)$:

$$\begin{aligned} & f_i(\mathbf{r}_\star + \mathbf{c}_i, t_\star + 1) \\ &= \epsilon^0 f_i \\ &+ \epsilon^1 \left(\partial_{1\beta}\, c_{i\beta} f_i + \partial_{t_1} f_i \right) \\ &+ \epsilon^2 \left(\frac{1}{2} \partial_{1\beta} \partial_{1\gamma}\, c_{i\beta} c_{i\gamma} f_i + \partial_{1\beta} \partial_{t_1}\, c_{i\beta} f_i + \frac{1}{2} \partial_{t_1}^2 f_i + \partial_{t_2} f_i \right) \\ &+ \mathcal{O}(\epsilon^3). \end{aligned} \tag{5.20}$$

Here, f_i and its derivatives are taken implicitly at node $\mathbf{r}_\star$ and time $t_\star$ (unless otherwise specified). To obtain this result, we have used Equation (5.8), which connects the derivatives with respect to the lattice space and time variables $(\mathbf{r}_\star, t_\star)$, and the derivatives with respect to the rescaled (macroscopic) space and time variables $(\mathbf{r}_1, t_1, t_2)$.

We now introduce the fact that the average populations f_i are close to their local equilibrium values; so they can also be expanded in powers of ϵ – see (5.2) – with a zeroth order term corresponding to the local equilibrium. Combining the two expansions, the l.h.s. of (5.16) becomes:

$$\begin{aligned}
&\epsilon^1\left(\partial_{1\beta} c_{i\beta}\, f_i^{(0)} + \partial_{t_1} f_i^{(0)}\right) \\
+\;& \epsilon^2\Big(\partial_{1\beta}\, c_{i\beta} f_i^{(1)} + \partial_{t_1} f_i^{(1)} + \tfrac{1}{2}\partial_{1\beta}\partial_{1\gamma}\, c_{i\beta}c_{i\gamma} f_i^{(0)} \\
&\quad + \partial_{1\beta}\partial_{t_1}\, c_{i\beta} f_i^{(0)} + \tfrac{1}{2}\partial_{t_1}^2 f_i^{(0)} + \partial_{t_2} f_i^{(0)}\Big) \\
+\;& \mathcal{O}(\epsilon^3).
\end{aligned} \tag{5.21}$$

The r.h.s. of (5.16) can also be formally expanded in powers of ϵ, using expression (5.14) for the averaged collision term. The resulting expansion is (argument $\mathbf{r}_\star$ omitted in the r.h.s. for simplicity):

$$\begin{aligned}
\Delta_i(\mathscr{P}, \mathbf{r}_\star) &= \epsilon^1\Big(\sum_{S\in\Gamma} L_{is}\, \mathscr{P}^{(1)}(S)\Big) \\
&+ \epsilon^2\Big(\sum_{S\in\Gamma} L_{is}\, \mathscr{P}^{(2)}(S)\Big) \\
&+ \mathcal{O}(\epsilon^3).
\end{aligned} \tag{5.22}$$

Using (5.21) and (5.22) as the respective expansions of the l.h.s. and r.h.s. of the averaged microdynamic equation (5.16), we obtain, to the first order in ϵ:

$$\sum_{S\in\Gamma} L_{is}\, \mathscr{P}^{(1)}(S) = \partial_{t_1} f_i^{(0)} + \partial_{1\beta}\, c_{i\beta} f_i^{(0)}, \tag{5.23}$$

and to second order:

$$\begin{aligned}
\sum_{S\in\Gamma} L_{is}\, \mathscr{P}^{(2)}(S) &= \partial_{t_1} f_i^{(1)} + \partial_{1\beta}\, c_{i\beta} f_i^{(1)} \\
&+ \partial_{t_2} f_i^{(0)} + \frac{1}{2}\partial_{1\beta}\partial_{1\gamma}\, c_{i\beta}c_{i\gamma} f_i^{(0)} \\
&+ \frac{1}{2}\partial_{t_1}^2 f_i^{(0)} + \partial_{1\beta}\partial_{t_1}\, c_{i\beta} f_i^{(0)}.
\end{aligned} \tag{5.24}$$

We shall now derive the first and second order macrodynamical equations of the lattice gas starting from Equations (5.23) and (5.24) which follow from the averaged microdynamic equation (5.16) expanded in powers of the scale separation parameter ϵ.

5.3 First order macrodynamics

Formally, the first order equation (5.23) is a set of b linear equations for $2^{b\cdot\mathcal{N}}$ unknown quantities, the probability perturbations $\mathscr{P}^{(1)}(S)$ for each lattice configuration S. Of course, these b equations are not sufficient to fully characterize $\mathscr{P}^{(1)}(S)$, which must also meet the requirements of the Liouville equation (see Chapter 4) governing the evolution of the probability distribution $\mathscr{P}$. However, (5.23) is a necessary condition for $\mathscr{P}^{(1)}(S)$, and it has to admit *at least* one solution. Because of the conservation laws, the rank of the linear system (5.23) is lower than b, the number of equations. So the r.h.s. of (5.23) must satisfy solvability conditions (Fredholm's criterion) which lead to the first order macrodynamics. We now see clearly the direct one-to-one relation between the conservation laws and the macrodynamics: the conservation of a microscopic observable q imposes a solvability condition on the r.h.s. of (5.23), and so yields a macrodynamic equation which is a continuity equation for the associated macroscopic variable.

By construction, the continuity equations so obtained are valid to first order in the scale parameter ϵ. They do not take into account diffusive and viscous effects that will come out of the second order equation (5.24).

5.3.1 Solvability conditions for the first order problem

We consider the first order equation (5.23) for $\mathscr{P}^{(1)}$; applying Fredholm's criterion of solvability (the l.h.s. must be orthogonal to all vectors in the kernel of the adjoint of the linear operator L_{is}), we find that $f_i^{(0)}$ must verify:

$$\partial_{t_1} \sum_{i=1}^{b} q_i^{[\kappa]} f_i^{(0)} + \partial_{1\beta} \sum_{i=1}^{b} q_i^{[\kappa]} c_{i\beta} f_i^{(0)} = 0, \qquad \kappa = 1, \ldots, \delta. \tag{5.25}$$

The sums over i are clearly the macroscopic (ensemble-averaged) density $\rho^{[\kappa]}$ and flux $\mathbf{\Phi}^{[\kappa]}$ of the observable $q^{[\kappa]}$,[4] and Equation (5.25) takes the familiar form:

$$\partial_{t_1} \rho^{[\kappa]} + \partial_{1\beta} \Phi_\beta^{[\kappa]} = 0, \qquad \kappa = 1, \ldots, \delta. \tag{5.26}$$

Under this form, the solvability condition (5.26) for the first order problem (5.23) appears clearly as a set of *macroscopic* continuity equations connecting densities $\rho^{[\kappa]}$ and fluxes $\mathbf{\Phi}^{[\kappa]}$ of the microscopically conserved observables (collisional invariants) $q^{[\kappa]}$.

We now need to evaluate these fluxes and to express them as functions of the macroscopic densities $\rho^{[\kappa]}$, so that the continuity equations (5.26) become a closed set of evolution equations for the macroscopic variables, i.e. the densities

[4] See Section 2.1.3 for the definition of macroscopic densities and fluxes, and Section 4.1.2 for the notion of ensemble-averaging.

$\rho^{[\kappa]}$ of the collisional invariants. To do so, we need information about the nature of these collisional invariants; and we also need to know the explicit form of the mean populations f_i. Thus, we now particularize the analysis to the class of thermal lattice gases (see Section 4.5.2) for which the collisional invariants are mass, momentum and energy. (The case of non-thermal models is treated in a less detailed manner in Section 5.7. Indeed, the analysis is fundamentally similar, albeit technically different.)

In Section 4.5.2, we derived expression (4.73) for the mean equilibrium populations in terms of the mass density ρ, the momentum density $\mathbf{j}$, and the effective energy density w. This expression, which is valid for nearly equally distributed states (low velocity, low effective energy) reads:

$$
\begin{aligned}
f_i^{(0)} &= \frac{\rho}{b} + \frac{1}{\xi_2} j_\alpha c_{i\alpha} + \frac{1}{\xi_4} w e_i \\
&+ \frac{b(b-2\rho)}{2\rho(b-\rho)} \left[\frac{1}{\xi_2^2} j_\alpha j_\beta \left(c_{i\alpha} c_{i\beta} - \frac{\mathbf{c}_i^2}{D} \delta_{\alpha\beta} \right) \right. \\
&\quad + \frac{2}{\xi_2 \xi_4} w j_\alpha \left(e_i - \frac{2\xi_4}{D\xi_2} \right) c_{i\alpha} \\
&\quad \left. + \frac{1}{\xi_4^2} w^2 \left(e_i^2 - \frac{\xi_6}{\xi_4} e_i - \frac{\xi_4}{b} \right) \right] + \mathcal{O}(\eta^3),
\end{aligned}
\tag{5.27}
$$

where e_i is the microscopic effective energy: $e_i = \frac{\mathbf{c}_i^2}{2} - \frac{D\xi_2}{2b}$ (see Section 4.5.2). In (5.27), the expansion parameter η is not a scale separation parameter as ϵ. It governs the proximity to the truly equally distributed macrostate with the same mass density. It is also the order of magnitude of both $\mathbf{j}$ and w (see Section 4.5.2).

Let us also recall the expressions defining the geometric coefficients ξ_2, ξ_4 and ξ_6 in (5.27):

$$
\begin{aligned}
\sum_{i=1}^{b} c_{i\alpha} c_{i\beta} &= \xi_2 \, \delta_{\alpha\beta}, \\
\sum_{i=1}^{b} c_{i\alpha} c_{i\beta} c_{i\gamma} c_{i\delta} &= \frac{1}{D(D+2)} \left(4\xi_4 + \frac{D^2 \xi_2^2}{b} \right) \\
&\quad \times \left(\delta_{\alpha\beta}\delta_{\gamma\delta} + \delta_{\alpha\gamma}\delta_{\beta\delta} + \delta_{\alpha\delta}\delta_{\beta\gamma} \right), \\
\sum_{i=1}^{b} \frac{\mathbf{c}_i^2}{2} c_{i\alpha} c_{i\beta} c_{i\gamma} c_{i\delta} &= \frac{1}{D(D+2)} \left(4\xi_6 + \frac{6D\xi_2\xi_4}{b} + \frac{D^3 \xi_2^3}{2b^2} \right) \\
&\quad \times \left(\delta_{\alpha\beta}\delta_{\gamma\delta} + \delta_{\alpha\gamma}\delta_{\beta\delta} + \delta_{\alpha\delta}\delta_{\beta\gamma} \right).
\end{aligned}
\tag{5.28}
$$

These coefficients depend only on the geometric structure of the lattice gas and not on the collision rules.[5]

We now use (5.27) to obtain the averaged mass, momentum and effective energy fluxes $\mathbf{\Phi}^{[0]}$, $\mathbf{\Phi}^{[\alpha]}$ and $\mathbf{\Phi}^{[D+1]}$, involved in the first order macrodynamic equations (5.25). The details of the calculations are given in the Appendix, Section A.9.

(i) The mass flux is just the momentum density; its components are:

$$\Phi_\beta^{[0]} \equiv \sum_{i=1}^{b} c_{i\beta} f_i^{(0)} = j_\beta = \rho u_\beta, \qquad \beta = 1, \ldots, D, \tag{5.29}$$

where $\mathbf{u} = \mathbf{j}/\rho$ denotes the macroscopic fluid velocity (see Section 4.5.2).

(ii) The momentum flux is a second order tensor whose components can also be expressed in terms of the macroscopic variables:

$$\begin{aligned} \Phi_\beta^{[\alpha]} &\equiv \sum_{i=1}^{b} c_{i\alpha} c_{i\beta} f_i^{(0)} \\ &= g(\rho) j_\alpha u_\beta + P(\rho, \mathbf{j}^2, w)\delta_{\alpha\beta} \\ &\quad + \mathcal{O}(\eta^3), \qquad \alpha, \beta = 1, \ldots, D, \end{aligned} \tag{5.30}$$

where

$$g(\rho) = \frac{b - 2\rho}{b - \rho} \frac{b}{D(D+2)} \left(\frac{4\xi_4}{\xi_2^2} + \frac{D^2}{b} \right), \tag{5.31}$$

and

$$P(\rho, \mathbf{j}^2, w) = \frac{\xi_2}{b}\rho - \frac{1}{D}\frac{g(\rho)}{\rho}\mathbf{j}^2 + \frac{2}{D}w. \tag{5.32}$$

$P(\rho, \mathbf{j}^2, w)$ appears as a diagonal part of the momentum flux tensor: it is interpreted as the hydrostatic pressure of the lattice gas.

(iii) The effective energy flux follows similarly; its components are:

$$\begin{aligned} \Phi_\beta^{[D+1]} &\equiv \sum_{i=1}^{b} e_i c_{i\beta} f_i^{(0)} \\ &= g'(\rho) w u_\beta + \frac{2\xi_4}{D\xi_2} j_\beta \\ &\quad + \mathcal{O}(\eta^3), \qquad \beta = 1, \ldots, D, \end{aligned} \tag{5.33}$$

where

$$g'(\rho) = \frac{b - 2\rho}{b - \rho}\left(1 + \frac{2b\xi_6}{D\xi_2\xi_4} - \frac{4b\xi_4}{D^2\xi_2^2}\right). \tag{5.34}$$

[5] See Section 4.5.2 and Table 4.1 for the values of these coefficients for some simple models.

The density-dependent coefficient $g(\rho)$ that appears in both (5.30) and (5.32) is called the 'non-Galilean factor', or, more precisely, the 'non-Galilean factor for momentum'.[6] This non-Galilean factor is very important both for theory and for applications of lattice gases. Indeed, Equation (5.30) is *not* Galilean invariant unless $g(\rho)$ is exactly equal to 1, which *a priori* is not the case. However, as we shall see in Chapter 9, Galilean invariance can be recovered, at least for non-thermal models, in incompressible regimes where the density ρ is nearly constant: $g(\rho)$ is then a constant factor that can be eliminated by proper time rescaling. The factor $g'(\rho)$ in Equation (5.33) is the 'non-Galilean factor for energy' because it plays, for the energy, the same role as the non-Galilean factor $g(\rho)$ in the momentum equation.

Somewhat in the same way as transport coefficients are expressed in terms of time-correlation functions, these non-Galilean factors may also be given a microscopic content: they can be expressed in terms of static correlation functions (for details see Brito, Bussemaker and Ernst, 1992). For LGAs satisfying detailed or semi-detailed balance – so possessing a universal equilibrium distribution – the non-Galilean factors are given in terms of the correlation functions $\kappa_i^{(2)}$ and $\kappa_i^{(3)}$ discussed in Section 4.7; so, in some way, they can be interpreted as a thermodynamic property of the lattice gas.

We now return to the first order continuity equations (5.26) which, with the above results, we can write in the standard Eulerian form:

$$\partial_{t_1}\rho + \partial_{1\beta}(\rho u_\beta) = 0, \tag{5.35}$$

$$\partial_{t_1} j_\alpha + \partial_{1\beta}\Big(g(\rho) j_\alpha u_\beta\Big) + \partial_{1\alpha} P(\rho, \mathbf{j}^2, w) = \mathcal{O}(\eta^3), \tag{5.36}$$

$$\partial_{t_1} w + \partial_{1\beta}\Big(g'(\rho) w u_\beta\Big) + \partial_{1\beta}\left(\frac{2\xi_4}{D\xi_2} j_\beta\right) = \mathcal{O}(\eta^3). \tag{5.37}$$

The mass continuity equation is valid *to any order* in $\mathbf{j}$, whereas the momentum and effective energy continuity equations are approximations neglecting terms $\mathcal{O}(\eta^3)$. The reason is that the mass flux is also the momentum density, but the momentum and effective energy fluxes are functions of the macroscopic variables, which can be obtained only through an asymptotic expansion in powers of $\mathbf{j}$ and w.

5.3.2 Solution of the first order problem

The first order macroscopic equations (5.35) to (5.37) appear as solvability conditions for the first order averaged microdynamic equation (5.23). But we have not solved this first order equation, and its solution $\mathcal{P}^{(1)}(S)$ should be necessary to compute the first order mean populations $f_i^{(1)}$ involved in the

[6] $g(\rho)$ and the factor $G(\rho)$ of Section 4.5.1 differ by a multiplicative constant.

second order problem (5.24). Nevertheless, it is possible to proceed further (to the second order) *without* actually solving this difficult problem, at least if one is interested in the general form of the second order macrodynamic equations, rather than in the explicit expression (and the numerical value) of the coefficients. Indeed, suitable symmetry arguments will provide sufficient information on the first order mean populations $f_i^{(1)}$ and on how they depend on the macroscopic variables. In this manner, linear response coefficients will be introduced, whose values can be either 'measured' by numerical simulations (see Chapter 10), or evaluated with the Boltzmann approximation (see Section 5.6).

Let us start from the first order equation (5.23), and rewrite the r.h.s. in terms of the gradients of the macroscopic variables, using the expression (5.27) for the equilibrium distribution $f_i^{(0)}$, and the solvability conditions (5.35) to (5.37). The resulting first order equation is:

$$\begin{aligned} \sum_{S\in\Gamma} L_{iS}\, \mathscr{P}^{(1)}(S) &= \frac{1}{\xi_2}\left(c_{i\gamma}c_{i\delta} - \frac{\mathbf{c}_i^2}{D}\delta_{\gamma\delta}\right)\frac{\partial_{1\gamma} j_\delta + \partial_{1\delta} j_\gamma}{2} \\ &+ \frac{1}{\xi_4}\left(e_i - \frac{2\xi_4}{D\xi_2}\right) c_{i\gamma}\partial_{1\gamma} w \\ &+ \mathcal{O}(\epsilon\eta^2). \end{aligned} \tag{5.38}$$

It is relatively easy to convince oneself that the solution $\mathscr{P}^{(1)}(S)$ of the first order problem must be a linear combination of the gradients $\partial_{1\gamma} j_\delta + \partial_{1\delta} j_\gamma$ and $\partial_{1\beta} w$. The first order corrections $f_i^{(1)}$ to the mean populations, that result from $\mathscr{P}^{(1)}(S)$, must thus also be a linear combination of these gradients. The most general form for this combination reads:

$$f_i^{(1)} = M_{i\gamma\delta}\frac{\partial_{1\gamma} j_\delta + \partial_{1\delta} j_\gamma}{2} + N_{i\gamma}\partial_{1\gamma} w, \tag{5.39}$$

where $\mathbf{N}_i$ is a set of i-indexed D-dimensional vectors, and where $\mathbf{M}_i$ is a set of i-indexed D-dimensional second order tensors that may depend on the macroscopic variables $(\rho, \mathbf{j}^2, w)$. Furthermore, as the second order tensors $\mathbf{M}_i$ are contracted with the symmetric expression $\partial_{1\gamma} j_\delta + \partial_{1\delta} j_\gamma$, they can be assumed, without loss of generality, to be symmetrical with respect to exchanges of their first and second Greek indices.

So far, we have made no hypothesis on the value of these vectors and tensors, and they can be regarded as a set of unknown quantities depending on the macroscopic variables. However, we can be more specific if the lattice gas is G-invariant and verifies the irreducibility property (see Sections 2.3.4 and 2.3.6). Then, we can apply the simplification procedure described in Section 2.3.6.3: the vectors $\mathbf{N}_i$ must be proportional to the velocity vectors $\mathbf{c}_i$, with equal proportionality coefficients for all channels belonging to the same class:[7]

$$\mathbf{N}_i = N_{(I)}\,\mathbf{c}_i, \qquad \forall i = 1, \ldots, b, \tag{5.40}$$

[7] For the notion of class, see Section 2.3.4.2.

where I is the index of the class to which channel i belongs. Along the same lines of reasoning, the second order tensors $\mathbf{M}_i$ are linear combinations of $\mathbf{c}_i \otimes \mathbf{c}_i$ and δ, the Kronecker tensor, with coefficients which depend only on the class of the corresponding channel:

$$\mathbf{M}_i = M_{(I)}\mathbf{c}_i \otimes \mathbf{c}_i + M'_{(I)}\delta, \qquad \forall i = 1, \dots, b. \tag{5.41}$$

This simplification is possible because the vectors and tensors $\mathbf{M}_i$ and $\mathbf{N}_i$ are such that for all isometry $\mathbf{g}$ in G and for all $i = 1, \dots, b$, we have $\mathbf{g}(\mathbf{M}_i) = \mathbf{M}_{g(i)}$ and $\mathbf{g}(\mathbf{N}_i) = \mathbf{N}_{g(i)}$ (for the definitions of $\mathbf{g}$, G and g, see Section 2.3.4).

The resulting form for $f_i^{(1)}$ reads:

$$\begin{aligned} f_i^{(1)} &= \left(M_{(I)}c_{i\gamma}c_{i\delta} + M'_{(I)}\delta_{\gamma\delta}\right)\frac{\partial_{1\gamma}j_\delta + \partial_{1\delta}j_\gamma}{2} \\ &\quad + N_{(I)}c_{i\gamma}\ \partial_{1\gamma}w. \end{aligned} \tag{5.42}$$

As emphasized in Section 5.2.2, the first order corrections $f_i^{(1)}$ to the mean populations must not contribute to the mass, momentum and effective energy densities. This severely constrains the possible values of the arbitrary set of parameters $M_{(I)}$, $M'_{(I)}$ and $N_{(I)}$. Indeed, the fact that $f_i^{(1)}$ does not contribute to the mass variable ($\sum_i f_i^{(1)} = 0$) implies that:

$$\sum_{i=1}^{b}\left(M_{(I)}c_{i\gamma}c_{i\delta} + M'_{(I)}\delta_{\gamma\delta}\right) = 0, \qquad \forall\, \gamma, \delta, \tag{5.43}$$

that $f_i^{(1)}$ does not contribute to the momentum variable imposes that:

$$\sum_{i=1}^{b} N_{(I)}c_{i\gamma}c_{i\delta} = 0, \qquad \forall\, \gamma, \delta, \tag{5.44}$$

and that $f_i^{(1)}$ does not contribute to the effective energy requires that:

$$\sum_{i=1}^{b}\left(M_{(I)}c_{i\gamma}c_{i\delta} + M'_{(I)}\delta_{\gamma\delta}\right)e_i = 0, \qquad \forall\, \gamma, \delta. \tag{5.45}$$

The $f_i^{(1)}$s contribute neither to the effective energy density nor to the mass density, thus, they do not contribute to the kinetic energy density either, since the latter is a linear combination of effective energy and mass densities. So, the following relation holds between M and M':

$$\sum_{i=1}^{b}\left(M_{(I)}c_{i\gamma}c_{i\delta} + M'_{(I)}\delta_{\gamma\delta}\right)\mathbf{c}_i^2 = 0, \qquad \forall\, \gamma, \delta. \tag{5.46}$$

This relation, taken for $\gamma = \delta$ and summed over γ gives the following simpler form:

$$\sum_{i=1}^{b} M_{(I)} \mathbf{c}_i^2 \mathbf{c}_i^2 = -D \sum_{i=1}^{b} M'_{(I)} \mathbf{c}_i^2, \tag{5.47}$$

which will lead to some simplification in the forthcoming algebra.

To write the solvability conditions for the second order problem, we need an expression for the contributions $\sum_i c_{i\alpha} c_{i\beta} f_i^{(1)}$ and $\sum_i e_i c_{i\beta} f_i^{(1)}$ of the first order mean populations $f_i^{(1)}$, to the momentum and effective energy fluxes. The contribution to the mass flux $\sum_i c_{i\beta} f_i^{(1)}$ is obviously zero, since the mass flux is nothing else but the momentum density, to which the first order populations $f_i^{(1)}$ cannot contribute. We show hereafter how to obtain the functional form of these contributions, without knowing the actual values of the coefficients involved.

The first order contribution $\sum_i c_{i\alpha} c_{i\beta} f_i^{(1)}$ to the momentum flux can be computed from (5.42), which gives the following expression:

$$\begin{aligned} &\sum_{i=1}^{b} c_{i\alpha} c_{i\beta} f_i^{(1)} = \\ &\left(\sum_{i=1}^{b} M_{(I)} c_{i\alpha} c_{i\beta} c_{i\gamma} c_{i\delta} + \sum_{i=1}^{b} M'_{(I)} c_{i\alpha} c_{i\beta} \delta_{\gamma\delta} \right) \frac{\partial_{1\gamma} j_\delta + \partial_{1\delta} j_\gamma}{2}. \end{aligned} \tag{5.48}$$

The lattice gas model is assumed to verify the fourth order crystallographic isotropy property (see Section 2.3.5). Thus, there exist two coefficients $\nu^{(c)}$ and $\tilde{\nu}^{(c)}$, depending on the macroscopic variables $(\rho, \mathbf{j}^2, w)$, such that:

$$\sum_{i=1}^{b} M_{(I)} c_{i\alpha} c_{i\beta} c_{i\gamma} c_{i\delta} = -\nu^{(c)} \left(\delta_{\alpha\beta} \delta_{\gamma\delta} + \delta_{\alpha\gamma} \delta_{\beta\delta} + \delta_{\alpha\delta} \delta_{\beta\gamma} \right),$$

and (5.49)

$$\sum_{i=1}^{b} M'_{(I)} c_{i\alpha} c_{i\beta} = -\tilde{\nu}^{(c)} \delta_{\alpha\beta}.$$

The superscript '(c)' is used to indicate that these coefficients come from the first order perturbation $f_i^{(1)}$ to the mean populations, which are solutions of the first order problem, and thus depend on the collision rules. They are not purely geometrical coefficients like ξ_2, ξ_4 and ξ_6.

Because of (5.47), the two coefficients $\nu^{(c)}$ and $\tilde{\nu}^{(c)}$ cannot be independent. Indeed, it follows from their definition (5.49) that:

$$\nu^{(c)} = -\frac{1}{D(D+2)} \sum_{i=1}^{b} M_{(I)} \mathbf{c}_i^2 \mathbf{c}_i^2,$$

and (5.50)

$$\tilde{\nu}^{(c)} = -\frac{1}{D} \sum_{i=1}^{b} M'_{(I)} \mathbf{c}_i^2.$$

So, Equation (5.47) implies that:

$$\tilde{\nu}^{(c)} + \frac{D+2}{D} \nu^{(c)} = 0. \tag{5.51}$$

With this result, one can express the first order contribution to the momentum flux tensor (5.48) in terms of the coefficient $\nu^{(c)}$:

$$\sum_{i=1}^{b} c_{i\alpha} c_{i\beta} f_i^{(1)} = -\nu^{(c)} \left(\delta_{\alpha\gamma}\delta_{\beta\delta} + \delta_{\alpha\delta}\delta_{\beta\gamma} - \frac{2}{D}\delta_{\alpha\beta}\delta_{\gamma\delta} \right) \frac{\partial_{1\gamma} j_\delta + \partial_{1\delta} j_\gamma}{2}, \tag{5.52}$$

which yields immediately:

$$\sum_{i=1}^{b} c_{i\alpha} c_{i\beta} f_i^{(1)} = -2\nu^{(c)} \left(\frac{\partial_{1\alpha} j_\beta + \partial_{1\beta} j_\alpha}{2} - \frac{1}{D} \partial_{1\gamma} j_\gamma \delta_{\alpha\beta} \right). \tag{5.53}$$

Thanks to (5.51) which connects $\nu^{(c)}$ and $\tilde{\nu}^{(c)}$, the first order contribution to the momentum flux tensor takes the simple form (5.53), involving only the symmetric traceless part of the momentum density gradient tensor $\partial\, j$.

In a similar manner, the first order contribution $\sum_i e_i c_{i\beta} f_i^{(1)}$ to the effective energy flux is given by:

$$\sum_{i=1}^{b} e_i c_{i\beta} f_i^{(1)} = \sum_{i=1}^{b} N_{(I)} e_i c_{i\beta} c_{i\gamma}\ \partial_{1\gamma} w. \tag{5.54}$$

Observing that e_i depends only on the class to which channel i belongs, and invoking again the crystallographic isotropy property, we see that there exists a coefficient $\zeta^{(c)}$, which depend on the macroscopic variables $(\rho, \mathbf{j}^2, w)$, such that:

$$\sum_{i=1}^{b} N_{(I)} e_i c_{i\beta} c_{i\gamma} = -\zeta^{(c)}\ \delta_{\beta\gamma}. \tag{5.55}$$

The resulting expression for the first order contribution to the effective energy flux vector reads:

$$\sum_{i=1}^{b} e_i c_{i\beta} f_i^{(1)} = -\zeta^{(c)} \, \partial_{1\beta} w. \tag{5.56}$$

We emphasize again that the actual values of $v^{(c)}$ and $\zeta^{(c)}$ are still unknown. At this point, all we can say – without the Boltzmann approximation – is that these coefficients do exist, and that the first order contributions to the momentum and effective energy fluxes are of the form (5.53) and (5.56). In Section 5.6, we shall use the Boltzmann approximation to compute these coefficients explicitly. Their numerical value can also be 'measured' by computer simulations of wave damping (see Chapter 10).

5.4 Second order macrodynamics

The second order problem (5.24) has exactly the same mathematical structure as the first order problem (5.23), except that the r.h.s. is more complicated. The analysis will also involve solvability conditions, that is, the r.h.s. must be orthogonal to all vectors in the kernel of the adjoint of the linear operator L_{is} (Fredholm's criterion).

5.4.1 Solvability conditions for the second order problem

We apply Fredholm's criterion to the second order problem (5.24): the r.h.s. must be orthogonal to the local invariants of the model $q^{[\kappa]}$, $\kappa = 1, \ldots, \delta$. We obtain that $f_i^{(0)}$ and $f_i^{(1)}$ must verify the following relations:

$$\begin{aligned} 0 \;=\;& \partial_{t_1} \sum_{i=1}^{b} q_i^{[\kappa]} f_i^{(1)} + \partial_{1\beta} \sum_{i=1}^{b} q_i^{[\kappa]} c_{i\beta} f_i^{(1)} \\ &+ \partial_{t_2} \sum_{i=1}^{b} q_i^{[\kappa]} f_i^{(0)} + \frac{1}{2} \partial_{1\beta} \partial_{1\gamma} \sum_{i=1}^{b} q_i^{[\kappa]} c_{i\beta} c_{i\gamma} f_i^{(0)} \\ &+ \frac{1}{2} \partial_{t_1}^2 \sum_{i=1}^{b} q_i^{[\kappa]} f_i^{(0)} + \partial_{1\beta} \partial_{t_1} \sum_{i=1}^{b} q_i^{[\kappa]} c_{i\beta} f_i^{(0)}, \end{aligned} \tag{5.57}$$

for all $\kappa = 1, \ldots, \delta$, that is, for all collisional invariants. Since we treat the case of single-species thermal lattice gases, the linear invariants are the mass, the momentum and the effective energy. The algebraic manipulations leading from (5.57) to equations (5.58), (5.59) and (5.61) hereafter are given in the Appendix, Section A.10. In these manipulations, we use the first order macrodynamical equations, (5.35) to (5.37), and the fact that the first order perturbations

$f_i^{(1)}$ do not contribute to the mass, momentum and effective energy densities (see Equation (5.4)).

For the mass invariant $q^{[0]} = (1, \ldots, 1)$, the solvability condition (5.57) leads to the simple relation:

$$\partial_{t_2} \rho = 0, \tag{5.58}$$

which shows that there is no mass diffusion. The first order perturbations $f_i^{(1)}$ do not appear in the second order solvability condition (5.58). Indeed, by definition, the $f_i^{(1)}$s contribute neither to the mass density, nor to the mass flux, since the latter is also the momentum density (see Equation (5.4)).

The case of the momentum invariant is more difficult, because the first order perturbations $f_i^{(1)}$ now contribute to the second order solvability condition through the momentum flux $\sum_i c_{i\alpha} c_{i\beta} f_i^{(1)}$ (see Equation (5.53)). The resulting solvability condition is:

$$\partial_{t_2} j_\alpha + \partial_{1\beta} \left(-\nu \left(\partial_{1\alpha} j_\beta + \partial_{1\beta} j_\alpha - \frac{2}{D} \partial_{1\gamma} j_\gamma \delta_{\alpha\beta} \right) \right) = \mathcal{O}(\eta^2), \tag{5.59}$$

where the coefficient ν, which *a priori* depends on $(\rho, \mathbf{j}^2, w)$, amounts to:

$$\nu = \nu^{(c)} - \frac{1}{2D(D+2)} \left(\frac{4\xi_4}{\xi_2} + \frac{D^2 \xi_2}{b} \right). \tag{5.60}$$

This coefficient ν is a transport coefficient that will be interpreted in Chapter 9 as a 'shear viscosity coefficient'.

The effective energy invariant leads to the same kind of expression:

$$\partial_{t_2} w + \partial_{1\beta} \left(-\zeta \, \partial_{1\beta} w \right) = \mathcal{O}(\eta^2), \tag{5.61}$$

where the coefficient ζ is:

$$\zeta = \zeta^{(c)} + \left(\frac{\xi_6}{D\xi_4} + \frac{\xi_2}{2b} - \frac{2\xi_4}{D^2 \xi_2} \right). \tag{5.62}$$

The coefficient ζ is also a transport coefficient that will be interpreted in Chapter 9 as a 'thermal conductivity coefficient'.

In both (5.60) and (5.62), the first term is the contribution of $f_i^{(1)}$, which depends on the collision rules, and the second term is a purely geometrical contribution, which depends only on the propagation rule and on the lattice geometry.

5.5 The macrodynamic equations

We can now merge the results of the first order problem, (5.35), (5.36) and (5.37), and of the second order problem, (5.58), (5.59) and (5.61), to obtain the macrodynamical equations governing the time-evolution of the macroscopic variables

ρ, $\mathbf{j}$ and w. To do this, we multiply (5.35), (5.36) and (5.37) by ϵ, and (5.58), (5.59) and (5.61) by ϵ^2. After summation, we use (5.8) to recover the real time- and space-derivatives ∂_t and ∂_α from their rescaled versions ∂_{t_1}, ∂_{t_2} and $\partial_{1\alpha}$. We obtain, for the mass invariant:

$$\partial_t \rho + \partial_\beta(\rho u_\beta) = 0, \tag{5.63}$$

for the momentum invariant:

$$\begin{aligned} \partial_t j_\alpha + \partial_\beta\Big(g j_\alpha u_\beta\Big) + \partial_\alpha P &= \partial_\beta\Big(\nu\left(\partial_\alpha j_\beta + \partial_\beta j_\alpha - \tfrac{2}{D}\partial_\gamma j_\gamma \delta_{\alpha\beta}\right)\Big) \\ &\quad + \mathcal{O}(\epsilon^3\eta) + \mathcal{O}(\epsilon^2\eta^2) + \mathcal{O}(\epsilon\eta^3), \end{aligned} \tag{5.64}$$

and for the energy invariant:

$$\begin{aligned} \partial_t w + \partial_\beta\Big(g' w u_\beta\Big) + \partial_\beta\Big(\tfrac{2\xi_4}{D\xi_2} j_\beta\Big) &= \partial_\beta\Big(\zeta\ \partial_\beta w\Big) \\ &\quad + \mathcal{O}(\epsilon^3\eta) + \mathcal{O}(\epsilon^2\eta^2) + \mathcal{O}(\epsilon\eta^3). \end{aligned} \tag{5.65}$$

The non-Galilean factors g and g', defined in (5.31) and (5.34) respectively, depend on the mass density ρ. The pressure P is defined in (5.32), and depends on all the macroscopic parameters $(\rho, \mathbf{j}^2, w)$. The transport coefficients ν and ζ are defined in (5.60) and (5.62). Their values also depend *a priori* on all the macroscopic parameters $(\rho, \mathbf{j}^2, w)$. However, any dependence of ν or ζ on $\mathbf{j}^2$ or w would lead to terms of order of $\epsilon^2\eta^2$ or higher in the macrodynamic equations. To be consistent with the order of approximation in the macrodynamic equations, these terms will be neglected; from now on, we shall use the short-hand notation ν and ζ for $\nu(\rho, 0, 0)$ and $\zeta(\rho, 0, 0)$.

Equations (5.63), (5.64) and (5.65) give the general form of the macrodynamic equations governing the long-time, long-wavelength behavior of the lattice gas.[8] These macrodynamic equations show close resemblance with the equations of continuous fluid dynamics (such as the Navier–Stokes equation). However, there is a difference: the non-Galilean factors $g(\rho)$ and $g'(\rho)$, in (5.64) and (5.65) respectively, do not appear in the usual equations of continuous fluid mechanics. The lack of Galilean invariance in the fluid-like behavior of the lattice gas is a direct consequence of the discrete nature of the microscopic velocity space, which itself results from the lattice structure.[9] In Chapter 9, we shall show how Galilean invariance can be recovered in hydrodynamic regimes such as incompressible flow and linear acoustic flow.

[8] In Chapter 8, we give an alternative derivation of the lattice gas hydrodynamic equations using the lattice Boltzmann approximation.

[9] A factor also appears in front of the advection term in the hydrodynamic equations of discrete kinetic models (Gatignol, 1975) where position space is continuous and velocity space is discrete.

5.6 Transport coefficients within the Boltzmann approximation

To compute theoretically the transport coefficients ν and ζ, we need a physical approximation: the Boltzmann ansatz by which we assume that particles entering a collision are statistically uncorrelated (of course they are correlated *after* the collision). Physically this means that the 'memory' that particles may have of a past collision is erased by the propagation phase, before these particles meet again for a new collision. Clearly this approximation is acceptable at sufficiently low density.

Under this strong physical hypothesis, the averaged collision operator $\Delta_i(\mathscr{P}, \mathbf{r}_\star, t_\star)$ defined in (5.12) is replaced by the 'Boltzmann collision operator' $\Delta^{(B)}$, where the average of the products of n_is is factorized into the product of the averages:

$$\Delta_i^{(B)}(f_i) \equiv \sum_{(s,s')\in\gamma^2} (s_i' - s_i) A(s \to s') \prod_{j=1}^{b} f_j^{s_j} \overline{f}_j^{\overline{s}_j}. \tag{5.66}$$

Combining the Boltzmann collision operator and the propagation operator, we obtain the 'lattice Boltzmann equation':[10]

$$f_i(\mathbf{r}_\star + \mathbf{c}_i, t_\star + 1) - f_i(\mathbf{r}_\star, t_\star) = \sum_{(s,s')\in\gamma^2} (s_i' - s_i) A(s \to s') \prod_{j=1}^{b} f_j^{s_j} \overline{f}_j^{\overline{s}_j}. \tag{5.67}$$

As for the averaged microdynamic equation in Section 5.2.5, we can again introduce the expansion of f_i in powers of the scale separation parameter ϵ. Following the same procedure, we obtain the first order lattice Boltzmann equation, which connects the as yet unknown first order mean populations $f_i^{(1)}$ to the local equilibrium mean populations $f_i^{(0)}$:

$$\sum_{j=1}^{b} \Omega_{ij} f_j^{(1)} = \partial_{t_1} f_i^{(0)} + \partial_{1\beta} c_{i\beta} f_i^{(0)}, \tag{5.68}$$

where Ω_{ij} is the Boltzmann collision matrix, that is, the Boltzmann collision operator linearized around the uniform, zero-speed, zero-effective energy equilibrium populations $f_i^{(0)} = d \equiv \rho/b$:

$$\Omega_{ij} \equiv \left. \frac{\partial \Delta_i^{(B)}}{\partial f_j} \right|_{f_i = d}. \tag{5.69}$$

One may wonder why the linearization is performed around the *zero-speed* and *zero-effective energy* uniform equilibrium $f_i = d$, rather than around $f_i = f_i^{(0)}$, as it logically should. The reason is that the difference between the Boltzmann

[10] The Boltzmann equation was discussed in detail in Section 4.2.2; we should keep in mind, it is an *approximate* equation: strictly, in (5.67) we should write '$\simeq$' rather than '$=$'.

collision operator linearized around $f_i = d$ and around $f_i = f_i^{(0)}$ is of order η. Taking this difference into account would result in $\epsilon^2\eta^2$ corrections to the macrodynamics that would not be consistent with the chosen approximation scheme (see Equations (5.64) and (5.65)).

From the Boltzmann collision operator (5.66), it can be shown that the Boltzmann collision matrix Ω_{ij} is given by:[11]

$$\Omega_{ij} = \sum_{s,s' \in \gamma} (s_i' - s_i) A(s \rightarrow s')\, s_j\, d^{p-1}(1-d)^{b-p-1}, \tag{5.70}$$

where p is the number of particles in the state s:

$$p = \sum_{i=1}^{b} s_i.$$

Expression (5.70) provides the basis of a procedure for the computation of the values of the matrix coefficients Ω_{ij}, once the collision rules $A(s \rightarrow s')$ of the lattice gas model are given.

The Boltzmann collision matrix is not regular, because of the existence of conserved quantities. Yet, Equation (5.68) has a family of solutions provided the first order solvability conditions, (5.35) to (5.37), are satisfied. The desired solution $f_i^{(1)}$ will be the only element in this family that also satisfies the condition to have zero contribution to the mass, momentum and effective energy (see Equation (5.4) and the discussion preceding (5.4)).

Using the solvability conditions together with the expression (5.27) for the local equilibria $f_i^{(0)}$, one can re-express the r.h.s. of (5.68) in terms of the gradients of the macroscopic variables ρ, $\mathbf{j}$ and w as:

$$\begin{aligned} \sum_{j=1}^{b} \Omega_{ij} f_j^{(1)} &= \frac{1}{\xi_2}\left(c_{i\gamma}c_{i\delta} - \frac{\mathbf{c}_i^2}{D}\delta_{\gamma\delta}\right)\frac{\partial_{1\gamma} j_\delta + \partial_{1\delta} j_\gamma}{2} \\ &+ \frac{1}{\xi_4}\left(e_i - \frac{2\xi_4}{D\xi_2}\right) c_{i\gamma}\partial_{1\gamma} w \\ &+ \mathcal{O}(\epsilon\eta^2). \end{aligned} \tag{5.71}$$

The standard procedure for solving (5.71) is to introduce $\Omega^{[-1]}$, the 'pseudo-inverse' of the Boltzmann collision matrix. More precisely, consider the restriction $\tilde{\Omega}$ of Boltzmann's collision matrix Ω, to the subspace of $\mathbb{R}^b$, orthogonal to its kernel. This restricted operator is regular and its inverse operator $\tilde{\Omega}^{-1}$ can be defined. The 'pseudo-inverse' $\Omega^{[-1]}$ of Ω is defined as an operator which acts as the null operator on the kernel of Ω, and as $\tilde{\Omega}^{-1}$, on the subspace of $\mathbb{R}^b$, orthogonal to the kernel.

[11] The linearized Boltzmann equation is discussed in detail in Section 6.1.

In terms of $\Omega^{[-1]}$, the first order corrections $f_i^{(1)}$ are:

$$f_i^{(1)} = \frac{1}{\xi_2}\sum_{j=1}^{b}\Omega_{ij}^{[-1]}\left(c_{j\gamma}c_{j\delta} - \frac{\mathbf{c}_j^2}{D}\delta_{\gamma\delta}\right)\frac{\partial_{1\gamma}j_\delta + \partial_{1\delta}j_\gamma}{2} + \frac{1}{\xi_4}\sum_{j=1}^{b}\Omega_{ij}^{[-1]}\left(e_j - \frac{2\xi_4}{D\xi_2}\right)c_{j\gamma}\partial_{1\gamma}w + \mathcal{O}(\epsilon\eta^2). \tag{5.72}$$

It is then straightforward to compute the contributions of the first order populations $f_i^{(1)}$ to the momentum and effective energy fluxes:

$$\sum_{i=1}^{b} c_{i\alpha}c_{i\beta}f_i^{(1)} = \frac{1}{\xi_2}\sum_{i,j=1}^{b} c_{i\alpha}c_{i\beta}\Omega_{ij}^{[-1]}\left(c_{j\gamma}c_{j\delta} - \frac{\mathbf{c}_j^2}{D}\delta_{\gamma\delta}\right)\frac{\partial_{1\gamma}j_\delta + \partial_{1\delta}j_\gamma}{2} + \mathcal{O}(\epsilon\eta^2), \tag{5.73}$$

and

$$\sum_{i=1}^{b} e_i c_{i\beta}f_i^{(1)} = \frac{1}{\xi_4}\sum_{i,j=1}^{b} e_i c_{i\beta}\Omega_{ij}^{[-1]}\left(e_j - \frac{2\xi_4}{D\xi_2}\right)c_{j\gamma}\,\partial_{1\gamma}w + \mathcal{O}(\epsilon\eta^2). \tag{5.74}$$

Considering Equations (5.52) and (5.56) as definitions for $\nu^{(c)}$ and $\zeta^{(c)}$, the collisional contributions to the transport coefficients ν and ζ, we can use the relations (5.73) and (5.74) to obtain the desired expressions for $\nu^{(c)}$ and $\zeta^{(c)}$:

$$\nu^{(c)} = -\frac{1}{(D-1)(D+2)}\frac{1}{\xi_2}\sum_{\alpha\beta}\sum_{i,j=1}^{b} c_{i\alpha}c_{i\beta}\Omega_{ij}^{[-1]}\left(c_{j\alpha}c_{j\beta} - \frac{\mathbf{c}_j^2}{D}\delta_{\alpha\beta}\right), \tag{5.75}$$

and

$$\zeta^{(c)} = -\frac{1}{D}\frac{1}{\xi_4}\sum_{\beta}\sum_{i,j=1}^{b} e_i c_{i\beta}\Omega_{ij}^{[-1]}\left(e_j - \frac{2\xi_4}{D\xi_2}\right)c_{j\beta}. \tag{5.76}$$

The resulting values for $\nu^{(c)}$ and $\zeta^{(c)}$ depend on the density per channel d through $\Omega^{[-1]}$; these are *approximate* results, since we used the Boltzmann approximation in the computation. They are important theoretical results, but in practice, it is more convenient – and also more accurate – to 'measure' the transport coefficients using a wave relaxation technique by lattice gas automaton simulations (see Chapter 10).

5.7 Non-thermal models

Non-thermal models (see Section 4.5.1) differ from thermal models by the conservation laws: non-thermal models conserve mass and momentum, but *do not* conserve a linearly independent kinetic energy observable. Examples are the FHP-II and FHP-III models. Consequently, the macrodynamics of non-thermal models is governed by only $D+1$ macroscopic continuity equations, while there are $D+2$ equations for thermal models. However, it would be erroneous to think that the macrodynamics of non-thermal models simply degenerates from the macrodynamics of thermal models, by 'omitting' the energy continuity equation and the contributions of the effective energy density w to the mass and momentum continuity equations. Indeed, there are substantial differences. The first difference already appears at the stage of the nearly equally distributed equilibrium distributions, which, for non-thermal models, read:

$$\begin{aligned} f_i &= \frac{\rho}{b} + \frac{1}{\xi_2} c_{i\alpha} j_\alpha \\ &+ \frac{b}{2\xi_2^2} \frac{b-2\rho}{\rho(b-\rho)} \left(c_{i\alpha} c_{i\beta} - \frac{\xi_2}{b} \delta_{\alpha\beta}\right) j_\alpha j_\beta \\ &+ \mathcal{O}(j^3). \end{aligned} \tag{5.77}$$

(see Equation (4.58) in Section 4.5.1). This expression *cannot* be obtained from the equivalent expression (4.73) for thermal models by setting $w=0$. Indeed, the non-linear $j_\alpha j_\beta$ term for non-thermal models involves the second order tensor $(c_{i\alpha} c_{i\alpha} - \frac{\xi_2}{b} \delta_{\alpha\beta})$, whereas the same term for thermal models involves the tensor $(c_{i\alpha} c_{i\alpha} - \frac{c_i^2}{D} \delta_{\alpha\beta})$. These tensors are different in the most general case, for example for models with rest particles. This slight difference in the equilibrium distribution produces modifications in the first and second order macrodynamical equations.

5.7.1 First order macrodynamics

For non-thermal models, the momentum flux tensor $\Phi_\beta^{[\alpha]} \equiv \sum_i c_{i\alpha} c_{i\beta} f_i^{(0)}$ takes a form that remains rather similar to the equivalent expression (5.30) for thermal models, except for the diagonal pressure term which is slightly modified. The resulting first order macrodynamical equations for the mass and momentum density are respectively:

$$\partial_{t_1} \rho + \partial_{1\beta}(\rho u_\beta) = 0, \tag{5.78}$$

$$\partial_{t_1} j_\alpha + \partial_{1\beta}\Big(g(\rho) j_\alpha u_\beta\Big) + \partial_{1\alpha} P(\rho, \mathbf{j}^2) = \mathcal{O}(\eta^3), \tag{5.79}$$

where the pressure term reads:

$$P(\rho, \mathbf{j}^2) = \frac{\xi_2}{b}\rho - \frac{\Lambda}{D}\frac{g(\rho)}{\rho}\mathbf{j}^2. \tag{5.80}$$

The coefficient Λ, which is equal to 1 for thermal models (see Equation (5.32)), is given by:

$$\Lambda = \frac{D^2\xi_2^2 - D2\xi_4 b}{D^2\xi_2^2 + 4\xi_4 b}. \tag{5.81}$$

The factor $g(\rho)$ is not modified:

$$g(\rho) = \frac{b-2\rho}{b-\rho}\frac{b}{D(D+2)}\left(\frac{4\xi_4}{\xi_2^2} + \frac{D^2}{b}\right).$$

As already emphasized, the non-thermal pressure (5.80) cannot be obtained from its thermal counterpart (5.32) by making w vanish, since the value of Λ, the coefficient of the $\mathbf{j}^2$ term, is *a priori* modified, at least in the most general case. However, there are lattice gas models, such as FHP-1, which do conserve kinetic energy, but do not enter the category of thermal models because the kinetic energy observable is proportional to the mass observable (thus they are linearly dependent). This is the case for the models where all velocity vectors $\mathbf{c}_i$ have equal moduli (homokinetic models). For these models, the coefficient ξ_4 vanishes, and so the coefficient Λ is equal to 1, as for thermal models. As far as only macrodynamical aspects are concerned, these homokinetic models can be viewed as special cases of non-thermal models (with $\Lambda = 1$); they can also be viewed as degenerated cases of thermal models (with vanishing effective energy density w).[12]

5.7.2 Second order macrodynamics

The lack of an independent kinetic energy invariant has more dramatic consequences on the second order macrodynamics. In the thermal case, the second order momentum continuity equation (5.59) involve only the symmetric traceless part of the momentum gradient tensor:

$$\left(\partial_{1\alpha} j_\beta + \partial_{1\beta} j_\alpha - \frac{2}{D}\partial_{1\gamma} j_\gamma \delta_{\alpha\beta}\right).$$

For non-thermal models, there is an additional diagonal term, proportional to the divergence of $\mathbf{j}$. There are two reasons for this: first, the equilibrium distribution is different (no energy term), second, the lack of a kinetic energy invariant releases the constraint (5.46) and its direct consequence (5.51).

[12] For details on homokinetic models, see Frisch, d'Humières, Hasslacher, Lallemand, Pomeau and Rivet, 1987, or Rivet, 1988.

The second order macroscopic equations for the mass and momentum densities read:

$$\partial_{t_2}\rho = 0, \tag{5.82}$$

and

$$\begin{aligned} \partial_{t_2} j_\alpha \quad &+ \quad \partial_{1\beta}\left(-\nu\left(\partial_{1\alpha} j_\beta + \partial_{1\beta} j_\alpha - \tfrac{2}{D}\partial_{1\gamma} j_\gamma \delta_{\alpha\beta}\right)\right) \\ &+ \quad \partial_{1\beta}\left(-\nu'\left(\partial_{1\gamma} j_\gamma \delta_{\alpha\beta}\right)\right) = \mathcal{O}(\eta^2), \end{aligned} \tag{5.83}$$

where the coefficients ν and ν', which *a priori* depend on ρ, are:

$$\nu = \nu^{(c)} - \frac{1}{2D(D+2)}\left(\frac{4\xi_4}{\xi_2} + \frac{D^2\xi_2}{b}\right), \tag{5.84}$$

and

$$\nu' = \nu'^{(c)} - \frac{2\xi_4}{D^2\xi_2}. \tag{5.85}$$

The coefficient ν' in the additional $\partial_{1\gamma} j_\gamma$ term will be interpreted in Chapter 9 as the 'bulk viscosity coefficient'; this coefficient is a specific feature of (non-thermal) lattice gases where rest particles are assigned an 'internal energy' (see the discussion in Section 3.3.2). The bulk viscosity coefficient vanishes for kinetic thermal models (such as the GBL lattice gas). The terms $\nu^{(c)}$ and $\nu'^{(c)}$ are the collisional contributions to the coefficients ν and ν'. They can be 'measured' numerically by sound wave damping simulations (see Chapter 10), and they can be evaluated theoretically within the Boltzmann approximation (see Section 5.7.4).

5.7.3 The macrodynamic equation

In the same way as we did for the thermal lattice gas (see Section 5.5), we merge the first and second order contributions to the macrodynamics to obtain the 'macrodynamic equations':

$$\partial_t\rho + \partial_\beta(\rho u_\beta) = 0, \tag{5.86}$$

$$\begin{aligned} &\partial_t j_\alpha + \partial_\beta\left(g j_\alpha u_\beta\right) + \partial_\alpha P = \\ &\partial_\beta\left(\nu\left(\partial_\alpha j_\beta + \partial_\beta j_\alpha - \tfrac{2}{D}\partial_\gamma j_\gamma \delta_{\alpha\beta}\right) + \nu'\left(\partial_\gamma j_\gamma \delta_{\alpha\beta}\right)\right) \\ &+\mathcal{O}(\epsilon^3\eta) + \mathcal{O}(\epsilon^2\eta^2) + \mathcal{O}(\epsilon\eta^3). \end{aligned} \tag{5.87}$$

The non-Galilean factor g is defined in (5.31), the pressure P, defined in (5.80), depends on the macroscopic quantities $(\rho, \mathbf{j}^2)$, and the transport coefficients ν

and ν' are given in (5.60) and (5.62), respectively. The values of ν and ν' also depend *a priori* on the macroscopic quantities $(\rho, \mathbf{j}^2)$. However, any dependence of ν or ν' on $\mathbf{j}^2$ would lead to terms at least of order $(\epsilon^2\eta^2)$ in the macrodynamic equations. For consistency with the order of approximation in the macrodynamic equations, these terms are neglected; for simplicity, we shall use the short-hand notation ν and ν' for $\nu(\rho, 0)$ and $\nu'(\rho, 0)$.

5.7.4 The transport coefficients

We proceed essentially along the lines of Section 5.6. The linearized Boltzmann equation for the first order perturbation $f_i^{(1)}$ reduces to:

$$\sum_{j=1}^{b} \Omega_{ij} f_j^{(1)} = \frac{1}{\xi_2}\left(c_{i\gamma}c_{i\delta} - \frac{\xi_2}{b}\delta_{\gamma\delta}\right)\frac{\partial_{1\gamma} j_\delta + \partial_{1\delta} j_\gamma}{2} + \mathcal{O}(\epsilon\eta^2). \tag{5.88}$$

As discussed earlier, for non-thermal models the tensor $c_{i\gamma}c_{i\delta} - \frac{\xi_2}{b}\delta_{\gamma\delta}$ replaces the tensor $c_{i\gamma}c_{i\delta} - \frac{c_i^2}{D}\delta_{\gamma\delta}$ of thermal models.

The collisional contributions $\tilde{\nu}^{(c)}$ and $\nu'^{(c)}$ to the shear and bulk viscosity coefficients can be computed once the pseudo-inverse $\Omega_{ij}^{[-1]}$ of the linearized Boltzmann collision operator is known. These contributions are:

$$\nu'^{(c)} = -\frac{1}{D^2}\frac{1}{\xi_2}\sum_{i,j=1}^{b} \mathbf{c}_i^2 \Omega_{ij}^{[-1]} \mathbf{c}_j^2, \tag{5.89}$$

and

$$\nu^{(c)} = -\frac{1}{(D-1)(D+2)}\left(\frac{1}{\xi_2}\sum_{\alpha\beta}\sum_{i,j=1}^{b} c_{i\alpha}c_{i\beta}\Omega_{ij}^{[-1]} c_{j\alpha}c_{j\beta} - D\nu'^{(c)}\right). \tag{5.90}$$

For homokinetic models (see Section 5.7.1), the bulk viscosity coefficient vanishes, and the Boltzmann expression for the shear viscosity coefficient is identical to the shear viscosity of thermal lattice gases.

5.8 Comments

In this chapter, we developed a procedure inspired from the Chapman–Enskog method (see Chapman and Cowling, 1970) to obtain the macroscopic equations of lattice gases, starting from the microscopic dynamics. In Chapter 8, we present a different and complementary approach to the 'micro- to macro-' derivation of lattice gas hydrodynamics. We close by summing up the characteristics of the LGA as a fluid model:

- The model is G-invariant (see Section 2.3.4).
- The model has crystallographic isotropy, at least up to fourth order (see Section 2.3.5).
- The model verifies the irreducibility condition (see Section 2.3.6).
- The model verifies the semi-detailed balance condition (see Section 2.3.1).
- All particles have unit mass.
- The collisional invariants of the model are the mass, the momentum, and the kinetic energy for thermal models, and the mass and momentum for non-thermal models (see Section 2.3.3).

Two additional assumptions define the regimes covered by the analysis:

- The scale separation hypothesis, which justifies the Chapman–Enskog procedure, expresses the hydrodynamic limit.
- The local equilibrium around which the Chapman–Enskog expansion is performed is 'nearly equally distributed', i.e. at any node, the mean populations of all channels are close to a common value, which means physically that the fluid velocity and the effective energy density are small.

The resulting macrodynamical equations describe the hydrodynamics of the lattice gas. These equations are quite *similar* – albeit not identical – to the equations of classical fluid mechanics. They appear clearly as continuity equations, they are translation-invariant in space and time, and they are isotropic; however, they are *not Galilean invariant* because of the presence of the factors $g(\rho)$ and $g'(\rho)$, a consequence of the underlying discreteness of the microscopic velocity space. In Chapter 9, we shall see that in certain hydrodynamic regimes, which from the physical viewpoint are most interesting, the lattice gas is governed by Galilean invariant equations.

Chapter 6

Linearized hydrodynamics

As LGAs are constructed as model systems where point particles undergo displacements in discrete time steps and where configurational transitions on the lattice nodes represent collisional processes, one can view the lattice gas as a discretized version of a hard sphere gas on a regular lattice where particles are subject to an exclusion principle instead of an excluded volume.[1] The advantage with LGAs is that, starting from exact microdynamical equations, statistical mechanical computations can be conducted rather straightforwardly in a logical fashion with well controlled assumptions to bypass the many-body problem. This is well exemplified by the development in Chapter 4 leading to the lattice Boltzmann equation. For the moment we shall consider the lattice gas automaton as a *bona fide* statistical mechanical model with extremely simplified dynamics. Nevertheless we may argue that the lattice gas exhibits two important features:

(i) it possesses a large number of degrees of freedom;
(ii) its Boolean microscopic nature combined with stochastic microdynamics results in intrinsic fluctuations.

Because of these spontaneous fluctuations and of its large number of degrees of freedom, the lattice gas can be considered as a 'reservoir of thermal excitations' in much the same way as a real fluid. Now the question must be raised – as for the hard sphere model in usual statistical mechanics – as to the validity of the lattice gas automaton to *represent* actual fluids. In Chapters 5 and 8 we consider full hydrodynamics and macroscopic phenomena. Here we address

[1] See also the discussion on the second virial coefficient in Section 4.6.

the question whether the intrinsic fluctuations in LGA capture the essentials of actual fluctuations in real fluids as described by *linearized hydrodynamics.*

6.1 The linearized Boltzmann equation

We are interested in the dynamics of fluctuations which create perturbations that are sufficiently weak to justify first order perturbation analysis. We consider *small* deviations $\delta f(\mathbf{r}_\star, \mathbf{c}; t_\star)$ from the equilibrium distribution $f^{(\mathrm{eq})}$:

$$\delta f(\mathbf{r}_\star, \mathbf{c}; t_\star) = f(\mathbf{r}_\star, \mathbf{c}; t_\star) - f^{(\mathrm{eq})}, \tag{6.1}$$

which are produced by internal (spontaneous) fluctuations in position – velocity space $(\mathbf{r}_\star, \mathbf{c})$.[2] Here $\delta f(\mathbf{r}_\star, \mathbf{c}; t_\star)$ is the *non-equilibrium* ensemble average of the fluctuations in the number of particles with velocity $\mathbf{c}$ at node $\mathbf{r}_\star$ at time $t_\star$:

$$\delta f(\mathbf{r}_\star, \mathbf{c}; t_\star) = \langle n(\mathbf{r}_\star, \mathbf{c}; t_\star) - \langle n(\mathbf{c}) \rangle \rangle_{\mathrm{NE}}, \tag{6.2}$$

where $\langle n(\mathbf{c}) \rangle = f^{(\mathrm{eq})}$ is the Fermi distribution (see Chapter 4) at *global* equilibrium:

$$f_i^{(\mathrm{eq})} \equiv \langle n(\mathbf{c}_i) \rangle = [1 + \exp(-\alpha + \beta\, e_i)]^{-1}. \tag{6.3}$$

The space- and time-evolution of $\delta f(\mathbf{r}_\star, \mathbf{c}; t_\star)$ is governed by the equation which is obtained by substituting $f(\mathbf{r}_\star, \mathbf{c}; t_\star) = f^{(\mathrm{eq})} + \delta f(\mathbf{r}_\star, \mathbf{c}; t_\star)$ in the lattice Boltzmann equation (4.7) and applying first order perturbation analysis. To first order we obtain the *linearized lattice Boltzmann equation*:

$$\delta f_i(\mathbf{r}_\star + \mathbf{c}_i; t_\star + 1) - \delta f_i(\mathbf{r}_\star; t_\star) = \sum_{j=1}^{b} \Omega_{ij}\, \delta f_j(\mathbf{r}_\star; t_\star), \tag{6.4}$$

with the notation $f_i(\mathbf{r}_\star; t_\star) \equiv f(\mathbf{r}_\star, \mathbf{c}_i; t_\star)$ and where we have used the property that the action of the collision operator on $f^{(\mathrm{eq})}$ is zero. The $b \times b$ matrix Ω_{ij} denotes the linearized collision operator (5.69):

$$\begin{aligned}
\Omega_{ij} &= \sum_{ss'} (s_i' - s_i)\, A(s \to s') \prod_{k \neq j} (f_k^{(\mathrm{eq})})^{s_k} (1 - f_k^{(\mathrm{eq})})^{(1-s_k)} \\
&\times \left[s_j (f_j^{(\mathrm{eq})})^{s_j - 1} (1 - f_j^{(\mathrm{eq})})^{(1-s_j)} - (1 - s_j)(f_j^{(\mathrm{eq})})^{s_j} (1 - f_j^{(\mathrm{eq})})^{-s_j} \right] \\
&= \sum_{ss'} (s_i' - s_i)\, A(s \to s') \prod_{k \neq j} (f_k^{(\mathrm{eq})})^{s_k} (1 - f_k^{(\mathrm{eq})})^{(1-s_k)} \\
&\times \left[s_j (f_j^{(\mathrm{eq})})^{s_j - 1} (1 - f_j^{(\mathrm{eq})})^{-s_j} - (1 - f_j^{(\mathrm{eq})})^{-1} f_j^{(\mathrm{eq})\, s_j} (1 - f_j^{(\mathrm{eq})})^{(1-s_j)} \right]
\end{aligned}$$

[2] The equilibrium state $f^{(\mathrm{eq})}$ is a *global* equilibrium $f^{(\mathrm{eq})}(\mathbf{c})$. Fluctuations can also be related to external (forced) perturbations when a constraint is imposed to the system; then $f^{(\mathrm{eq})}$ must be taken as *local* equilibrium $f^{(\mathrm{eq})}(\mathbf{r}_\star, \mathbf{c})$.

$$= \sum_{s s'} (s'_i - s_i) s_j \, A(s \to s') \, \frac{\prod_k (f_k^{(\mathrm{eq})})^{s_k} (1 - f_k^{(\mathrm{eq})})^{(1-s_k)}}{(f_j^{(\mathrm{eq})})(1 - f_j^{(\mathrm{eq})})}$$

$$- \sum_{s s'} (s'_i - s_i) \, A(s \to s') \, \frac{\prod_k (f_k^{(\mathrm{eq})})^{s_k} (1 - f_k^{(\mathrm{eq})})^{(1-s_k)}}{(1 - f_j^{(\mathrm{eq})})} ,$$

where the last line on the r.h.s. vanishes because of the equilibrium condition (4.36).

For *non-thermal* LGAs (models with single non-zero velocity modulus) the equilibrium distribution function (6.3) is independent of the velocity: $f^{(\mathrm{eq})} = d$, the average particle density per channel. Then the linearized collision operator reduces to (see 5.70)

$$\Omega_{ij} = \sum_{s s'} (s'_i - s_i) s_j \, A(s \to s') \, d^{p-1} (1 - d)^{b-p-1} , \tag{6.5}$$

where $p = \sum_k s_k$ is the number of particles in configuration s. When detailed balance $A(s \to s') = A(s' \to s)$ is satisfied, (6.5) can be written in symmetrical form:

$$\Omega_{ij} = \Omega_{ji} = \frac{1}{2} \sum_{s s'} (s'_i - s_i)(s_j - s'_j) \, A(s \to s') \, d^{p-1} (1 - d)^{b-p-1} . \tag{6.6}$$

For multi-speed models (like *thermal* LGAs) this symmetry property does not hold. Nevertheless a symmetrical expression can be obtained (Ernst and Das, 1992) by considering the modified operator $\Omega_{ij} \kappa_j^{(2)}$ where $\kappa_j^{(2)} = f_j^{\mathrm{eq}} (1 - f_j^{\mathrm{eq}})$ is related to the static correlation function $S(k)$ (see Section 4.7). Defining the factorized equilibrium distribution function:

$$f^{(\mathrm{eq})}(s) = \prod_{k=1}^{b} \left[(f_k^{(\mathrm{eq})})^{s_k} \, (1 - f_k^{(\mathrm{eq})})^{(1-s_k)} \right] , \tag{6.7}$$

(which is α- and β-dependent) and using the expression obtained above for Ω_{ij}, we have:

$$\begin{aligned} \Omega_{ij} \kappa_j^{(2)} &= \sum_{s s'} (s'_i - s_i) s_j \, A(s \to s') \, f^{(\mathrm{eq})}(s) \\ &= \sum_{s s'} (s'_i - s_i) s_j \, A(s \to s') \prod_k \left(\frac{f_k^{(\mathrm{eq})}}{1 - f_k^{(\mathrm{eq})}} \right)^{s_k} (1 - f_k^{(\mathrm{eq})}) , \end{aligned} \tag{6.8}$$

where, with the explicit expression for the equilibrium distribution function (6.3),

$$\prod_k \left(\frac{f_k^{(\mathrm{eq})}}{1 - f_k^{(\mathrm{eq})}} \right)^{s_k} = e^{\alpha \, \rho(s) - \beta \, e(s)} . \tag{6.9}$$

Here $\rho(s) = \sum_k s_k$ and $e(s) = \sum_k e(c_k) \, s_k$ are respectively the number of particles in configuration s and the corresponding energy; these quantities are conserved during collisions, i.e.

$$\rho(s) = \sum_k s_k = \sum_k s'_k, \tag{6.10}$$

$$e(s) = \sum_k e(c_k)\, s_k = \sum_k e(c_k)\, s'_k. \tag{6.11}$$

So, we obtain an explicitly symmetrical expression for the modified linearized Boltzmann operator:

$$\Omega_{ij}\kappa_j^{(2)} = \frac{1}{2}\sum_{ss'} (s'_i - s_i)(s_j - s'_j)\, A(s \to s')\, e^{\alpha\,\rho(s) - \beta\, e(s)} \prod_k (1 - f^{(\text{eq})k}), \tag{6.12}$$

which is valid for any model satisfying detailed balance.

We close this section with a technical point. For later use we introduce the *thermal* scalar product:[3]

$$\langle A|B\rangle = \sum_i A(\mathbf{c}_i)\, B(\mathbf{c}_i)\, \kappa_i\,. \tag{6.13}$$

With this notation, the symmetry of $\Omega_{ij}\kappa_j^{(2)}$ implies:

$$\langle A|\mathbf{\Omega}|B\rangle = \sum_{ij} A_i \Omega_{ij}\kappa_j^{(2)} B_j = \langle B|\mathbf{\Omega}|A\rangle\,, \tag{6.14}$$

a property that will prove to be quite useful.

6.2 Slow and fast variables

In the present chapter we are interested in describing the space and time dynamics of fluctuating quantities, such as the number density, the current density, and the energy density, in the lattice gas. The relevant objects for evaluating the dynamics of these quantities are their autocorrelation functions, which involve the value of the fluctuating quantity at some arbitrary initial time, provided the system is in a stationary state. Therefore the computation of the correlation functions can be formulated as an initial value problem, and the projection operator technique is quite appropriate (Zwanzig, 1965; Mori, 1965a, 1965b). It provides a convenient separation of the dynamical variables into *slow variables* and *fast variables*: the latter correspond to the kinetic modes and they decay practically instantaneously in comparison with those variables which vary slowly in space and time. These slow variables can be identified when the constants of motion of the system are known; we shall see that they correspond to the hydrodynamic modes.

[3] The weight κ_i is β dependent.

It is convenient to rewrite the deviation $\delta f(\mathbf{r}_\star, \mathbf{c}_i; t_\star)$ from the equilibrium distribution (see (6.1)) as:

$$\delta f(\mathbf{r}_\star, \mathbf{c}_i; t_\star) = \kappa^{(2)}(\mathbf{c}_i)\, \phi_i(\mathbf{r}; t), \tag{6.15}$$

and consequently the density fluctuations, the current density fluctuations, and the energy density fluctuations respectively as:

$$\delta\rho(\mathbf{r}_\star; t_\star) = \sum_j \kappa_j^{(2)}\, \phi_i(\mathbf{r}; t), \tag{6.16}$$

$$\delta\mathbf{j}(\mathbf{r}_\star; t_\star) = \sum_j \kappa_j^{(2)}\, \mathbf{c}_j\, \phi_i(\mathbf{r}; t), \tag{6.17}$$

$$\delta e(\mathbf{r}_\star; t_\star) = \sum_j \kappa_j^{(2)}\, \frac{1}{2} c_j^2\, \phi_i(\mathbf{r}; t). \tag{6.18}$$

We shall describe the *long-time* behavior of the lattice gas in terms of $\phi_i(\mathbf{r}; t)$. Long times are large compared to the microscopic time Δt over which collision and propagation take place. So far we have omitted to write explicitly the time interval Δt because it is just the automaton unitary time, but here it is important to do so for the clarity of mathematical manipulations; then the discrete time variable is $t_\star = n\Delta t$, where n is an integer. So the linearized Boltzmann equation (6.4) now reads:

$$\phi_i(\mathbf{r}_\star + \mathbf{c}_i\Delta t; t_\star + \Delta t) - \phi_i(\mathbf{r}_\star; t_\star) = \sum_j \bar{\Omega}_{ij}\, \phi_j(\mathbf{r}_\star; t_\star), \tag{6.19}$$

where the linearized collision operator is redefined as:

$$\bar{\Omega}_{ij} = \kappa_i^{-1}\, \Omega_{ij}\, \kappa_j, \tag{6.20}$$

with $\kappa_i = \kappa^{(2)}(\mathbf{c}_i) = \kappa_i^{(2)}$.

In order to obtain the large-scale description that we are seeking, we eliminate the fast variables by means of a projection operator $\mathscr{P}$, which when operating on a function $g(\mathbf{c})$ projects the function onto the subspace $\mathscr{V}_0$ spanned by the eigenvectors of $\bar{\Omega}_{ij}$ with zero eigenvalues:

$$\mathscr{P}\, g(\mathbf{c}) = \sum_{\alpha=1}^{4} |a_\alpha\rangle\, N_\alpha\, \langle a_\alpha | g\rangle, \tag{6.21}$$

where $a_\alpha(\mathbf{c}) = \{1, \mathbf{c}, e(c) - c_T^2\}$ is a set of orthogonal vectors and N_α is a normalization factor defined by $\langle a_\alpha | a_\beta\rangle = N_\alpha^{-1}\, \delta_{\alpha\beta}$. The constant c_T is defined such that:

$$\langle a_0 | a_3\rangle = \langle 1 | (\frac{1}{2}c^2 - c_T^2)\rangle = 0\,; \tag{6.22}$$

its physical interpretation will be given subsequently. It is easy to check that the operator $\mathscr{P}$, as defined by (6.21), has the properties of a projector, i.e.

$$\mathscr{P}^2 = \mathscr{P}, \qquad \text{and} \qquad \mathscr{P} g = g, \qquad \forall g \in \mathscr{V}_0\,; \tag{6.23}$$

the complementary operator $\mathscr{Q} \equiv 1 - \mathscr{P}$ projects onto the subspace orthogonal to $\mathscr{V}_0$.

It will be most convenient to perform the subsequent developments using Laplace–Fourier transformations; therefore we define the double transform of the discrete dynamical variable $a(\mathbf{r}_\star; t_\star)$ as:

$$\tilde{a}(\mathbf{k};s) = \sum_{n=0}^{\infty} \sum_{\mathbf{r}_\star \in \mathscr{L}} a(\mathbf{r}_\star; t_\star)\, e^{-\imath \mathbf{k}\cdot\mathbf{r}_\star}\, e^{-s n \Delta t}\, \Delta t\,. \tag{6.24}$$

Applying this definition to $\phi_i(\mathbf{r}_\star; t_\star)$, we project $\tilde{\phi}(\mathbf{k}, \mathbf{c}; s)$ onto $\mathscr{V}_0$ to obtain the Laplace–Fourier transformed hydrodynamic variables, i.e.

$$\mathscr{P}\tilde{\phi}(\mathbf{k}, \mathbf{c}; s) = \sum_{\alpha=1}^{4} |a_\alpha\rangle\, N_\alpha\, \tilde{A}_\alpha(\mathbf{k}; s)\,, \tag{6.25}$$

with

$$\tilde{A}_1(\mathbf{k};s) = \tilde{\delta}\rho(\mathbf{k};s) = \sum_i \kappa_i^{(2)}\, \tilde{\phi}(\mathbf{k}, \mathbf{c}; s), \tag{6.26}$$

$$\tilde{A}_2(\mathbf{k};s) = \tilde{\delta} j_x(\mathbf{k};s) = \sum_i \kappa_i^{(2)}\, c_{i,x}\, \tilde{\phi}(\mathbf{k}, \mathbf{c}; s), \tag{6.27}$$

$$\tilde{A}_3(\mathbf{k};s) = \tilde{\delta} j_y(\mathbf{k};s) = \sum_i \kappa_i^{(2)}\, c_{i,y}\, \tilde{\phi}(\mathbf{k}, \mathbf{c}; s), \tag{6.28}$$

$$\begin{aligned} \tilde{A}_4(\mathbf{k};s) &= \tilde{\delta} e(\mathbf{k};s) - c_T^2\, \delta\rho(\mathbf{k};s) \\ &= \sum_i \kappa_i^{(2)}\, (\frac{1}{2}c_i^2 - c_T^2)\, \tilde{\phi}(\mathbf{k}, \mathbf{c}; s). \end{aligned} \tag{6.29}$$

Now Laplace–Fourier transformation of the linearized Boltzmann equation (6.19) yields:

$$e^{s\Delta t} e^{\imath \mathbf{k}\cdot\mathbf{c}_i \Delta t} \tilde{\phi}_i(\mathbf{k};s) - \tilde{\phi}_i(\mathbf{k};s) - \sum_j \bar{\Omega}_{ij}\, \tilde{\phi}_j(\mathbf{k};s) = e^{s\Delta t} \Delta t\, e^{\imath \mathbf{k}\cdot\mathbf{c}_i \Delta t} \phi_i(\mathbf{k}), \tag{6.30}$$

where $\phi_i(\mathbf{k}) = \phi_i(\mathbf{k}; t_\star = 0)$ is the space-Fourier transform of $\phi_i(\mathbf{r}; t)$ at initial time.[4]

We introduce the following notation $\tilde{\phi}_i(\mathbf{k};s) \equiv |\phi(\mathbf{k};s)\rangle$, and we define the operator $\mathscr{L}$ by:

$$\mathscr{L}\, |\phi(\mathbf{k};s)\rangle = e^{-\imath \mathbf{k}\cdot\mathbf{c}_i \Delta t} \sum_j (\, \delta_{ij} + \bar{\Omega}_{ij}\,)\, \tilde{\phi}_j(\mathbf{k};s)\,, \tag{6.31}$$

[4] The factor $\exp(s\,\Delta t)$ on the l.h.s. of (6.30) is a consequence of discreteness. Note that by taking the limit

$$\lim_{\Delta t \to 0} \frac{1}{\Delta t} \left[e^{s\Delta t} \tilde{\phi}(s) - \tilde{\phi}(s) - e^{s\Delta t} \Delta t \phi(0) \right] = s\,\tilde{\phi}(s) - \phi(0)\,,$$

one retrieves the usual expression of the continuous transform

$$\int_0^\infty e^{-st}\, \partial_t\, \phi(t) = s\,\tilde{\phi}(s) - \phi(0).$$

to rewrite Equation (6.30) as:

$$(e^{s\Delta t} - \mathscr{L})\,|\phi(\mathbf{k};s)\rangle = e^{s\Delta t}\Delta t\,|\phi(\mathbf{k})\rangle . \tag{6.32}$$

This result exhibits an appropriate form for applying the projectors $\mathscr{P}$ and $\mathscr{Q}$ to separate the equation into a 'hydrodynamic equation':

$$e^{s\Delta t}\mathscr{P}\,|\phi(\mathbf{k};s)\rangle - \mathscr{P}\mathscr{L}\mathscr{P}\,|\phi(\mathbf{k};s)\rangle - \mathscr{P}\mathscr{L}\mathscr{Q}\,|\phi(\mathbf{k};s)\rangle = e^{s\Delta t}\Delta t\,\mathscr{P}\,|\phi(\mathbf{k})\rangle , \tag{6.33}$$

and a 'transport equation':

$$e^{s\Delta t}\mathscr{Q}\,|\phi(\mathbf{k};s)\rangle - \mathscr{Q}\mathscr{L}\mathscr{Q}\,|\phi(\mathbf{k};s)\rangle - \mathscr{Q}\mathscr{L}\mathscr{P}\,|\phi(\mathbf{k};s)\rangle = e^{s\Delta t}\Delta t\,\mathscr{Q}\,|\phi(\mathbf{k})\rangle , \tag{6.34}$$

where we have used the property $(\mathscr{P}+\mathscr{Q})\,|\phi(\mathbf{k};s)\rangle = |\phi(\mathbf{k};s)\rangle$.

Our goal is to solve the hydrodynamic equation, but we observe that the third term on the l.h.s. of (6.33) requires the knowledge of the perturbation $\mathscr{Q}\,|\phi(\mathbf{k};s)\rangle$ (which is orthogonal to the slow variables), that is we must first obtain the solution to the transport equation (6.34). Therefore we consider the resolvent operator:

$$R = [e^{s\Delta t} - \mathscr{Q}\mathscr{L}]^{-1}, \tag{6.35}$$

with the property:

$$R\,\mathscr{Q} = \left[e^{s\Delta t} - \mathscr{Q}\mathscr{L}\right]^{-1}\mathscr{Q} = \left[e^{s\Delta t} - \mathscr{Q}\mathscr{L}\mathscr{Q}\right]^{-1}\mathscr{Q}, \tag{6.36}$$

which follows from the operator identity: $R = S + R\,(S^{-1} - R^{-1})\,S$, with $S = [\exp(s\,\Delta t) - \mathscr{Q}\mathscr{L}\,Q]^{-1}$, and $(S^{-1} - R^{-1}) = \mathscr{Q}\mathscr{L}\mathscr{P}$. We can then write the formal solution to Equation (6.34) as:

$$\mathscr{Q}|\phi(\mathbf{k};s)\rangle = R\left(e^{s\Delta t}\Delta t\,\mathscr{Q}\,|\phi(\mathbf{k})\rangle + \mathscr{Q}\mathscr{L}\mathscr{P}\,|\phi(\mathbf{k};s)\rangle\right), \tag{6.37}$$

and inject this result into Equation (6.33) to obtain:

$$(e^{s\Delta t} - \mathscr{P}\mathscr{L})\mathscr{P}\,|\phi(\mathbf{k};s)\rangle - \mathscr{P}\mathscr{L}\,R\,\mathscr{Q}\mathscr{L}\mathscr{P}\,|\phi(\mathbf{k};s)\rangle$$

$$= e^{s\Delta t}\Delta t\,(\mathscr{P}\,|\phi(\mathbf{k})\rangle + \mathscr{P}\mathscr{L}\,R\,\mathscr{Q}\,|\phi(\mathbf{k})\rangle), \tag{6.38}$$

where there remains a dependency on $\mathscr{Q}|\phi(\mathbf{k};t_\star = 0)\rangle$, the initial value of a quantity orthogonal to the slow variables.

We are now in a position to write, formally at least, the set of equations for the hydrodynamic variables, by using explicitly (6.25) and taking the scalar product of (6.38) with the vectors $\langle a_\alpha|$, $(\alpha = 1, 2, 3, 4)$. The result is the *relaxation equation*:

$$e^{s\Delta t}\,\tilde{A}_\alpha(\mathbf{k};s) - \sum_{\sigma=1}^{4} \mathcal{L}_{\alpha\sigma}\,\tilde{A}_\sigma(\mathbf{k};s) - \sum_{\sigma=1}^{4} K_{\alpha\sigma}\,\tilde{A}_\sigma(\mathbf{k};s)$$
$$= e^{s\Delta t}\Delta t\, A_\alpha(\mathbf{k};0) + I_\alpha(\mathbf{k};s). \tag{6.39}$$

where (i) the second term on the l.h.s. represents the coupling between the hydrodynamic variables, and the matrix elements $\mathcal{L}_{\alpha\sigma}$ are given by:

$$\mathcal{L}_{\alpha\sigma} = \langle a_\alpha|\,\mathcal{L}\,|a_\sigma\rangle\, N_\sigma\,, \tag{6.40}$$

or, with (6.31) and noting that $\bar{\mathbf{\Omega}}\mathcal{P}\,|\phi\rangle = 0$:

$$\mathcal{L}_{\alpha\sigma} = \langle a_\alpha|\,e^{-\imath\mathbf{k}\cdot\mathbf{c}\Delta t}\,|a_\sigma\rangle\, N_\sigma\,; \tag{6.41}$$

(ii) the third term on the l.h.s. expresses collisional effects, and the matrix

$$K_{\alpha\sigma} = \langle a_\alpha|\,\mathcal{L}\,\mathcal{Q}\,R\,\mathcal{Q}\,\mathcal{L}|a_\sigma\rangle\, N_\sigma \tag{6.42}$$

is referred to as the *memory function*;
(iii) the term $I_\alpha(\mathbf{k};s)$ on the r.h.s.

$$I_\alpha(\mathbf{k};s) = \langle a_\alpha|\,\mathcal{L}\mathcal{Q}\,R\,e^{s\Delta t}\,\Delta t\mathcal{Q}\,|\phi(\mathbf{k};0)\rangle\,, \tag{6.43}$$

contains the contributions from the initial state and also includes collision effects. Note that to obtain (6.42) and (6.43), we have used the property $R\,\mathcal{Q} = \mathcal{Q}\,R\,\mathcal{Q}$.[5]

The set of equations (6.39) describes the dynamics of the hydrodynamic quantities $\tilde{A}_\alpha(\mathbf{k};s)$; these equations are exact within the limits of validity of the linearized Boltzmann equation.

6.3 The hydrodynamic limit

We now want to obtain an explicit analytical formulation of the dynamics of the lattice gas automaton fluctuations in the hydrodynamic limit, that is for long times and large spatial scales (t much larger than Δt and $k = |\mathbf{k}|$ much smaller than the reciprocal lattice unit length). Therefore we consider the limit of small Δt by expanding our results in s and k (to second order); for instance the factor $e^{s\Delta t}$ expands as:

$$e^{s\Delta t} \simeq 1 + s\,\Delta t + \frac{1}{2}\,(s\,\Delta t)^2. \tag{6.45}$$

[5] This property follows from the definition $\mathcal{Q}\,R\,\mathcal{Q} = R\,\mathcal{Q} - \mathcal{P}\,R\,\mathcal{Q}$, and the observation that the action of the second term vanishes; indeed, using the operator identity given after Equation (6.36) and with the notation $T = [e^{s\Delta t} - \mathcal{L}]^{-1}$, one has

$$\mathcal{P}\,R\,\mathcal{Q} = \mathcal{P}[T + R(\mathcal{Q}\mathcal{L} - \mathcal{L})T]\mathcal{Q} = \mathcal{P}T\mathcal{Q} - \mathcal{P}R\mathcal{P}\mathcal{L}T\mathcal{Q} = 0, \tag{6.44}$$

because $\mathcal{P}$ commutes with $\mathcal{L}$ and with T.

The expansion of the other quantities in (6.39) is not so trivial, and they will now be considered separately.

6.3.1 The coupling function

For the matrix elements $\mathscr{L}_{\alpha\sigma}$, (6.41), we have:

$$\begin{aligned}\mathscr{L}_{\alpha\sigma} &= \langle a_\alpha|\, e^{-\imath \mathbf{k}\cdot\mathbf{c}\Delta t}\, |a_\sigma\rangle\, N_\sigma \\ &\simeq \langle a_\alpha|\, (1 - \imath\, \mathbf{k}\cdot\mathbf{c}\Delta t - \frac{1}{2}(\mathbf{k}\cdot\mathbf{c}\Delta t)^2)\, |a_\sigma\rangle\, N_\sigma \\ &\simeq \delta_{\alpha\sigma} - \imath k_l\, \Delta t\, \langle a_\alpha|J^l_\sigma\rangle\, N_\sigma - k_l k_m \langle J^l_\alpha| \frac{\Delta t^2}{2} |J^m_\sigma\rangle\, N_\sigma\,. \end{aligned} \tag{6.46}$$

Here k_l ($l = \{x, y\}$) denotes the components of the wave vector $\mathbf{k}$,[6] and $|\mathbf{J}_\alpha\rangle$ the current associate to $|a_\alpha\rangle$, with:

$$\begin{aligned}\imath k_l\, \langle a_\alpha|J^l_\sigma\rangle\, N_\sigma &= \begin{pmatrix} 0 & \imath k_x & \imath k_y & 0 \\ \imath k_x\, c_T^2 & 0 & 0 & \imath k_x \\ \imath k_y\, c_T^2 & 0 & 0 & \imath k_y \\ 0 & \imath k_x\, (c_s^2 - c_T^2) & \imath k_y\, (c_s^2 - c_T^2) & 0 \end{pmatrix} \\ &\equiv L_{\alpha\sigma}\,, \end{aligned} \tag{6.47}$$

where the constants c_s^2 and c_T^2 are given by:

$$c_s^2 = \frac{1}{2} \frac{\langle c^2|c^2\rangle}{\langle c^2|1\rangle}\,, \tag{6.48}$$

$$c_T^2 = \frac{1}{2} \frac{\langle c^2|1\rangle}{\langle 1|1\rangle}\,. \tag{6.49}$$

6.3.2 The memory function

The memory function matrix (6.42) is more complicated: its exact computation would require the inversion of the resolvent matrix R, which is analytically not possible; so we shall use a perturbation procedure.

We first consider the *right* vector:

$$\mathscr{Q}\mathscr{L}\, |a_\sigma\rangle = \mathscr{Q} \left[e^{-\imath \mathbf{k}\cdot\mathbf{c}\Delta t}(\mathbf{1} + \bar{\Omega}) \right] |a_\sigma\rangle\,; \tag{6.50}$$

expansion to first order gives

$$\mathscr{Q}\mathscr{L}\, |a_\sigma\rangle \simeq -\imath k_l\, \Delta t\, \mathscr{Q}\, |J^l_\sigma\rangle\,. \tag{6.51}$$

Similarly for the *left* vector:

$$\langle a_\sigma|\, \mathscr{L}\mathscr{Q} = \langle a_\sigma| \left[e^{-\imath \mathbf{k}\cdot\mathbf{c}\Delta t}(1 + \bar{\Omega}) \right] \mathscr{Q}\,, \tag{6.52}$$

[6] In (6.46), the terms $\mathcal{O}(k^2)$ are related to the propagation step in the lattice dynamics; such terms are not present in continuous theory; here these terms will be incorporated in the transport matrix yielding an extra contribution to the transport coefficients.

we obtain:

$$\langle a_\sigma | \mathscr{L}\mathscr{Q} \simeq -\imath k_l \, \Delta t \, \langle J^l_\sigma | (1 + \bar{\Omega}) \mathscr{Q} . \tag{6.53}$$

We observe that the current density appears as:

$$\mathscr{Q} \, |\mathbf{J}_\sigma\rangle = |\mathbf{J}_\sigma\rangle - \sum_{\alpha=1}^{4} N_\alpha \, |a_\alpha\rangle \, \langle a_\alpha | \mathbf{J}_\sigma\rangle \equiv |\hat{\mathbf{J}}_\sigma\rangle \, ; \tag{6.54}$$

this quantity is called the *subtracted current*, and is orthogonal to the hydrodynamic eigenvectors.

We now expand the resolvent operator:

$$\begin{aligned} R &= \left[e^{s\Delta t} - \mathscr{Q}\mathscr{L}\mathscr{Q} \right]^{-1} \\ &\simeq \left[1 + s\Delta t - \mathscr{Q}^2 - \mathscr{Q}\bar{\Omega}\mathscr{Q} + \mathscr{Q}(\imath \mathbf{k} \cdot \mathbf{c}\Delta t)(1 + \bar{\Omega})\mathscr{Q} \right]^{-1} . \end{aligned} \tag{6.55}$$

One can easily verify that $\mathscr{Q}\bar{\mathbf{\Omega}}\mathscr{Q} \, |\phi\rangle = \bar{\mathbf{\Omega}} \, |\phi\rangle$ because $\mathscr{Q}$ projects any vector onto the subspace orthogonal to the subspace spanned by the vectors with zero eigenvalue; so in practice we can substitute $\mathscr{Q}\bar{\mathbf{\Omega}}\mathscr{Q}$ by $\bar{\mathbf{\Omega}}$. On the other hand, we notice from Equation (6.34) that $R^{-1} = (\exp(s\,\Delta t) - \mathscr{Q}\,\mathscr{L})$ operates on the projected vector $\mathscr{Q} \, |\phi\rangle$, and since $\mathscr{Q}^2 \, \mathscr{Q} \, |\phi\rangle = \mathscr{Q} \, |\phi\rangle$, we may therefore, in (6.55), replace operationally $\mathscr{Q}^2$ by the identity operator. With these simplifications, we have

$$R \simeq \left[s\Delta t - \bar{\Omega} + \mathscr{Q}(\imath \mathbf{k} \cdot \mathbf{c}\Delta t)(1 + \bar{\Omega})\mathscr{Q} \right]^{-1} . \tag{6.56}$$

Combining (6.51) and (6.53) into (6.42), we have

$$K_{\alpha\sigma} \simeq -k_l k_m \, (\Delta t)^2 \, \langle \hat{J}^l_\alpha | (1 + \bar{\Omega}) \, R \, | \hat{J}^m_\sigma \rangle \, N_\sigma \, , \tag{6.57}$$

with R given by (6.56).[7] We now treat the resolvent by applying the operator identity

$$\frac{1}{A} = \frac{1}{B} - \frac{1}{B}(A - B)\frac{1}{A} , \tag{6.58}$$

to (6.56); with the notation

$$H(\mathbf{k}) = \mathscr{Q}(\imath \mathbf{k} \cdot \mathbf{c}\Delta t)(1 + \bar{\Omega})\mathscr{Q} , \tag{6.59}$$

we obtain

$$R = \frac{1}{s\Delta t - \bar{\Omega}} - \frac{1}{s\Delta t - \bar{\Omega}} H(\mathbf{k}) \, R , \tag{6.60}$$

[7] Expression (6.57) has a structure similar to the memory function in continuous theory (see e.g. Boon and Yip, 1980, Chapter 3).

which is iterated to yield

$$R = \frac{1}{s\Delta t - \bar{\Omega}} - \frac{1}{s\Delta t - \bar{\Omega}} H(\mathbf{k}) \frac{1}{s\Delta t - \bar{\Omega}} + \frac{1}{s\Delta t - \bar{\Omega}} H(\mathbf{k}) \frac{1}{s\Delta t - \bar{\Omega}} H(\mathbf{k}) - \cdots . \tag{6.61}$$

We take (6.61) as a perturbative expansion in powers of k, so that the memory function can be written as:

$$K_{\alpha\sigma} = K^{(2)}_{\alpha\sigma} + K^{(3)}_{\alpha\sigma} + \cdots , \tag{6.62}$$

where $K^{(n)}_{\alpha\sigma}$ is proportional to k^n, that is to lowest order in k:

$$K_{\alpha\sigma} \simeq K^{(2)}_{\alpha\sigma} = -k_l k_m (\Delta t)^2 \langle \hat{J}^l_\alpha | (1 + \bar{\Omega}) \frac{1}{s\Delta t - \bar{\Omega}} | \hat{J}^m_\sigma \rangle N_\sigma . \tag{6.63}$$

This memory function is well behaved in the limit $s \to 0$ and for all values of k, provided the non-zero eigenvalues of $\bar{\mathbf{\Omega}}$ be sufficiently different from zero at $k = 0$. This 'correct' long-time behavior is a consequence of the combination of the operator $(s\,\Delta t - \bar{\Omega})^{-1}$ with the projector $\mathcal{Q}$ which prevents singularities $(\sim s^{-1})$ that would result from the action of $\bar{\mathbf{\Omega}}$ on the hydrodynamic variables (which have zero eigenvalues at $k = 0$).

6.3.3 The random force term

The term $I_\alpha(\mathbf{k}; s)$ in (6.39) and given by (6.43), expresses the effect produced by the fast variables from the initial state, and is usually referred to as the 'random force'.[8] Since we are interested in the long-time behavior, we use (6.53) and the resolvent expansion (6.56) in (6.43), to obtain:

$$I_\alpha(\mathbf{k}; s) \simeq -i k_l (\Delta t)^2 \langle \hat{J}^l_\alpha | (1 + \bar{\Omega}) \frac{1}{\left[s\Delta t - \bar{\Omega} + H(\mathbf{k}) \right]} \mathcal{Q} | \phi(\mathbf{k}; 0) \rangle . \tag{6.64}$$

If the initial distribution function has the form (see (6.15)):

$$f(\mathbf{r}_\star, \mathbf{c}; 0) = f^{(\mathrm{eq})}(\mathbf{c}) + \kappa^{(2)}_j(\mathbf{c}) \left(C_1(\mathbf{r}_\star) + \mathbf{C}_2(\mathbf{r}_\star) \cdot \mathbf{c} + C_3(\mathbf{r}_\star) \left(\frac{1}{2} c^2 - c_T^2 \right) \right) , \tag{6.65}$$

where C_γ depends only on the space variable, then $\mathcal{Q}|\phi(\mathbf{k}; 0)\rangle = 0$ (because $\mathcal{Q}$ projects on the subspace orthogonal to the subspace spanned by $\{1, \mathbf{c}, \frac{1}{2}c^2 - c_T^2\}$); this result holds $\forall\, t > 0$, and $I_\alpha(\mathbf{k}; s)$ vanishes in the long time limit.

What about initial states with a distribution function different from (6.65)? From (6.64), we observe that $I_\alpha(\mathbf{k}; s)$ is governed by the application of the

[8] In the Langevin formulation of relaxation dynamics, the random force is often inserted 'by hand' in the evolution equation in order to account for the effect of the microscopic fluctuations on the dynamics of the macroscopic variables; here we have an analytical – although implicit – expression in terms of the kinetic variables.

resolvent operator on $\mathscr{Q}|\phi(\mathbf{k};0)\rangle$; the result of the application depends on the eigenvalue spectrum of the collision operator $\bar{\mathbf{\Omega}}$, and we may expect that, for small values of k, the operator $\left[H(\mathbf{k}) - \bar{\Omega}\right]$ exhibits an eigenvalue spectrum equal to that of $\bar{\mathbf{\Omega}}$ plus a term proportional to k. Let $\lambda_n(k) < 0$ be the non-zero eigenvalues of the eigenvectors corresponding to the fast variables; then the time behavior of $I_\alpha(\mathbf{k};t_\star)$ should be of the type:

$$I_\alpha(\mathbf{k};t_\star) = \sum_{n\geq 5}^{b} e^{\lambda_n(k)\,t}\,\alpha_n(\mathbf{k};0), \tag{6.66}$$

where $\alpha_n(\mathbf{k};0), (n = 5,\ldots,b)$, corresponds to the fast modes. The negative eigenvalues (distinct from the zero eigenvalues, $n = 1,\ldots,4$) are responsible for the fast decay and eventual vanishing of $I_\alpha(\mathbf{k};t_\star)$.[9] So we may safely assume that in the hydrodynamic limit:

$$\lim_{t\to\infty}\lim_{k\to 0} I_\alpha(\mathbf{k};t_\star) = 0. \tag{6.67}$$

6.3.4 The long-wavelength, long-time limit

We can now collect our results (6.46), (6.63), and (6.67), to write the *relaxation equation* (6.39) in the long-wavelength, long-time limit, i.e. for $k|\mathbf{c}|\Delta t \ll 1$ and $s\Delta t \ll 1$. Noting that:

$$\mathscr{L}_{\alpha\sigma} = \delta_{\alpha\sigma} - i\,k_l\,\Delta t\,\langle a_\alpha|J^l_\sigma\rangle\,N_\sigma + \mathcal{O}(k^2), \tag{6.68}$$

and that $K_{\alpha\sigma} = \mathcal{O}(k^2)$, we obtain to first order, i.e. $\mathcal{O}(s)$ and $\mathcal{O}(k)$:

$$\begin{aligned}&(1 + s\Delta t)\,\tilde{A}_\alpha(\mathbf{k};s) - (1 + s\Delta t)\,\Delta t\,A_\alpha(\mathbf{k};0) - \sum_\sigma \delta_{\alpha\sigma}\,\tilde{A}_\sigma(\mathbf{k};s)\\ &+ i\,k_l\,\Delta t \sum_\sigma \langle a_\alpha|J^l_\sigma\rangle\,N_\sigma\,\tilde{A}_\sigma(\mathbf{k};s) = 0.\end{aligned} \tag{6.69}$$

This result is divided by Δt to yield, in the limit $s\Delta t \ll 1$:

$$s\,\tilde{A}_\alpha(\mathbf{k};s) + i\,k_l \sum_\sigma \langle a_\sigma|J^l_\sigma\rangle\,N_\sigma\,\tilde{A}_\sigma(\mathbf{k};s) = A_\alpha(\mathbf{k};0), \tag{6.70}$$

which corresponds to the Euler equations.

To second order, $\mathcal{O}(s^2)$ and $\mathcal{O}(k^2)$, we have:

$$\begin{aligned}&(1 + s\Delta t + \frac{1}{2}s^2\Delta t^2)\,\tilde{A}_\alpha(\mathbf{k};s)\\ &-(1 + s\Delta t + \frac{1}{2}s^2\Delta t^2)\,\Delta t\,A_\alpha(\mathbf{k};0)\\ &-\sum_\sigma \delta_{\alpha\sigma}\,\tilde{A}_\sigma(\mathbf{k};s) + \sum_\sigma L_{\alpha\sigma}(\mathbf{k})\,\Delta t\,\tilde{A}_\sigma(\mathbf{k};s)\end{aligned}$$

[9] The decay rate is model dependent; this point and the related question of the separation of eigenvalues at low k will be examined in the next chapter in the light of specific LGA model computations.

$$+ k_l k_m \sum_\sigma \left[\langle J^l_\alpha | \frac{\Delta t^2}{2} | J^m_\sigma \rangle + \langle \hat{J}^l_\alpha | \frac{\Delta t^2}{s\Delta t - \bar{\Omega}} | \hat{J}^m_\sigma \rangle \right.$$

$$\left. + \langle \hat{J}^l_\alpha | \frac{\Delta t^2 \bar{\Omega}}{s\Delta t - \bar{\Omega}} | \hat{J}^m_\sigma \rangle \right] N_\sigma \tilde{A}_\sigma(\mathbf{k}; s) = 0 . \tag{6.71}$$

Again we divide this equation by Δt, and take the limit $s\Delta t \ll 1$, to obtain:

$$s\tilde{A}_\alpha(\mathbf{k}; s) + \frac{1}{2} s^2 \Delta t \, \tilde{A}_\alpha(\mathbf{k}; s) + \sum_\sigma L_{\alpha\sigma}(\mathbf{k}) \tilde{A}_\sigma(\mathbf{k}; s)$$

$$+ k_l k_m \sum_\sigma \left[\Delta t \, \langle J^l_\alpha | \frac{1}{2} | J^m_\sigma \rangle + \langle \hat{J}^l_\alpha | \frac{1}{s - \tilde{\bar{\Omega}}} | \hat{J}^m_\sigma \rangle \right.$$

$$\left. + \Delta t \, \langle \hat{J}^l_\alpha | \frac{\tilde{\bar{\Omega}}}{s - \tilde{\bar{\Omega}}} | \hat{J}^m_\sigma \rangle \right] N_\sigma \tilde{A}_\sigma(\mathbf{k}; s) = A_\alpha(\mathbf{k}; 0), \tag{6.72}$$

where [10]

$$\tilde{\bar{\Omega}} = \bar{\Omega}/\Delta t. \tag{6.73}$$

We now use the first order result, Equation (6.70), to evaluate the second term on the l.h.s. of (6.72), i.e.

$$s^2 \tilde{A}_\alpha(\mathbf{k}; s) = s \, [s\tilde{A}_\alpha(\mathbf{k}; s)] = s \, [-i k_l \sum_\beta \langle a_\alpha | J^l_\beta \rangle N_\beta \tilde{A}_\beta] , \tag{6.74}$$

where we have omitted the term on the r.h.s. of (6.70) because when inserted in (6.72), it will give $\frac{1}{2} s\Delta t \, A_\alpha(\mathbf{k}; 0)$ which vanishes in the limit $s\Delta t \ll 1$. Iteration of (6.74) gives:

$$\begin{aligned} s^2 \tilde{A}_\alpha(\mathbf{k}; s) &= -k_l k_m \sum_\beta \sum_\sigma \langle J^l_\alpha | a_\beta \rangle \langle a_\beta | J^m_\sigma \rangle N_\beta N_\sigma \tilde{A}_\sigma \\ &= -k_l k_m \sum_\sigma \langle J^l_\alpha | \mathscr{P} | J^m_\sigma \rangle N_\sigma \tilde{A}_\sigma , \end{aligned} \tag{6.75}$$

where

$$\begin{aligned} \langle J^l_\alpha | \mathscr{P} | J^m_\sigma \rangle &= \langle J^l_\alpha | (1 - \mathscr{Q}) | J^m_\sigma \rangle \\ &= \langle J^l_\alpha | 1 | J^m_\sigma \rangle - \langle \hat{J}^l_\alpha | \hat{J}^m_\sigma \rangle . \end{aligned} \tag{6.76}$$

Substitution of these results into (6.72) yields:

$$\left[s + \sum_\sigma L_{\alpha\sigma}(\mathbf{k}) + k_l k_m \sum_\sigma \Phi_{\alpha\sigma}(\mathbf{k}; s) \right] \tilde{A}_\sigma(\mathbf{k}; s) = A_\alpha(\mathbf{k}; 0), \tag{6.77}$$

with the memory function:

$$\Phi_{\alpha\sigma}(\mathbf{k}; s) = \left[\langle \hat{J}^l_\alpha | \frac{1}{s - \tilde{\bar{\Omega}}} | \hat{J}^m_\sigma \rangle + \Delta t \, \langle \hat{J}^l_\alpha | \frac{1}{2} + \frac{\tilde{\bar{\Omega}}}{s - \tilde{\bar{\Omega}}} | \hat{J}^m_\sigma \rangle \right] N_\sigma . \tag{6.78}$$

[10] The difference between $\tilde{\bar{\Omega}}$ and $\bar{\Omega}$ is immaterial in practice (since Δt is the unitary time interval). However, in the continuous Boltzmann equation the collision operator has the dimension of a frequency, while in the discrete linearized Boltzmann equation (6.19), the operator $\bar{\Omega}$ is dimensionless. So (6.73) re-establishes the correct dimensional content of the collision operator.

Using (6.73) in the second term:

$$\Delta t\left(\frac{1}{2}+\frac{\tilde{\bar{\Omega}}}{s-\tilde{\bar{\Omega}}}\right) = \Delta t\left(\frac{1}{2}+\frac{\bar{\Omega}}{s\Delta t-\bar{\Omega}}\right), \tag{6.79}$$

then taking the limit $s\Delta t \to 0$ in (6.78), we obtain:[11]

$$\Phi_{\alpha\sigma}(\mathbf{k};s) \simeq \left[\langle \hat{J}^l_\alpha | \frac{1}{s-\tilde{\bar{\Omega}}} - \frac{\Delta t}{2} | \hat{J}^m_\sigma \rangle\right] N_\sigma . \tag{6.80}$$

These results, (6.77) with (6.80), conclude the derivation of the linearized hydrodynamic equations.

6.4 The transport matrix

The results obtained by computation in Laplace space are naturally equivalent to the results obtained by computation with the time variable expressions. However, some physical insight can be gained from the memory function (6.78) when expressed in real time. In the linear equations (6.77), the memory term $\Phi_{\alpha\sigma}(\mathbf{k};s)\tilde{A}_\sigma(\mathbf{k};s)$ is the Laplace transform of the convolution:

$$\sum_{\tau=0}^{t=n\Delta t} \Phi_{\alpha\sigma}(\mathbf{k};\tau)\, A_\sigma(\mathbf{k};t_\star-\tau)\Delta_\tau , \tag{6.81}$$

with the kernel:

$$\Phi_{\alpha\sigma}(\mathbf{k};\tau) = \left[\langle \hat{J}^l_\alpha | e^{(\tau+\Delta t)\tilde{\bar{\Omega}}} | \hat{J}^m_\sigma \rangle + \delta_{\tau 0}\langle \hat{J}^l_\alpha | \frac{1}{2} | \hat{J}^m_\sigma \rangle \Delta_\tau\right] N_\sigma .$$

The memory function expresses the effect of the past history of the dynamics related to the fast variables. But here we are interested in the dynamics of the slow variables $A_\sigma(\mathbf{k};t_\star)$ which relax over times ($\mathcal{O}(k^2)$) very long compared to the characteristic decay rate of $\Phi_{\alpha\sigma}(\mathbf{k};\tau) \sim \exp(\lambda\tau)$, where λ denotes the non-zero (negative) eigenvalues of the collision operator. Consequently, in (6.81), we may consider, to good approximation, the following simplifications: (i) $A_\sigma(\mathbf{k};t_\star)$ does not vary significantly over the summation time, and can therefore be taken out of the sum; and (ii) $\Phi_{\alpha\sigma}(\mathbf{k};\tau)$ decays sufficiently rapidly that it is practically zero for times larger than $\tau \sim \mathcal{O}(\lambda^{-1})$, and the summation can be extended to ∞. As a result the convolution (6.81) becomes $\Lambda_{\alpha\sigma}A_\sigma(\mathbf{k};t_\star)$ with:

$$\begin{aligned}\Lambda_{\alpha\sigma} &= \sum_{\tau=0}^{\infty}\Phi_{\alpha\sigma}(\mathbf{k};\tau)\,\Delta_\tau \\ &= \sum_{n=0}^{\infty}\left[\langle \hat{J}^l_\alpha | e^{(n+1)\Delta_\tau\tilde{\bar{\Omega}}} | \hat{J}^m_\sigma \rangle + \Delta t\,\delta_{n0}\langle \hat{J}^l_\alpha | \frac{1}{2} | \hat{J}^m_\sigma \rangle\right] N_\sigma\,\Delta_\tau\end{aligned}$$

[11] The limit $s \to 0$ in Laplace space corresponds to the limit $t \to \infty$ in real time.

$$= \sum_{n=1}^{\infty} \left[\langle \hat{J}^l_\alpha | e^{n\Delta_\tau \tilde{\tilde{\Omega}}} | \hat{J}^m_\sigma \rangle + \Delta t \, \delta_{n0} \langle \hat{J}^l_\alpha | \frac{1}{2} | \hat{J}^m_\sigma \rangle \right] N_\sigma \Delta_\tau$$

$$= \sum_{n=0}^{\infty} \left[\langle \hat{J}^l_\alpha | e^{n\Delta_\tau \tilde{\tilde{\Omega}}} | \hat{J}^m_\sigma \rangle - \Delta t \, \delta_{n0} \langle \hat{J}^l_\alpha | \frac{1}{2} | \hat{J}^m_\sigma \rangle \right] N_\sigma \Delta_\tau \,, \tag{6.82}$$

which defines the *transport matrix* $\Lambda_{\alpha\sigma}$. Note that the Laplace transform of the bottom line in (6.82) gives exactly (6.80).

The transport matrix (6.82) can also be written formally as:

$$\Lambda_{\alpha\sigma}(s) = N_\sigma \sum_{t=0}^{\infty} e^{-st} \langle \hat{J}^l_\alpha | \hat{J}(t)^l_\sigma \rangle \, \Delta t \,. \tag{6.83}$$

This expression is the analogue of the Green–Kubo integral in continuous theory; at zero frequency ($s = 0$) and in the continuous limit ($\Delta t \to 0$), one retrieves the classical formulation of the transport coefficients:

$$\Lambda_{\alpha\sigma} = N_\sigma \int_0^{\infty} dt \, \langle \hat{J}^l_\alpha | \hat{J}(t)^l_\sigma \rangle \,. \tag{6.84}$$

There are two important distinctions between the continuous expression and the lattice gas result: (i) in the latter, instead of a time integral, we have a summation, which is a consequence of the persistence at the macroscopic scale of the space and time discretization at the microscopic level; (ii) the transport coefficients contain an additional (negative) term $-\frac{1}{2} N_\sigma \langle \hat{J}^l_\alpha | \hat{J}^m_\sigma \rangle \Delta t$, (see (6.80) and (6.82)), which is another consequence of the discrete nature of the dynamics.[12] The transport coefficients were discussed in Section 5.6; they will be further discussed and evaluated explicitly in the subsequent chapters.

6.5 Comments

In this chapter, we derived the linearized equations describing the space-time evolution of the fluctuating quantities in the lattice gas at the hydrodynamic scale. Using the projection operator method, we obtained a logical separation into slow and fast variables; the ensuing set of relaxation equations for the hydrodynamic variables follows from the lattice Boltzmann equation. There is no analytical procedure by which, starting from a microscopic formulation, one could derive the hydrodynamic equations in an exact manner, but at least the approximations used in the derivation can be justified physically on the basis of scales separation. The resulting description of the dynamical behavior gives a picture of the system as viewed through a space-time window whose width is

[12] In the literature, the negative contribution to the transport coefficient has often been called a *propagative part* invoking the representation of the lattice gas in terms of particles undergoing, during the propagation step, free flight from node to node over the time interval Δt.

set by the observer in the range of sufficiently large wavelengths. In the next chapter we shall see that this picture, as provided by a simplified system such as the lattice gas automaton, is in accordance with the behavior of spontaneous fluctuations as observed in real fluids.

Chapter 7

Hydrodynamic fluctuations

The object of the present chapter is small deviations from local equilibrium which are triggered by spontaneous fluctuations. In real fluids these fluctuations which temporarily disturb the system from local equilibrium are such that a fluid at global equilibrium can be viewed as a reservoir of excitations extending over a broad range of wavelengths and frequencies from the hydrodynamic scale down to the range of the intermolecular potential. Non-intrusive scattering techniques are used to probe these fluctuations at the molecular level (neutron scattering spectroscopy) and at the level of collective excitations (light scattering spectroscopy) (Boon and Yip, 1980). The quantity measured by these scattering methods is the power spectrum of density fluctuations, i.e. the dynamic structure factor $S(\mathbf{k}, \omega)$ which is the space- and time-Fourier transform of the correlation function of the density fluctuations. The spectral function $S(\mathbf{k}, \omega)$ is important because it provides insight into the dynamical behavior of spontaneous fluctuations (or forced fluctuations in non-equilibrium systems). Whereas the fluctuations extend continuously from the molecular level to the hydrodynamic scale, there are experimental and theoretical limitations to the ranges where they can be probed and computed. Indeed, no theory provides a fully explicit analytical description of space-time dynamics establishing the bridge between kinetic theory and hydrodynamic theory. Scattering techniques have limited ranges of wavelengths over which fluctuation correlations can be probed. Numerical computational techniques can realize molecular dynamics simulations (Boon and Yip, 1980) covering in principle the whole desired range, but in practice there are computation time and memory requirement limitations

(Mareschal and Holian, 1992). So lattice gas automata provide an interesting alternative from both computational and theoretical viewpoints.

7.1 The dynamic structure factor

The dynamic structure factor is defined as the double Fourier transform with respect to space and time of the van Hove function $G(\mathbf{r}, t_\star)$, the correlation function of density fluctuations $\delta\rho(\mathbf{r}_\star, t_\star)$ around the equilibrium state (Boon and Yip, 1980):

$$\begin{aligned} \rho S(\mathbf{k}, \omega) &= \sum_{\mathbf{r}_\star} \sum_{t_\star=-\infty}^{\infty} e^{-\imath\omega t_\star - \imath\mathbf{k}\cdot\mathbf{r}_\star} \langle \delta\rho(\mathbf{r}_\star, |t_\star|)\delta\rho(0,0)\rangle \\ &= \frac{1}{V} \sum_{t_\star=-\infty}^{\infty} e^{-\imath\omega t_\star} \langle \delta\rho(\mathbf{k}, |t_\star|)\delta\rho^*(\mathbf{k}, 0)\rangle, \end{aligned} \tag{7.1}$$

where

$$h(\mathbf{k}) = \sum_{\mathbf{r}\in\mathscr{L}} \exp(-\imath\mathbf{k}\cdot\mathbf{r}_\star) h(\mathbf{r}_\star) \tag{7.2}$$

denotes the discrete spatial Fourier transform, and $V = L^D = \mathcal{N}$ is the number of nodes in the D-dimensional lattice $\mathscr{L}$. In the case of deterministic and invertible dynamics, time reversal invariance of the microscopic equations of motion guarantees that the van Hove function is *even* in time. In the case of deterministic but non-invertible, or stochastic dynamics, $S(\mathbf{k}, \omega)$ is simply defined through (7.1), as twice the half-sided temporal Fourier cosine transform, where the time evolution is defined for positive time argument only. The density fluctuation can be expressed in terms of the fluctuations of the channel occupation variables, n_i $(i = 0, \ldots, b-1)$:

$$\delta\rho(\mathbf{r}_\star, t_\star) = \sum_i \delta n_i(\mathbf{r}_\star, t_\star) = \sum_i \left[n_i(\mathbf{r}_\star, t_\star) - \langle n_i(\mathbf{r}_\star, t_\star)\rangle\right]. \tag{7.3}$$

The average is taken over an equilibrium ensemble and the equilibrium distribution has the form of the Fermi–Dirac distribution (see Chapter 4):

$$\begin{aligned} f_i^{(\mathrm{eq})} &\equiv \langle n_i(\mathbf{r}_\star, t_\star)\rangle = \left[1 + e^{-\alpha+\beta e_i - \boldsymbol{\gamma}\cdot\mathbf{c}_i}\right]^{-1} \\ &= \left[1 + \exp\{-\sum_n b_n a_n(\mathbf{c}_i)\}\right]^{-1}. \end{aligned} \tag{7.4}$$

Here we consider *global* equilibrium, where the system is macroscopically at rest $(\mathbf{u} = \boldsymbol{\gamma} = \mathbf{0})$.

The most important quantity in the non-equilibrium description of fluids and LGAs is the *kinetic propagator* or Green's function:

$$\Gamma_{ij}(\mathbf{k}, t_\star)\kappa_j^{(2)} = \langle \delta n_i(\mathbf{k}, t_\star)\delta n_j^*(\mathbf{k}, 0)\rangle \tag{7.5}$$

where $\delta n_i(\mathbf{k}, t_\star)$ is the spatial Fourier transform of $\delta n_i(\mathbf{r}_\star, t_\star)$, and $\kappa_j^{(2)}$ is the equal time correlation function of the δn_is (see Section 4.7). With (7.3), the dynamic structure factor (7.1) is expressed in terms of the propagator (7.5) as:

$$\rho S(\mathbf{k}, \omega) = \sum_{t_\star=-\infty}^{\infty} e^{-\iota\omega t_\star} \sum_{ij} \Gamma_{ij}(\mathbf{k}, t_\star)\, \kappa_j^{(2)} = \sum_{ij} \Gamma_{ij}(\mathbf{k}, \omega)\, \kappa_j^{(2)}, \tag{7.6}$$

and the static structure factor $S(\mathbf{k})$ – the Fourier transform of the equal time density correlation function, discussed in Section 4.7 – is obtained from (7.5) taken at $t = 0$:

$$\sum_{ij} \Gamma_{ij}(\mathbf{k}, 0)\kappa_j^{(2)} = \sum_{ij} \delta_{ij}\kappa_j^{(2)} = \sum_j \kappa_j^{(2)} = \rho S(\mathbf{k}). \tag{7.7}$$

7.2 Fluctuation correlations

To study the fluctuations dynamics, the main quantity to be computed is the kinetic propagator $\Gamma_{ij}(\mathbf{k}, t_\star)$ which we will calculate in the mean field or Boltzmann approximation. Starting from the lattice Boltzmann equation (see Chapter 4):

$$f_i(\mathbf{r}_\star + \mathbf{c}_i, t_\star + 1) = f_i(\mathbf{r}_\star, t_\star) + \Delta_i^{(B)}(\{n_j\}), \tag{7.8}$$

we want to study fluctuations which disturb the gas locally from its equilibrium state. We proceed as in Chapter 6 by expanding the collision term $\Delta_i^{(B)}$ around the stationary distribution f, (7.4):

$$\Delta_i^{(B)}(f + \delta n) = \Delta_i^{(B)}(f) + \sum_j \Omega_{ij}\delta n_j + \sum_{j,k} \mathcal{O}(\delta n_j \delta n_k). \tag{7.9}$$

Here f satisfies $\Delta_i^{(B)}(f) = 0$ (because of mass conservation), the $b \times b$ matrix $\mathbf{\Omega}$ is the linearized Boltzmann collision operator, and the definition of δn_i is contained in (7.3). Combination of (7.8) and (7.9) gives the linearized lattice Boltzmann equation:

$$\delta n_i(\mathbf{r}_\star + \mathbf{c}_i, t_\star + 1) = \delta n_i(\mathbf{r}_\star, t_\star) + \sum_j \Omega_{ij}\delta n_j(\mathbf{r}_\star, t_\star). \tag{7.10}$$

This equation is equivalent to that obtained in Section 6.1. By space Fourier transformation, we rewrite Equation (7.10) as:

$$\delta n_i(\mathbf{k}, t_\star + 1) = \sum_j e^{-\iota \mathbf{k}\cdot\mathbf{c}_i} \left(\delta_{ij} + \Omega_{ij}\right) \delta n_j(\mathbf{k}, t_\star), \tag{7.11}$$

which expresses the fluctuation δn_i in terms of its value at the previous time step. Solving Equation (7.11) by iteration, we obtain $\delta n_i(\mathbf{k}, t)$ from the knowledge of its initial time value:

$$\delta n_i(\mathbf{k}, t_\star) = \left[e^{-\imath \mathbf{k}\cdot\mathbf{c}} (\mathbb{1} + \mathbf{\Omega}) \right]_{ij}^{t_\star} \delta n_j(\mathbf{k}, 0), \tag{7.12}$$

which is valid for $t_\star \geq 0$. Here $[\exp(-\imath\mathbf{k}\cdot\mathbf{c})]_{ij} = \delta_{ij}\exp(-\imath\mathbf{k}\cdot\mathbf{c}_i)$ is a diagonal matrix. Substituting (7.12) in (7.5) and noting from (7.5) that $\Gamma_{ij}(\mathbf{k}, 0) = \delta_{ij}$, we obtain the expression for the propagator:

$$\Gamma_{ij}(\mathbf{k}, t_\star)\kappa_j^{(2)} = \left[e^{-\imath\mathbf{k}\cdot\mathbf{c}} (\mathbb{1} - \mathbf{\Omega}) \right]_{ij}^{t_\star} \kappa_j^{(2)}. \tag{7.13}$$

We now introduce the resolvent, i.e. the Laplace transform of (7.13):

$$\begin{aligned} \tilde{\Gamma}_{ij}(\mathbf{k}, s) &= \sum_{t=0}^{\infty} e^{-st}\,\Gamma_{ij}(\mathbf{k}, t_\star) \\ &= \left[\frac{1}{e^{s+\imath\mathbf{k}\cdot\mathbf{c}} - \mathbb{1} - \mathbf{\Omega}} \right]_{ij} e^{s+\imath\mathbf{k}\cdot\mathbf{c}}, \end{aligned} \tag{7.14}$$

where the second equality follows from $\sum_{\alpha=0}^{\infty} x^\alpha = (1-x)^{-1}, |x| < 1$. After summation over (i, j), we rewrite (7.14) equivalently as:

$$\begin{aligned} \sum_{ij} \tilde{\Gamma}_{ij}(\mathbf{k}, s)\kappa_j^{(2)} &= \sum_{ijl} \left[\frac{1}{e^{s+\imath\mathbf{k}\cdot\mathbf{c}} - \mathbb{1} - \mathbf{\Omega}} \right]_{il} (e^{s+\imath\mathbf{k}\cdot\mathbf{c}} - \mathbb{1} - \mathbf{\Omega})_{lj}\,\kappa_j^{(2)} \\ &+ \sum_{ijl} \left[\frac{1}{e^{s+\imath\mathbf{k}\cdot\mathbf{c}} - \mathbb{1} - \mathbf{\Omega}} \right]_{il} (\mathbb{1} + \mathbf{\Omega})_{lj}\kappa_j^{(2)}\,; \end{aligned} \tag{7.15}$$

then we make use of the relation $\sum_j \Omega_{lj}\kappa_j^{(2)} = 0$ (which follows from mass conservation) to obtain:

$$\sum_{ij} \tilde{\Gamma}_{ij}(\mathbf{k}, s)\kappa_j^{(2)} = \sum_j \kappa_j^{(2)} + \sum_{ij} \left[\frac{1}{e^{s+\imath\mathbf{k}\cdot\mathbf{c}} - \mathbb{1} - \mathbf{\Omega}} \right]_{ij} \kappa_j^{(2)}. \tag{7.16}$$

The dynamic structure factor $S(\mathbf{k}, \omega)$ is now readily obtained from (7.6) by setting $s = \imath\omega$ in (7.16):

$$\rho S(\mathbf{k}, \omega) \equiv 2\,\mathrm{Re}F(\mathbf{k}, \omega), \tag{7.17}$$

$$F(\mathbf{k}, \omega) = \sum_{ij} \left\{ \left[e^{\imath\omega+\imath\mathbf{k}\cdot\mathbf{c}} - \mathbb{1} - \mathbf{\Omega} \right]^{-1} + \tfrac{1}{2} \right\}_{ij} \kappa_j^{(2)} \tag{7.18}$$

where Re denotes the real part.

This is the general expression for the dynamic structure factor within the Boltzmann approximation. The inverse matrix in (7.18) is b-dimensional and depends on ω and $\mathbf{k}$: it is a complicated mathematical object which in all generality resists analytical inversion. Nevertheless (7.18) can be used for direct numerical evaluation once the explicit collision matrix of the LGA model is known, as will be illustrated later. There is one extreme case where the explicit computation of $S(\mathbf{k},\omega)$ is straightforward, that is for very short wavelengths (or when the particles are in free flight motion); then from (7.6) and (7.13) with $\mathbf{\Omega}=0$, we obtain

$$\begin{aligned} \rho S(\mathbf{k},\omega) &= \mathrm{Re}\sum_{t=-\infty}^{\infty} e^{-\imath\omega t}\sum_{ij}\Gamma_{ij}(\mathbf{k},t_\star)\,\kappa_j^{(2)} \\ &= b\sum_i \delta_{\omega,\mathbf{ik}\cdot\mathbf{c}}\,\kappa_i^{(2)} + \rho S(\mathbf{k}); \end{aligned} \tag{7.19}$$

this is the spectrum for the *ballistic* regime. On the other hand, when one is interested in the hydrodynamic regime, i.e. when the system is probed with a wavelength large compared to the microscopic scale (the inter-particle distance in real fluids; the lattice unit length in LGAs) and large compared to the mean free path between collisions (this condition prevails for dilute fluids), then one can compute $S(\mathbf{k},\omega)$ analytically from (7.18) in the limit of weak spatial and temporal variations ($|\mathbf{k}|\to 0$, $\omega\to 0$).[1] The hydrodynamic structure factor is the subject of Section 7.4.

7.3 The hydrodynamic modes

7.3.1 The spectral decomposition

To compute $S(\mathbf{k},\omega)$ analytically from (7.18) in the hydrodynamic regime,[2] – the domain of *small* values of $|\mathbf{k}|$ and ω – a perturbation method can be used. We shall search the solutions of the linearized Boltzmann equation in the hydrodynamic limit by an eigenvalue problem technique (Résibois and de Leener, 1977) on the basis that the hydrodynamic modes correspond to the eigenvectors whose eigenvalues approach zero when $|\mathbf{k}|\to 0$. We shall find that, up to second order in k, these eigenvectors yield the searched propagative and diffusive modes characterizing the dynamics of the fluid in the linearized hydrodynamic regime.

In Section 7.2, to obtain (7.16), we have used one of the *right* eigenfunctions of Ω_{ij} with zero eigenvalue, i.e.

$$\sum_j \Omega_{ij}\frac{df_j}{db_n} = \sum_j \Omega_{ij}\kappa_j^{(2)} a_n(\mathbf{c}_j) = 0, \tag{7.20}$$

with f_j given by (7.4), and where

$$a_n(\mathbf{c}_i) = \{1, \mathbf{c}_i, e_i = \tfrac{1}{2}c_i^2\} \tag{7.21}$$

[1] Physically $\left[|\mathbf{k}|\to 0,\ \omega\to 0\right]$ means $\left[|\mathbf{k}|\ell_f \ll 1,\ \omega\tau_f \ll 1\right]$, where ℓ_f is the mean free path and τ_f is the mean duration between collisions, that is we consider large wavelength ($\gg \ell_f$), long time ($\gg \tau_f$) phenomena.

[2] The regime considered here is for the fluid at global equilibrium where linearized hydrodynamics holds.

are the collisional invariants.[3] We will also need the *left* eigenvectors with zero eigenvalue of the non-symmetric matrix $\mathbf{\Omega}$. They follow from the *conservation laws* for the total number of particles, the total energy and the total momentum: $\sum_i a_n(\mathbf{c}_i)\Delta_i^{(B)}(\{n_j\}) = 0$, or

$$\sum_i a_n(\mathbf{c}_i)\Omega_{ij} = 0. \tag{7.22}$$

We now consider the eigenvectors and eigenvalues of the matrix $[\exp(-\imath\mathbf{k}\cdot\mathbf{c})(\mathbb{1}+\mathbf{\Omega})]$ which is non-symmetrical and non-Hermitian. The right eigenvectors are defined through the relation:

$$e^{-\imath\mathbf{k}\cdot\mathbf{c}}(\mathbb{1}+\mathbf{\Omega})|\psi_\mu(\mathbf{k})\rangle = e^{z_\mu(\mathbf{k})}|\psi_\mu(\mathbf{k})\rangle, \tag{7.23}$$

where $|\psi_\mu\rangle$ is the b-component vector $|\psi_\mu\rangle_i = \kappa_i^{(2)}\psi_\mu(\mathbf{c}_i)$ $(i = 1,\ldots,b)$, and $z_\mu(\mathbf{k})$ denotes the associated eigenvalue. The zero eigenvectors $a_n(\mathbf{c})$ of $\mathbf{\Omega}$ represent the right eigenvectors ψ_μ in the long wavelength limit ($\mathbf{k} \to 0$). Denoting the eigenvectors by:

$$|a_n\rangle_i = \{|\rho\rangle_i, |e\rangle_i, |\vec{g}\,\rangle_i\} = \kappa_i^{(2)}\{1, \tfrac{1}{2}c_i^2, \mathbf{c}_i\}, \tag{7.24}$$

the eigenvalue equations (7.20) and (7.22) can be written as:

$$\mathbf{\Omega}|a_n\rangle = 0, \qquad \langle a_n|\mathbf{\Omega} = 0. \tag{7.25}$$

The zero eigenvectors $a_n(\mathbf{c})$ of $\mathbf{\Omega}$ represent the right eigenvectors ψ_μ in the long wavelength limit ($\mathbf{k} \to 0$).

The left eigenvectors, $\langle\phi_\mu|$, defined through the relation:

$$\langle\phi_\mu(\mathbf{k})|e^{-\imath\mathbf{k}\cdot\mathbf{c}}(\mathbb{1}+\mathbf{\Omega}) = e^{z_\mu(\mathbf{k})}\langle\phi_\mu(\mathbf{k})|, \tag{7.26}$$

differ from the right eigenvectors $|\psi_\mu(\mathbf{k})\rangle$. Taking the transpose of (7.26) and using the symmetry of the *thermal* scalar product introduced in Section 6.1 (see Equations (6.13) and (6.14)):

$$\langle A|\mathbf{\Omega}|B\rangle = \langle B|\mathbf{\Omega}|A\rangle = \sum_{ij} A_i\Omega_{ij}\kappa_j^{(2)}B_j, \tag{7.27}$$

and comparison with (7.23) yields:

$$\phi_\mu(\mathbf{k}) = e^{-\imath\mathbf{k}\cdot\mathbf{c}}\psi_\mu(\mathbf{k})/\mathcal{M}_\mu(\mathbf{k}), \tag{7.28}$$

where $\mathcal{M}_\mu(\mathbf{k})$ is a normalization constant. Neither the right nor left eigenvectors are orthogonal, but they form a *complete bi-orthonormal* set satisfying the relations:

$$\begin{aligned}\sum_\mu |\psi_\mu\rangle\langle\phi_\mu| &= \mathbb{1}, \\ \langle\phi_\mu|\psi_\lambda\rangle &= \langle\psi_\mu|e^{\imath\mathbf{k}\cdot\mathbf{c}}|\psi_\lambda\rangle/\mathcal{M}_\mu = \delta_{\lambda\mu}.\end{aligned} \tag{7.29}$$

[3] (7.20) follows directly from the observation that for a stationary distribution with Lagrange multipliers b_n (and also with $b_n + \delta b_n$), $\Delta_i^{(B)}(f(b_n)) = 0$.

With the help of these eigenfunctions, we make the following spectral decomposition of the Boltzmann propagator (7.13):

$$\Gamma(\mathbf{k}, t_\star)\kappa^{(2)} = \sum_\mu |\psi_\mu(\mathbf{k})\rangle e^{z_\mu(\mathbf{k})t} \langle \phi_\mu(\mathbf{k})|; \quad t \geq 0. \tag{7.30}$$

With the notation introduced above, we rewrite the spectral function $F(\mathbf{k}, \omega)$ as:

$$\begin{aligned} F(\mathbf{k}, \omega) &= \sum_{t=-\infty}^{\infty} e^{-\imath\omega t} \sum_{ij} \Gamma_{ij}(\mathbf{k}, t_\star)\, \kappa_j^{(2)} \\ &= \langle \rho |\, \left[e^{\imath\omega + \imath\mathbf{k}\cdot\mathbf{c}} - \mathbb{1} - \mathbf{\Omega}\right]^{-1} + \tfrac{1}{2} |\rho\rangle \\ &= \sum_\mu \langle \rho|\psi_\mu\rangle\langle\phi_\mu|\rho\rangle \left\{ [e^{\imath\omega - z_\mu(\mathbf{k})} - 1]^{-1} + \tfrac{1}{2} \right\} \\ &\equiv \sum_\mu \mathcal{N}_\mu \mathcal{D}_\mu(\omega), \end{aligned} \tag{7.31}$$

where, for later convenience, we have introduced the quantities:

$$\begin{aligned} \mathcal{N}_\mu &= \langle \rho|\psi_\mu\rangle\langle\phi_\mu|\rho\rangle, \\ \mathcal{D}_\mu(\omega) &= \frac{1}{e^{\imath\omega - z_\mu(\mathbf{k})} - 1} + \tfrac{1}{2}. \end{aligned} \tag{7.32}$$

At this point we can verify that the first *sum rule* is satisfied, that is the integrated intensity of $S(\mathbf{k}, \omega)$ (at fixed $\mathbf{k}$-value) yields the static structure factor $S(\mathbf{k})$ (7.7). This is easily seen as follows: we evaluate the integral (by contour integration):

$$2\mathrm{Re} \int_{-\pi}^{\pi} d\omega \mathcal{D}_\mu(\omega) = 2\mathrm{Re} \int_{-\pi}^{\pi} d\omega \left\{ [e^{\imath\omega - z_\mu(\mathbf{k})} - 1]^{-1} + \tfrac{1}{2} \right\} = 2\pi, \tag{7.33}$$

and we use the relations (7.7), (7.24) and (7.25) to obtain $\langle\rho|\rho\rangle = \rho S(\mathbf{k})$ and hence:

$$\int_{-\pi}^{\pi} \frac{d\omega}{2\pi} \frac{S(\mathbf{k}, \omega)}{S(\mathbf{k})} = \mathrm{Re} \sum_\mu \frac{\mathcal{N}_\mu}{\langle\rho|\rho\rangle} = 1. \tag{7.34}$$

Each mode contributes a spectral line in (7.34), and their total weight factors add up to unity.[4]

The dominant contributions to $F(\mathbf{k}, \omega)$ come from small values of the denominators in (7.32) which are obtained for the eigenvalues corresponding to the hydrodynamic modes since in the hydrodynamic regime these eigenvalues $z_\mu(\mathbf{k})$ are either $\mathcal{O}(k)$ or $\mathcal{O}(k^2)$ and $k(=|\mathbf{k}|)$ is small. So the hydrodynamic modes with $z_\mu(k) \to 0$ as $k \to 0$ are *slow* modes (that is they exhibit slow decay in space and time). In the long wavelength limit, there are also kinetic modes, with $\mathrm{Re}\, z_\mu(0) < 0$, which give terms with *fast* exponential decay in (7.30) and

4 At finite $|\mathbf{k}|$-values, 'weight factors' are not necessarily positive, because the matrices are complex, but not Hermitian.

whose contribution to $F(\mathbf{k},\omega)$ is therefore negligible with respect to the hydrodynamic modes. Consequently, when we are interested in the hydrodynamic behavior, we can restrict the summation in (7.31) to the μs corresponding to the hydrodynamic modes whose effect is dominant.

7.3.2 The eigenvalues

We start from the eigenvalue equation:

$$\left[e^{z_\mu(\mathbf{k})+\imath\mathbf{k}\cdot\mathbf{c}} - \mathbb{1} - \mathbf{\Omega}\right] |\psi_\mu(\mathbf{k})\rangle = 0. \tag{7.35}$$

In the long wavelength limit the hydrodynamic modes $\psi_\mu(\mathbf{k})$ and eigenvalues $z_\mu(\mathbf{k})$ can be determined by a Taylor series expansion:

$$\begin{aligned} \psi_\mu(\mathbf{k}) &= \psi_\mu^{(0)} + \imath k\psi_\mu^{(1)} + (\imath k)^2\psi_\mu^{(2)} + \cdots \\ z_\mu(\mathbf{k}) &= \imath k z_\mu^{(1)} + (\imath k)^2 z_\mu^{(2)} + \cdots. \end{aligned} \tag{7.36}$$

Substitution of (7.36) into (7.35) gives the set of equations:

$$\begin{aligned} \Omega|\psi_\mu^{(0)}\rangle &= 0 \\ \Omega|\psi_\mu^{(1)}\rangle &= (c_l + z_\mu^{(1)})|\psi_\mu^{(0)}\rangle \\ \Omega|\psi_\mu^{(2)}\rangle &= (c_l + z_\mu^{(1)})|\psi_\mu^{(1)}\rangle + [z_\mu^{(2)} + \tfrac{1}{2}(c_l + z_\mu^{(1)})^2]|\psi_\mu^{(0)}\rangle, \end{aligned} \tag{7.37}$$

with $c_l = \widehat{\mathbf{k}}\cdot\mathbf{c}$, where $\widehat{\mathbf{k}}$ is the unit vector along the $\mathbf{k}$-direction. The general solution of the zeroth order equation is an arbitrary linear combination of the collisional invariants (7.24):

$$\begin{aligned} |\psi_\mu^{(0)}\rangle &= \sum_n B_n|a_n\rangle \\ &= B_T|s\rangle + B_p|p\rangle + B_l|c_l\rangle + B_\perp|c_\perp\rangle. \end{aligned} \tag{7.38}$$

Here c_l and $c_\perp = \widehat{\mathbf{k}}_\perp\cdot\mathbf{c}$ are respectively the longitudinal and transverse components of the microscopic momentum $\mathbf{g}(\mathbf{c}) = \mathbf{c}$.[5] The microscopic *entropy*, $s(\mathbf{c}) = e(\mathbf{c}) - h\rho(\mathbf{c}) = \frac{1}{2}c^2 - c_s^2$, and the microscopic *pressure*, $p(\mathbf{c}) = \frac{1}{2}c^2$, are linear combinations of the collisional invariants $\rho(\mathbf{c}) = 1$ and $e(\mathbf{c}) = \frac{1}{2}c^2$ in (7.24). The constant h:

$$h \equiv c_s^2 = \frac{\langle p|p\rangle}{\langle p|\rho\rangle} = \frac{\sum_i \kappa_i c_i^4}{2\sum_i \kappa_i c_i^2}, \tag{7.39}$$

is chosen such that the microscopic pressure and entropy are orthogonal,[6] i.e. $\langle s|p\rangle = 0$. We have also introduced the quantity c_s which will appear to be the 'adiabatic' speed of sound.

[5] This decomposition will appear quite logical in terms of the longitudinal and transverse modes.

[6] The choice of the couple of thermodynamic variables s and p is convenient here, but density and temperature would also be appropriate.

The coefficients B_n in (7.38) and eigenvalues $z_\mu^{(1)}$ are determined by solving exactly the eigenvalue problem to order k in the subspace spanned by $\{s, p, c_l, c_\perp\}$. Multiplication of the second equation in (7.37) to the left with $\langle a_m|$ yields:

$$\sum_n \langle a_m | c_l + z_\mu^{(1)} | a_n \rangle B_n = 0, \tag{7.40}$$

with $n, m = \{s, p, l, \perp\}$. The matrix $\langle a_m | c_l + z_\mu^{(1)} | a_n \rangle$ reads:

$$\begin{array}{c} \\ s \\ p \\ l \\ \perp \end{array} \begin{array}{cccc} s & p & l & \perp \\ \left(\begin{array}{c} z_\mu^{(1)} \langle s|s \rangle \\ 0 \\ 0 \\ 0 \end{array} \right. & \begin{array}{c} 0 \\ z_\mu^{(1)} \langle p|p \rangle \\ \langle p|c_l^2 \rangle \\ 0 \end{array} & \begin{array}{c} 0 \\ \langle p|c_l^2 \rangle \\ z_\mu^{(1)} \langle c_l|c_l \rangle \\ 0 \end{array} & \left. \begin{array}{c} 0 \\ 0 \\ 0 \\ z_\mu^{(1)} \langle c_\perp|c_\perp \rangle \end{array} \right) \end{array}, \tag{7.41}$$

and is diagonal in the labels s and $\perp$, yielding $z_s^{(1)} = z_\perp^{(1)} = 0$. In the subspace $\{p, l\}$:

$$(z_\sigma^{(1)})^2 = \frac{(\langle p|c_l^2 \rangle)^2}{\langle p|p \rangle \langle c_l|c_l \rangle} = c_s^2, \tag{7.42}$$

where we have used (7.39) and the relations $\langle p|c_l^2 \rangle = \langle p|p \rangle$ and $\langle c_l|c_l \rangle = \langle p|\rho \rangle$. The eigenvalues $z_\mu^{(1)}$ to first order, the hydrodynamic modes $\psi_\mu^{(0)}$ to zeroth order, and the currents $j_\mu \equiv (c_l + z_\mu^{(1)})\psi_\mu^{(0)}$ are then given respectively by:

$$\begin{array}{lll} z_s^{(1)} = 0; & \psi_s^{(0)} = s; & j_s = c_l s; \\ z_\perp^{(1)} = 0; & \psi_\perp^{(0)} = c_\perp; & j_\perp = \tau_{xy} = c_l c_\perp; \\ z_\sigma^{(1)} = -\sigma c_s; & \psi_\sigma^{(0)} = p + \sigma c_s c_l; & j_\sigma = j_T + \sigma c_s \tau_{xx}. \end{array} \tag{7.43}$$

with $\tau_{xx} = \frac{1}{2}(c_l^2 - c_\perp^2)$ and $\sigma = \pm$; here s labels the entropy (or heat) mode, $\perp$ the shear (or transverse momentum) mode, and σ the two sound modes.

The solution to the second equation of (7.37) is:

$$|\psi_\mu^{(1)}\rangle = \frac{1}{\Omega}|j_\mu\rangle + \sum_\nu B_{\mu\nu}|\psi_\nu^{(0)}\rangle. \tag{7.44}$$

The first term on the right hand side belongs to the orthogonal complement of the null space of $\mathbf{\Omega}$. The coefficients $B_{\mu\nu}$ remain undetermined. By inserting (7.44) into the third equation of (7.37) and multiplying it on the left by $\langle \psi_\lambda^{(0)}|$ (with $\lambda = s, \perp$ and σ), we obtain:

$$\begin{aligned} 0 \;=\; & \sum_\nu B_{\mu\nu} \langle \psi_\lambda^{(0)} | (c_l + z_\mu^{(1)}) | \psi_\nu^{(0)} \rangle \\ & + \langle \psi_\lambda^{(0)} (c_l + z_\mu^{(0)}) | \frac{1}{\Omega} | (c_l + z_\mu^{(1)}) \, \psi_\mu^{(0)} \rangle \\ & + \langle \psi_\lambda^{(0)} | z_\mu^{(2)} + \tfrac{1}{2}(c_l + z_\mu^{(1)})^2 | \, \psi_\mu^{(0)} \rangle. \end{aligned} \tag{7.45}$$

For $\lambda = \mu$, the first term on the r.h.s. vanishes and we obtain the eigenvalues to second order:

$$\begin{aligned} z_s^{(2)} &\equiv D_{\mathrm{T}} = -\langle j_s|\frac{1}{\Omega} + \frac{1}{2}|j_s\rangle\langle s|s\rangle^{-1} \\ z_\perp^{(2)} &\equiv \nu = -\langle \tau_{xy}|\frac{1}{\Omega} + \frac{1}{2}|\tau_{xy}\rangle\langle c_\perp|c_\perp\rangle^{-1} \\ z_\sigma^{(2)} &\equiv \Gamma = -\langle j_s + \sigma c_s \tau_{xx}|\frac{1}{\Omega} + \frac{1}{2}|j_s + \sigma c_s \tau_{xx}\rangle\langle \psi_\sigma^{(0)}|\psi_\sigma^{(0)}\rangle^{-1}. \end{aligned} \tag{7.46}$$

Here D_{T} is the thermal diffusivity, ν the kinematic viscosity, and Γ the sound damping coefficient. To further simplify the expression for Γ, we use the relation:

$$\langle \psi_\sigma^{(0)}|\psi_\sigma^{(0)}\rangle = 2\langle p|p\rangle = 2c_s^2\langle c_l|c_l\rangle = 2c_s^2\langle p|\rho\rangle, \tag{7.47}$$

which follows from (7.39), (7.42) and (7.43). For lattice gas models with sufficient symmetry (like the triangular lattice with hexagonal symmetry, see Chapter 3), a fourth rank tensor is isotropic, yielding the equalities:

$$\nu\langle c_\perp|c_\perp\rangle = -\langle \tau_{xy}|\frac{1}{\Omega} + \frac{1}{2}|\tau_{xy}\rangle = -\langle \tau_{xx}|\frac{1}{\Omega} + \frac{1}{2}|\tau_{xx}\rangle. \tag{7.48}$$

Combination of the above results provides an expression for the sound damping coefficient in terms of the viscosity and the thermal diffusivity,

$$\Gamma = \tfrac{1}{2}\nu + \tfrac{1}{2}\frac{\langle s|s\rangle}{\langle p|p\rangle}D_{\mathrm{T}}. \tag{7.49}$$

Now by writing $s = p - \rho\langle p|p\rangle/\langle p|\rho\rangle$ and introducing the isothermal speed of sound c_{T} through the relation:

$$\frac{\langle p|\rho\rangle}{\langle \rho|\rho\rangle} = \frac{(\partial p/\partial \alpha)_\beta}{(\partial \rho/\partial \alpha)_\beta} = \left(\frac{\partial p}{\partial \rho}\right)_\beta \equiv c_{\mathrm{T}}^2, \tag{7.50}$$

we have:

$$\frac{\langle s|s\rangle}{\langle p|p\rangle} + 1 = \frac{\langle p|p\rangle\langle \rho|\rho\rangle}{\langle p|\rho\rangle^2} = \frac{c_s^2\langle \rho|\rho\rangle}{\langle p|\rho\rangle} = \frac{c_s^2}{c_{\mathrm{T}}^2} \equiv \gamma. \tag{7.51}$$

So we can rewrite the sound damping coefficient as:

$$\Gamma = \tfrac{1}{2}\nu + \tfrac{1}{2}(\gamma - 1)D_{\mathrm{T}}, \tag{7.52}$$

with $\gamma \geq 1$, as seen from (7.51).

It remains to evaluate the coefficients $B_{\mu\nu}$ which determine the first order eigenvectors through (7.44). From (7.45) with $\mu \neq \nu$, and after some rearrangements, we find:

$$B_{\mu\nu}\langle \psi_\lambda^{(0)}|\psi_\lambda^{(0)}\rangle(z_\mu^{(1)} - z_\lambda^{(1)}) = -\langle j_\lambda|\frac{1}{\Omega} + \frac{1}{2}|j_\mu\rangle, \tag{7.53}$$

from which we obtain the *non-vanishing* coefficients, i.e.

$$\begin{aligned} B_{\sigma,-\sigma} &= (\nu - \Gamma)/[2\sigma c_s] \\ B_{\sigma,\mathrm{T}} &= -D_{\mathrm{T}}/[\sigma c_s] \\ B_{\mathrm{T},\sigma} &= (\gamma - 1)D_{\mathrm{T}}/[2\sigma c_s]. \end{aligned} \tag{7.54}$$

The coefficients $B_{\mu\mu}(\mu = T, \sigma, \perp)$ remain undetermined, but they are unimportant as they will not appear in the computation of the dynamic structure factor $S(\mathbf{k}, \omega)$.

In summary, starting from the linearized lattice Boltzmann equation, we solved the eigenvalue equation (7.35) to obtain explicit expressions for the hydrodynamic modes: the shear mode, the heat mode, and the two sound modes. The eigenvectors contain the transverse ($\perp$) and longitudinal (l) components of the momentum fluctuation $\mathbf{g}(\mathbf{c}) = \mathbf{c}$, the entropy fluctuation $s(c) = \frac{1}{2}c^2 - c_s^2$, and the pressure fluctuation $p(c) = \frac{1}{2}c^2$. The eigenvalues contain the speed of sound c_s and the transport coefficients: kinematic viscosity ν, thermal diffusivity D_{T} and sound damping coefficient Γ, whose explicit expressions were obtained in the Boltzmann approximation.

7.4 The hydrodynamic spectrum

In order to carry out a consistent perturbation expansion of $F(\mathbf{k}, \omega)$ in (7.31) for small k, we observe that we are interested in $\omega \sim \mathcal{O}(k^\alpha)$, with α up to 2. The dominant non-vanishing contribution ($\mathcal{O}(k)$) yields the line-shifts, and the next contribution ($\mathcal{O}(k^2)$) yields the line-shape (width) of the spectrum. From (7.31), using the relation $(e^x - 1)^{-1} + \frac{1}{2} \simeq x^{-1} + \mathcal{O}(x)$, valid for small x, we observe that:

$$F(\mathbf{k}, \omega) = \sum_{\mu}{}^{*} \mathcal{N}_\mu \left[\imath\omega - z_\mu(\mathbf{k})\right]^{-1} \left(1 + \mathcal{O}(k^2)\right), \tag{7.55}$$

where the asterisk on the summation sign indicates restriction to the hydrodynamic modes. The coefficients $\mathcal{N}_\mu$ in (7.32) are evaluated in Section 7.3.2 for small k, and yield:

$$\begin{aligned} \mathcal{N}_\sigma &\simeq \frac{\langle\rho|\rho\rangle}{2\gamma}\left\{1 + \frac{\imath k}{\sigma c_s}[\Gamma + (\gamma - 1)D_{\mathrm{T}}]\right\} + \mathcal{O}(k^2), \\ \mathcal{N}_{\mathrm{T}} &\simeq \langle\rho|\rho\rangle\left(\frac{\gamma - 1}{\gamma}\right) + \mathcal{O}(k^2). \end{aligned} \tag{7.56}$$

We have also introduced the ratio, Equation (7.51):

$$\gamma = \frac{c_s^2}{c_{\mathrm{T}}^2} = \frac{\langle p|p\rangle\langle\rho|\rho\rangle}{\langle p|\rho\rangle^2}, \tag{7.57}$$

with c_s and c_T the adiabatic and isothermal speed of sound respectively.

Combination of (7.31), (7.55) and (7.56) yields the dynamic structure factor in the Landau–Placzek approximation:

$$\frac{S(\mathbf{k},\omega)}{S(\mathbf{k})} = \left(\frac{\gamma-1}{\gamma}\right)\frac{2D_{\mathrm{T}}k^2}{\omega^2+(D_{\mathrm{T}}k^2)^2}+\frac{1}{\gamma}\sum_{\pm}\frac{\Gamma k^2}{(\omega\pm c_{\mathrm{s}}k)^2+(\Gamma k^2)^2}$$
$$+\frac{1}{\gamma}\left[\Gamma+(\gamma-1)D_T\right]\frac{k}{c_{\mathrm{s}}}\sum_{\pm}\frac{(c_{\mathrm{s}}k\pm\omega)}{(\omega\pm c_{\mathrm{s}}k)^2+(\Gamma k^2)^2}. \qquad (7.58)$$

For small k-values and $\omega \sim \mathcal{O}(k)$ this expression is consistent up to relative $\mathcal{O}(k^2)$.

The power spectrum (7.58) has the standard form as obtained from the linearized hydrodynamic equations in continuous theory (Boon and Yip, 1980). The first line on the r.h.s. of (7.58) contains three symmetrical Lorentzians: the first one is the Rayleigh peak with a width $D_{\mathrm{T}}k^2$ and the other two Doppler shifted ($\pm c_{\mathrm{s}}k$) lines are the Brillouin peaks with a width Γk^2. The last term in (7.58) describes asymmetric contributions to the Brillouin lines. On the relevant scale $\omega \sim \mathcal{O}(k)$, the Brillouin shift of $\mathcal{O}(k)$ is the dominant feature that determines the structure of $S(\mathbf{k},\omega)$. The line-shape of the Brillouin line is determined by the line-width and the asymmetric terms, which are of relative order $\mathcal{O}(k)$ with respect to the frequency shift. The power spectrum (7.58) consisting of the Rayleigh line and the asymmetric Brillouin lines, derived here from the Boltzmann equation for the lattice gas, are in complete agreement with the results of the Landau–Placzek theory for continuous isotropic fluids (Boon and Yip, 1980). All further corrections to the theory are at least of relative order $\mathcal{O}(k^2)$ and would involve higher order transport coefficients, such as Burnett coefficients. Finally we note that the sum rule for the integrated intensity in the Landau–Placzek theory follows from the integration:

$$\frac{1}{2\pi}\int_{-\infty}^{\infty} d\omega\, S(\mathbf{k},\omega) = S(\mathbf{k}); \qquad (7.59)$$

the same result was obtained for the lattice gas in Section 7.3.1.[7]

7.5 The eigenvalue spectrum

In order to analyze the dynamic structure factor $S(\mathbf{k},\omega)$ over the full spectral range of wavelengths and thereby to assess the limits of validity of the Landau–Placzek theory, we must know both the eigenvalues, $z_\mu(\mathbf{k}) = \mathrm{Re}\, z_\mu(\mathbf{k}) + \imath\, \mathrm{Im}\, z_\mu(\mathbf{k})$, and the eigenfunctions of the propagator (7.13). At fixed wavelength $\mathbf{k}$, each eigenmode contributes, according to (7.31), a spectral line with a maximum located approximately at $\mathrm{Im}\, z_\mu(\mathbf{k})$ ($\neq 0$ if the mode is propagating) and a width

[7] Notice that the small-k limit and the ω-integration cannot be interchanged, because the ω-integral does not converge uniformly near $\omega = 0$.

determined by $\text{Re}\, z_\mu(\mathbf{k})$; the eigenfunctions enter the weight factors $\langle\rho|\psi\rangle\langle\phi|\rho\rangle$ in (7.31), and determine predominantly the amplitude of the spectral lines. In the previous section, we obtained these quantities by a perturbative calculation in the limit of small values of k. Beyond this limit, analytical calculation is no longer possible, and one must have recourse to numerical computation to obtain the eigenvalues of the matrix $\left[e^{-\imath\mathbf{k}\cdot\mathbf{c}}(\mathbb{1}-\mathbf{\Omega})\right]$. The collision operator $\mathbf{\Omega}$ should then be known explicitly, that is the lattice gas model must be specified: we shall consider the GBL model described in Section 3.6 which, as will be seen, exhibits all the general features observed in an actual mono-atomic fluid.

Figure 7.1 shows the real and imaginary parts of the eigenvalues. The total number of eigenvalues equals the number b of allowed velocity states ($b = 19$ in the GBL model), and their values depend on the details of the collision rules (see Section 3.6).

(i) There are fast *kinetic* modes with $\text{Re}z_\mu(\mathbf{k}) \neq 0$ at $k = 0$, which may be purely diffusive or may become propagating at non-zero k-values.

(ii) There are slow *hydrodynamic* modes with a damping $\text{Re}z_\mu(\mathbf{k}) \sim \mathcal{O}(k^2)$ as $k \to 0$; these modes may be propagating with a (sound) speed $\text{Im}z_\mu(\mathbf{k})/k \equiv c_\mu(\mathbf{k})$, or they may be purely diffusive (if $\text{Im}z_\mu(\mathbf{k}) = 0$). At small k–values the four hydrodynamic modes can easily be identified as they are well separated from the kinetic modes.

The eigenvalue spectra of the lattice gas automaton have a number of universal features in common with those found in the continuous theory of liquids and gases; one interesting aspect of lattice gases is that there is a finite number of eigenvalues which can all be computed explicitly (at least numerically). These features are important in the analysis of the dynamic structure factor and they are most appropriately interpreted by considering the ratio of the mean free path, $\ell_f \sim 1/\rho$, to the wavelength, $\Lambda = 2\pi/k$, of the probe used to analyze the dynamics. In laboratory experiments, Λ is the characteristic wavelength determined by the geometry of the scattering set-up used in neutron or photon spectroscopy (Boon and Yip, 1980); in lattice gas automata, the same role is played by the reciprocal wavenumber used in the analysis of the simulation data. So here the appropriate quantity is the product $k\ell_f$.

7.5.1 Hydrodynamic regime: $k\ell_f \ll 1$

In the hydrodynamic regime, that is $k \lesssim 0.6$ (or $\Lambda \gtrsim 10$ lattice units) in Figure 7.1, there is a clear separation between the kinetic eigenvalues and the hydrodynamic ones which correspond to the shear mode ($\mu = \perp$), the heat mode ($\mu = s$), and the two sound modes ($\mu = \sigma = \pm$). The values $z_\mu(\mathbf{k})$ of the hydrodynamic modes are given by (7.46) and the macroscopic transport coefficients are independent of the wavenumber k. In this regime the hypothesis of the Landau–Placzek theory is satisfied: long wavelength fluctuations decay essentially according to the linearized macroscopic hydrodynamic equations; we shall see indeed, in Section 7.6, that the hydrodynamic power spectrum of the lattice gas is accurately described by Equation (7.58). This is the domain of validity of classical hydrodynamics which gives a *local* description of the response of the system: energy and momentum fluxes are linear in the temperature and velocity gradients respectively.

7.5.2 Generalized hydrodynamic regime: $k\ell_f < 1$

When we look at a smaller scale, that is if we probe the dynamics of the fluid at a wavelength Λ which is not *much larger* than ℓ_f, we observe that the classical hydrodynamic description with constant transport coefficients breaks down: the *non-local* response of the system to spatial inhomogeneities renders

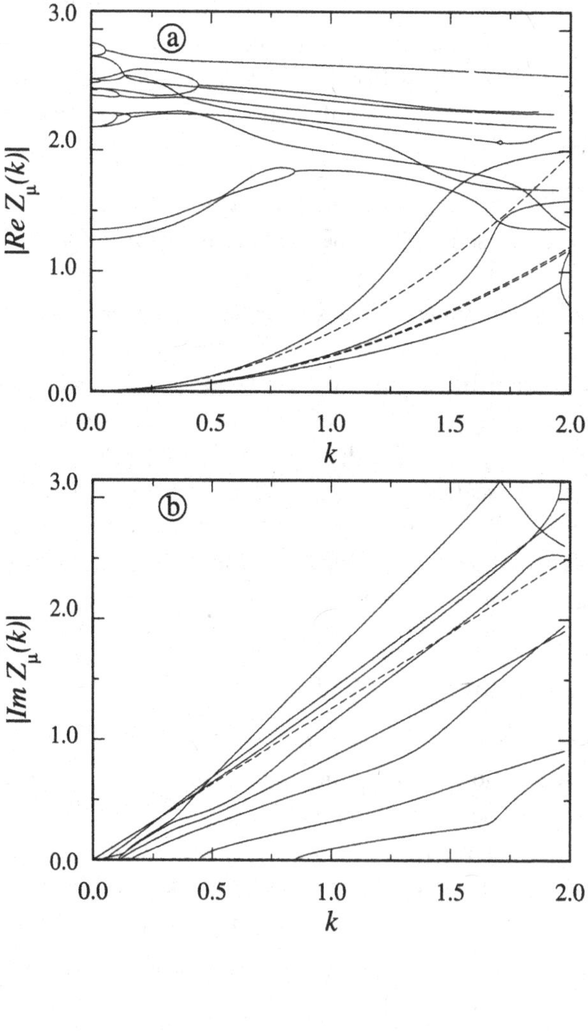

Figure 7.1 Eigenvalue spectrum of the GBL model: $z_\mu(k)$ as function of k. (a) Real eigenvalue $\mathrm{Re}\, z_\mu(k)$; the dotted lines represent the hydrodynamic relaxation rates: the upper curve corresponds to the kinematic viscosity, and the lower double curve to the thermal diffusivity and to the sound damping ($D_T = 0.302$, $\Gamma = 0.297$); (b) Imaginary eigenvalue $\mathrm{Im}\, z_\mu(k)$; the slope of the dotted line is the hydrodynamic speed of sound ($c_s = 1.26$). Density and energy per node: $\rho = 6.0$ and $e = 6.7$ respectively; **k** is oriented along a lattice axis and k is given in $2\pi \times$ reciprocal lattice units; $\mathrm{Re}\, z_\mu(k)$ and $\mathrm{Im}\, z_\mu(k)$ are frequencies expressed in units of $2\pi \times$ reciprocal automaton time steps.

the transport coefficients **k**-dependent We see indeed, in Figure 7.1, that the discrepancy between the macroscopic relaxation rates (dashed lines) and the exact ones (full lines) becomes larger and larger as k increases, and that the speed of sound $c_s(\mathbf{k})$ exhibits dispersion. The exact diffusivities, as we shall see in Section 7.6, can be described in terms of generalized **k**-dependent transport coefficients: $D_\mu(\mathbf{k}) = \operatorname{Re} z_\mu(\mathbf{k})/k^2 = \{\nu(\mathbf{k}), \Gamma(\mathbf{k}), D_T(\mathbf{k})\}$.

However, we observe that, in the range $0.6 \lesssim k \lesssim 1.25$, there still is a clear separation between hydrodynamic and kinetic real eigenvalues; so the contributions of the fast modes may be ignored in the analysis of the hydrodynamic spectrum. This defines the domain of *generalized hydrodynamics* where the general structure of the hydrodynamic spectrum should be preserved, but where important discrepancies will be observed between the Landau–Placzek theory and the Boltzmann theory for $S(\mathbf{k}, \omega)$. Closer inspection of Figure 7.1 leads to the following observation: for $\mathbf{k}$ parallel to a basis vector, say $\mathbf{1}_x$, the matrix in (7.31) is invariant under the reflection $c_y \leftrightarrow -c_y$. Consequently the matrix in (7.30) can be decomposed in two subspaces with even and odd parity in c_y. Therefore $S(\mathbf{k}, \omega)$ in (7.31) only couples to the even subspace containing sound ($\mu = \sigma = \pm$) and heat ($\mu = s$) modes. It does not couple to the shear mode ($\mu = \perp$) with odd c_y-parity. Therefore, when considering $S(\mathbf{k}, \omega)$, the domain of generalized hydrodynamics extends to about $k \simeq 1.5$, which corresponds to a wavelength of four lattice units.

In Section 7.6, we shall test these concepts of generalized hydrodynamics by substituting the **k**-dependent parameters $c_s(\mathbf{k})$, $\Gamma(\mathbf{k})$ and $D_T(\mathbf{k})$ (as obtained from the numerical evaluation of the eigenvalues) into the Landau–Placzek expression (7.58) and compare the result with the Boltzmann prediction (7.31) which does not involve any long wavelength or slow mode approximation.

7.5.3 Kinetic regime: $k\ell_f \gtrsim 1$

For large k-values ($k > 1.5$), that is for very short wavelengths ($\Lambda < 4$ lattice units), we can no longer make a distinction between fast and slow modes, as all decay rates have the same magnitude (see Figure 7.1). In the kinetic regime, it is physically meaningless to introduce any parameterization to reconstruct a hydrodynamic type spectrum, even with **k**-dependent coefficients: all modes of even c_y-parity couple to the dynamic structure factor and contribute a spectral line (Doppler shifted if the mode is propagating).

To illustrate the importance of the mean free path, $\ell_f \sim 1/\rho$, as the relevant length scale parameter for physical interpretation, let us mention that the eigenvalue spectrum at low density ($\rho \sim 1$) as compared to the spectrum of Figure 7.1 ($\rho = 6.0$) has a classical hydrodynamic regime ($k \lesssim 0.15$ or $\Lambda \gtrsim 42$) about a factor of 6 narrower, and a generalized hydrodynamic range about a factor of 8 smaller. For $k \gtrsim 0.4$ one observes the so-called Knudsen regime

(Das, Bussemaker and Ernst, 1993), where the collision matrix in (7.31) is only a small perturbation to the propagation term which constitutes the dominant contribution to the spectral density (see Equation (7.19)).

7.6 Power spectrum

We shall now examine the spontaneous density fluctuations in the microscopic simulations of the GBL lattice gas automaton. The lattice gas is at global equilibrium (i.e. the system is initialized with zero total momentum) and is subject to periodic boundary conditions. The GBL model has the interesting feature that its equilibrium thermodynamic state can be specified by the mass density $\rho = \sum_i f_i$, *and* the energy density $e = \sum_i e_i f_i$, which are set by the fugacity $z = \exp(\alpha)$ and the reduced temperature $\theta = \exp(-\beta/2)$ introduced in Chapter 4. We shall consider two states, one of high density ($\rho = 6.0, e = 6.7$) and one of low density ($\rho = 1.1, e = 1.1$).

The 'experimental' $S(\mathbf{k}, \omega)$ is obtained by space- and time-Fourier transformation of the data of the local density fluctuations $\delta\rho(\mathbf{r}_\star, t_\star)$; in the analysis, $S(\mathbf{k}, \omega)$ is treated as a spectral function, i.e. as a function of ω taken at fixed values of the wavenumber $k = |\mathbf{k}|$, and spectra will be analyzed for different typical values of k. We shall compare these 'experimental' spectra with the theoretical predictions obtained in the previous sections for the dynamic structure factor: the Boltzmann expression for $S(\mathbf{k}, \omega)$ in (7.17) and (7.31), and, for small $|\mathbf{k}|$, the Landau–Placzek expression (7.58) with the thermodynamic quantities γ and c_s computed from (7.57) and (7.39), and the Boltzmann transport coefficients ν and D_T, computed from (7.45). So the comparative analysis also provides a test of validity for the evaluation of the transport coefficients as they can be obtained from the measurement of the spectral line-widths.

In principle, measurements could be performed down to a value of the wavenumber equal to $k_0 = 2\pi/L$, where L is the lattice dimension. But we should note that the accuracy on the line-width measurement is set by the frequency resolution $\Delta\omega = 2\pi/T_0$, where T_0 is the total simulation time (usually of the order of a few thousand time steps), and so it decreases with the width of the spectral lines, i.e. when k decreases. Therefore in practice there is a lower bound to the k-range where measurements are reliable; for instance in the results discussed below, this lower limit is set at $k = 4k_0 \simeq 0.05$, which corresponds to a wavelength of 130 lattice units while the lattice has 512×512 nodes.

7.6.1 High density

Spectra $S(\mathbf{k}, \omega)$ measured at small wavenumbers are characteristic of the hydrodynamic behavior at long wavelengths: a typical example for the lattice gas at high density is shown in Figure 7.2. According to the discussion in Section 7.5.1 (see also Figure 7.1), this spectrum is measured at a k-value well inside the hydrodynamic regime where the predictions of the Boltzmann theory (7.18) and of the Landau–Placzek theory (7.58) coincide, and are in excellent agreement with the 'experimental' spectrum as Figure 7.2 shows. The spectrum consists of a doublet of spectral lines located at $\omega = \pm c_s k$, corresponding to the acoustic modes, and a peak centered at $\omega = 0$, which arises from fluctuations at constant pressure and corresponds to the thermal diffusivity mode (for non-thermal LGA models, this central peak is absent). Thus we observe that in the small k-domain the spectral density of lattice gas fluctuations is very well reproduced by the Landau–Placzek theory.[8] Most important is the observation that the dynamical structure factor of the lattice gas exhibits the typical structure and line-shapes of the Rayleigh–Brillouin spectrum measured in real fluids (Boon and Yip, 1980); it was precisely to explain how spontaneous hydrodynamic fluctuations in continuous fluids give rise to this type of spectrum that the Landau–Placzek

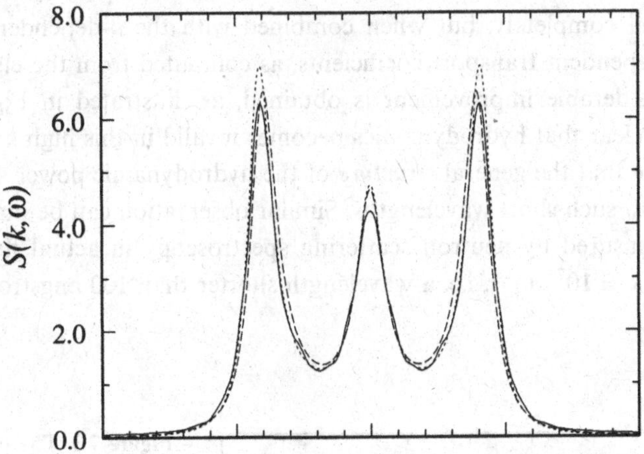

Figure 7.2 The LGA dynamic structure factor in the hydrodynamic regime. $S(\mathbf{k}, \omega)$ as a function of the frequency ω for the GBL lattice gas at high density and low k. Comparison between the simulation results (solid line) and the theoretical predictions: Landau–Placzek theory, Equation (7.58), (dashed curve), and Boltzmann theory, Equation (7.18), (dotted curve). Thermodynamic state: $\rho = 6.0$; $e = 6.7$. Automaton universe: $L \times L = (512)^2$. $k = 20k_0 \simeq 0.245$ reciprocal lattice units. ω is expressed in units of $2\pi/T_0$, where T_0 is the total number of time steps, and $S(k,\omega)/S(k)$ is expressed in reciprocal ω units.

[8] The slight discrepancy in the height of the doublet peaks (see Figure 7.2) is probably due to the weaker resolution at small k-values of the peaks where fluctuations in the simulation data are most pronounced.

theory was initially developed. The comparative inspection of Figure 7.2 and Figure 7.3 is instructive as the latter shows the power spectrum measured by light scattering spectroscopy in a real fluid at $k \simeq 2 \times 10^5$ cm^{-1}, i.e. a wavelength of a few thousands angstroms.

When the value of the wavenumber increases, the Landau–Placzek theory progressively fails to reproduce the measured spectra, because, as we discussed in Section 7.5.2, linearized hydrodynamics with constant transport coefficients becomes invalid at short wavelengths. For the GBL lattice gas, at $k = 50k_0 \simeq 0.6$ one enters the regime of generalized hydrodynamics where the Boltzmann theory gives very good account of the simulation data.

When k increases further, the deviations from classical hydrodynamics become larger, as exemplified in Figure 7.4, which shows a comparison between the measured and computed dynamical structure factors at $k = 136k_0 \simeq 1.67$ for the GBL model. The figure shows that the Boltzmann prediction remains valid down to quite short wavelengths (here $\Lambda \simeq 4$ lattice units). Here we are at the border between generalized hydrodynamics and the kinetic regime (see Figure 7.1): the Landau–Placzek theory with constant transport coefficients is expected to fail completely, but when combined with the k-dependent sound speed and k-dependent transport coefficients, as computed from the eigenvalue spectrum, considerable improvement is obtained, as illustrated in Figure 7.4. Although it is clear that hydrodynamics becomes invalid in this high k domain, it is remarkable that the general *structure* of the hydrodynamic power spectrum persists down to such short wavelengths. Similar observation can be made from the spectra measured by neutron scattering spectroscopy in actual fluids, e.g. liquid neon at $k \simeq 10^7$ cm^{-1}, i.e. a wavelength shorter than 100 angstroms (Bell *et al.*, 1975).

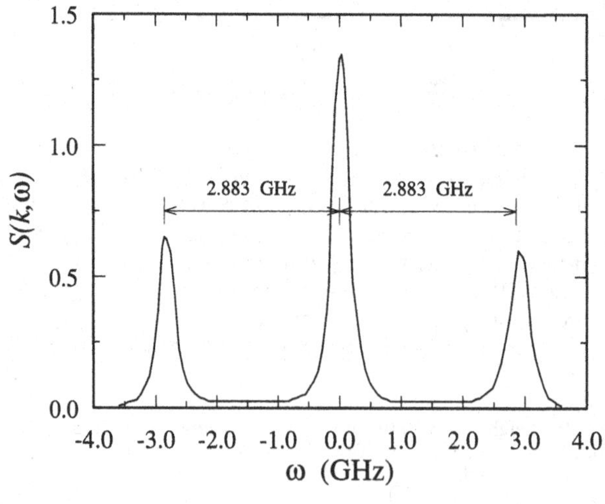

Figure 7.3 The power spectrum of liquid argon in its normal liquid range measured by light scattering spectroscopy (Fleury and Boon, 1969). ω is given in GHz and $S(k,\omega)$ in arbitrary units.

7.6.2 Low density

The various types of behavior discussed above are also found from the analysis of the spectra of the lattice gas at low density ($\rho \sim 1$), with the restriction that the domain where the hydrodynamic approximation is valid and even the generalized hydrodynamic domain are narrower in k, as can be inferred from the discussion in Section 7.5. In Figure 7.5, the density fluctuation spectrum is taken at $k = 12k_0 \simeq 0.15$ which is really the border between classical and generalized hydrodynamics for the low density lattice gas; here the Landau–Placzek and Boltzmann predictions are in good agreement with the spectrum obtained from the simulation data. On the other hand, when $k = 52k_0 \simeq 0.64$, the dilute gas is inside the kinetic regime, close to the Knudsen regime (Das, Bussemaker, and Ernst, 1993); the simulation data become very noisy and it is difficult to separate spurious oscillations from the actual kinetic spectral lines,

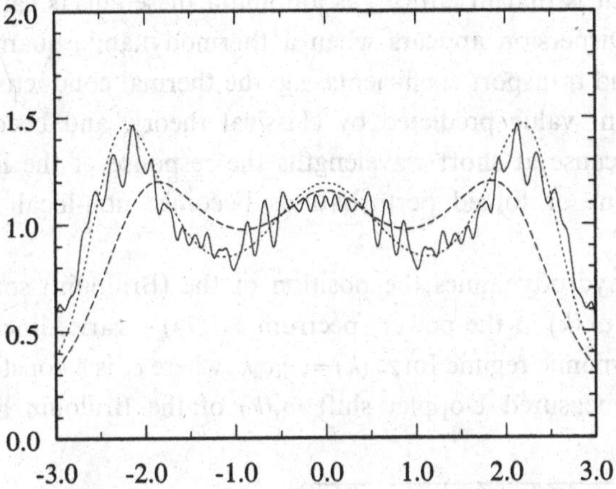

Figure 7.4 The LGA dynamic structure factor in the generalized hydrodynamic regime. Same conditions and symbols as for Figure 7.2; here $k = 20k_0 \simeq 0.245$ reciprocal lattice units.

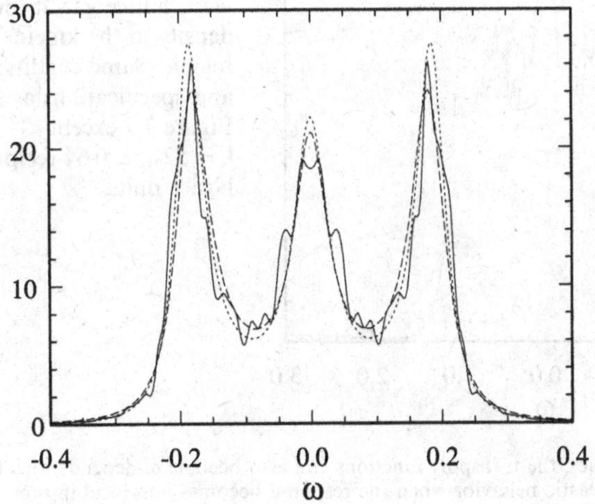

Figure 7.5 The dynamic structure factor of the GBL lattice gas at low density at the border of the classical and generalized hydrodynamic regimes. Thermodynamic state: $\rho = 1.1$; $e = 1.1$. $k = 12k_0 \simeq 0.15$ reciprocal lattice units. Other specifications as for Figure 7.2.

see Figure 7.6. In this almost collisionless regime, correlation effects through the periodic boundary conditions might be quite substantial. Nevertheless the Boltzmann theory yields predictions which follow more or less the general trends of the spectral features. In fact, from the eigenvalue spectrum and a numerical analysis of the eigenfunctions, one can see that $S(\mathbf{k}, \omega)$ is strongly coupled to six modes with $\mathrm{Re}\, z_\mu(\mathbf{k}) \sim 2$, four of which are propagating, which explains qualitatively why the 'Brillouin lines' in Figure 7.6 have a complex structure resulting from the superposition of several modes.

7.6.3 Dispersion effects

The results presented in the previous sections offer the possibility to examine the question, often discussed in statistical mechanics, of dispersion effects in simple fluids. The interesting point is that in lattice gas automata these effects can be analyzed quantitatively. Dispersion appears when a thermodynamic quantity, like the speed of sound, and transport coefficients, e.g. the thermal conductivity, deviate from their constant value predicted by classical theory, and become k-dependent functions, because at short wavelengths the response of the fluid to spontaneous fluctuations or forced perturbations becomes non-local (see Section 7.5.2).[9]

According to classical hydrodynamics the position of the (Brillouin) sound peaks – the Doppler shift $\omega_s(k)$ in the power spectrum $S(k, \omega)$ – varies linearly with k since in the hydrodynamic regime $\mathrm{Im}\, z_\pm(k) = \pm c_s k$, where c_s is a constant. In Figure 7.7, the LGA measured Doppler shift $\omega_s(k)$ of the Brillouin lines

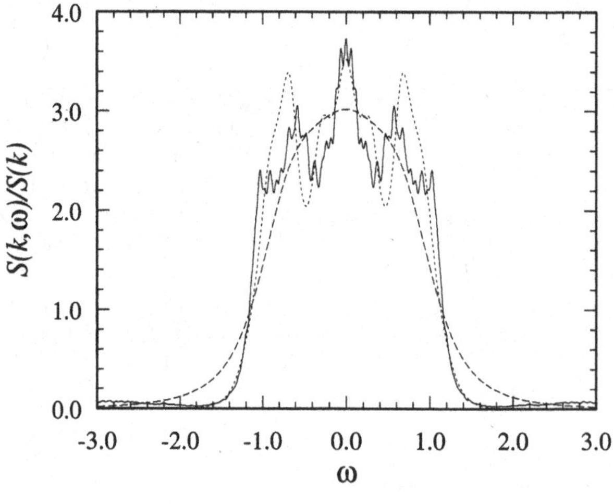

Figure 7.6 The dynamic structure factor of the GBL lattice gas at low density in the kinetic regime. Same conditions and specifications as for Figure 7.5 except $k = 52k_0 \simeq 0.64$ reciprocal lattice units.

[9] At very high frequencies, the transport functions can also become ω-dependent as the system exhibits viscoelastic behavior when the response becomes non-local in time.

(open circles) is compared with $\mathrm{Im}\, z_{\pm}(k) = \pm c_s(k)k$ (solid line) computed from the lattice Boltzmann equation; the classical behavior is shown by the dotted line, the thermodynamic speed of sound c_s being calculated from (7.39). The measured dispersion in the sound speed is seen to be remarkably well accounted for (up to $k \simeq 1.5$) by the Boltzmann theory for $c_s(k)$ confirming the validity of generalized hydrodynamics.

From the expression for the classical hydrodynamic spectrum (7.58), we can compute the ratio of the integrated intensity of the central Rayleigh peak to those of the Brillouin peaks, and obtain the Landau–Placzek ratio, which is a thermodynamic quantity equal to $(\gamma - 1)$. For the lattice gas, the measured value $\gamma_{\mathrm{exp}} \simeq 1.34$ is in very good agreement with the theoretical value 1.32, calculated from (7.51): $\gamma = c_s^2/c_T^2$. In the regime of generalized hydrodynamics, the Landau–Placzek ratio then becomes k-dependent, but it is very hard to measure quantitatively the dispersion effects on γ, and thereby make a comparison with theoretical predictions, because of the difficulty of evaluating the integrated intensities of the spectral peaks in a quantitative manner at high k-values (as it is in real fluids).

In the long wavelength limit, the value of the transport coefficients can be obtained from the line-widths of the spectral lines of $S(k, \omega)$ since the Landau–Placzek theory predicts three Lorentzians of width $D_T k^2$ (for the central Rayleigh peak) and Γk^2 (for the two shifted Brillouin peaks). In Figure 7.8, the simulation data for Rayleigh width (black dots) and for the Brillouin width (open circles) are shown as functions of k^2. The dashed line represents the classical values $D_T k^2$ and Γk^2 and corresponds in fact to two coinciding lines because here (by mere chance) $D_T \simeq \Gamma$ whose values are given by the Boltzmann transport coefficients in the limit $k \to 0$, derived in (7.46). The solid curves $D_T(k)k^2$ and $\Gamma(k)k^2$ computed from the Boltzmann eigenvalues (see Section 7.5) indicate that

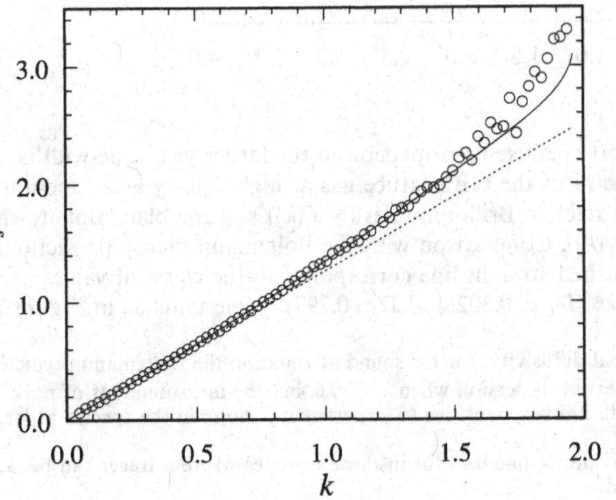

Figure 7.7 Sound dispersion in the lattice gas. Doppler shift $\omega_s(k)$ of the Brillouin lines measured from the dynamic structure factor of the GBL lattice gas at high density (open circles), classical hydrodynamic value $c_s \times k$ (dotted line), and Boltzmann theory prediction (solid curve). Same units as in Figure 7.5.

the dispersion observed in the generalized hydrodynamic domain ($k \lesssim 1.5$) is well accounted for by the lattice Boltzmann theory.[10]

In conclusion, the spectral analysis of the spontaneous fluctuations in the lattice gas automaton shows, theoretically as well as 'experimentally', that one finds in the LGA dynamic structure factor the essential features of the scattering function measured in real fluids.

7.7 Diffusion and correlations

In a fluid, spontaneous fluctuations also induce particle diffusion, but since particles diffuse in their own medium, diffusion cannot be observed unless one can, in one way or another, distinguish particles amongst themselves. This can be realized in a two-species fluid with say red and blue particles, which ideally do not differ from each other except by their color.[11] In technical terms, this means that there is an additional quantity which is necessary to characterize

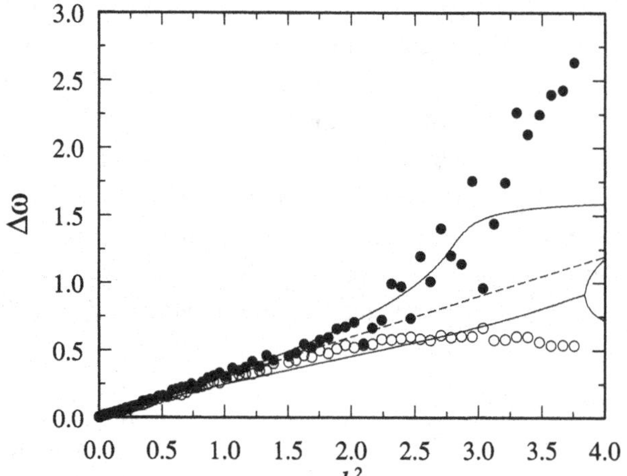

Figure 7.8 Transport coefficients dispersion in the lattice gas. Line-widths measured from spectra of the GBL lattice gas at high density as a function of k^2: the open circles refer to Brillouin lines ($\sim \Gamma(k)$) and the black dots to the Rayleigh line ($\sim D_T(k)$). Comparison with the Boltzmann theory predictions (solid lines). The dashed straight line corresponds to the classical value $|\mathrm{Re}\, z(k \to 0)| = 0.3k^2$ ($D_T \simeq 0.302$ and $\Gamma \simeq 0.297$). Same units as in Figure 7.5.

10 At least for the thermal diffusivity; for the sound attenuation the Boltzmann prediction deviates from the observed dispersion when $k \gtrsim 0.6$, but the measurements of peak position and line-width are very sensitive to 'experimental' noise in the spectra at large k-values.

11 In experimental observations, one uses for instance isotopes where a tracer can be conveniently identified.

the fluid, the *concentration* or *species density*, and which will couple to the other thermodynamic and mechanical variables.

The FHP model has been generalized to study diffusive phenomena in binary fluids using 'macroscopic' experiments (Bernardin and Sero-Guillaume, 1990; McNamara, 1990; Noullez, 1990). Typically the measurement would be performed in a lattice gas composed of red and blue particles (where the color is a passive property used to distinguish species which otherwise do not differ from one another) starting from an initial condition with a step function color density profile. The observer would then measure the evolution of the density profile, and evaluate the diffusion coefficient by fitting the 'experimental' profile, obtained after some time, to the solution of the diffusion equation subject to the appropriate boundary conditions (McNamara, 1990; Noullez, 1990).

Here we shall consider a more microscopic measure along the lines of the spectral analysis developed in this chapter. But the mixture of two real fluids exhibits a power spectrum where the central peak is not a simple Lorentzian, even in the long wavelength limit (Boon and Yip, 1980): it has a spectral structure where it is difficult to separate the contributions from entropy fluctuations and from concentration fluctuations which, in general, are not decoupled.[12] In this respect, the lattice gas offers an interesting approach to probe diffusion in a simple fluid mixture, because we can consider a non-thermal LGA and define an observable – the weighted difference of the species densities – whose fluctuation correlations yield the diffusive mode independently of the other modes, so that the corresponding power spectrum provides a measure of diffusion dynamics solely (Hanon and Boon, 1997).

7.7.1 The two-species lattice gas

The two-species lattice gas is the CFHP model described in Section 3.5: the particles are tagged as either 'red' or 'blue', and the proportion of red versus blue particles can be tuned to any preset value of concentration. Color is conserved by the dynamics, and, during the collision phase, is redistributed randomly amongst the channels, independently of the mass redistribution.

The technical analysis of the 'colored' lattice gas follows essentially the lines of the development presented in the previous sections for the single component lattice gas. The major distinction is the introduction of a 'colored' density, as we will be interested in the dynamics of the mass density fluctuations of one of the species to probe diffusive transport in the gas mixture. So we consider (i) the red mass density $\rho^{\text{red}}(\mathbf{r}_\star, t_\star)$, which is the number of red particles at node $\mathbf{r}_\star$ at time $t_\star$, and (ii) the corresponding fluctuations $\delta\rho^{\text{red}}(\mathbf{r}_\star, t_\star)$, defined in terms of the red channel occupations n_i ($i \in \{1, \ldots, \frac{1}{2}b\}$, with b the total number of

[12] Unless one of the two components is in trace amounts, in which case the two modes can be identified as they produce two independent central Lorentzians.

channels per node; $b = 14$ in the CFHP model):

$$\delta\rho^{\text{red}}(\mathbf{r}_\star, t_\star) = \sum_{i=1}^{b/2} \delta n_i(\mathbf{r}_\star, t_\star) = \sum_{i=1}^{b/2} \left[n_i(\mathbf{r}_\star, t_\star) - \langle n_i(\mathbf{r}_\star, t_\star) \rangle \right], \tag{7.60}$$

where $\langle n_i(\mathbf{r}_\star, t_\star)\rangle$ is given by:

$$\langle n_i \rangle \equiv f_i = \begin{cases} d\,\chi_{\text{r}} & \text{for } i = 1, \ldots, \frac{1}{2}b \quad \text{(red channels)}, \\ d\,(1-\chi_{\text{r}}) & \text{for } i = \frac{1}{2}b + 1, \ldots, b \text{ (blue channels)}, \end{cases} \tag{7.61}$$

with d the average density per channel, and χ_{r} the concentration of red particles. Note that:

$$\sum_{i=1}^{b} \langle n_i(\mathbf{r}_\star, t_\star)\rangle = \frac{1}{2} b d \chi_{\text{r}} + \frac{1}{2} b d (1 - \chi_{\text{r}}) = \rho^{\text{red}} + \rho^{\text{blue}} = \rho \tag{7.62}$$

defines the respective average densities per node.

The 'red mass' dynamic structure factor $S^{\text{red}}(\mathbf{k}, \omega)$, defined as the space- and time-Fourier transform of the van Hove correlation function:

$$G^{\text{red}}(\mathbf{r}_\star, t_\star) = \langle \delta\rho^{\text{red}}(\mathbf{r}_\star, t_\star)\, \delta\rho^{\text{red}}(0,0) \rangle \tag{7.63}$$

is given by:

$$\rho^{\text{red}}\, S^{\text{red}}(\mathbf{k}, \omega) = \sum_{\mathbf{r}_\star \in \mathscr{L}} \sum_{t=-\infty}^{\infty} e^{-\imath\omega t - \imath \mathbf{k}\cdot\mathbf{r}} G^{\text{red}}(\mathbf{r}_\star, |t_\star|) \tag{7.64}$$

$$= \frac{1}{V} \sum_{t=-\infty}^{\infty} e^{-\imath\omega t} \langle \delta\rho^{\text{red}}(\mathbf{k}, |t_\star|)\, \delta\rho^{\text{red}\,*}(\mathbf{k}, 0) \rangle, \tag{7.65}$$

where $\delta\rho^{\text{red}}(\mathbf{k}, t_\star)$ is the spatial Fourier transform of $\delta\rho^{\text{red}}(\mathbf{r}_\star, t_\star)$, and $V \equiv \mathscr{N}$ is the total number of nodes.[13] As for the single component lattice gas, the dynamic structure factor is expressed in terms of the kinetic propagator defined by:

$$\Gamma_{ij}(\mathbf{k}, t)\, \kappa_j = \langle \delta n_i(\mathbf{k}, t_\star)\, \delta n_j^*(\mathbf{k}, 0) \rangle, \tag{7.66}$$

where $\delta n_i(\mathbf{k}, t_\star)$ is the spatial Fourier transform of $\delta n_i(\mathbf{r}_\star, t_\star)$ and $\kappa_j = f_j^{(\text{eq})}(1 - f_j^{(\text{eq})})$. Using (7.66), we rewrite (7.64) as:

$$\rho^{\text{red}}\, S^{\text{red}}(\mathbf{k}, \omega) = \sum_{t=-\infty}^{\infty} e^{-\imath\omega t} \sum_{i=1}^{b/2} \sum_{j=1}^{b/2} \Gamma_{ij}(\mathbf{k}, t_\star)\, \kappa_j, \tag{7.67}$$

and the static structure factor (the Fourier transform of the equal-time van Hove

[13] Remember that V is also interpreted as the volume of the lattice $\mathscr{L}$ universe, and here the lattice is finite and has periodic boundary conditions.

function; see Section 4.7) as:

$$\rho^{\rm red}\, S^{\rm red}(\mathbf{k}) \;=\; \sum_{i=1}^{b/2}\sum_{j=1}^{b/2} \Gamma_{ij}(\mathbf{k},0)\,\kappa_j \tag{7.68}$$

$$\;=\; \sum_{i=1}^{b/2}\sum_{j=1}^{b/2} \delta_{ij}\,\kappa_j = \sum_{j=1}^{b/2} \kappa_j\,, \tag{7.69}$$

or

$$S^{\rm red}(\mathbf{k}) = 1 - d\,\chi_{\rm r}\,. \tag{7.70}$$

The kinetic propagator is then evaluated in the Boltzmann approximation following the same lines as in Section 7.2, and we obtain:

$$\rho^{\rm red}\, S^{\rm red}(\mathbf{k},\omega) \;\equiv\; 2\,{\rm Re}\, F^{\rm red}(\mathbf{k},\omega), \tag{7.71}$$

$$F^{\rm red}(\mathbf{k},\omega) \;=\; \sum_{i=1}^{b/2}\sum_{j=1}^{b/2} \left[\frac{1}{e^{\iota\omega+\iota\mathbf{k}\cdot\mathbf{c}} - \delta + \mathbf{\Omega}} + \frac{1}{2}\right]_{ij} \kappa_j\,. \tag{7.72}$$

This expression for the dynamic structure factor is similar to the result given in (7.17) and (7.18), and is exact within the Boltzmann approximation, but again the explicit analytical inversion of the $b \times b$ matrix in (7.72) cannot be performed in all generality. However, perturbation methods can be used to compute analytically $S^{\rm red}(\mathbf{k},\omega)$ in the hydrodynamic limit: $|\mathbf{k}| \longrightarrow 0$ and $\omega \sim \mathcal{O}(|\mathbf{k}|)$, $\mathcal{O}(|\mathbf{k}|^2)$; beyond the long-wavelength, long-time domain, one has recourse to numerical evaluation of (7.72) to compute the Boltzmann power spectrum. Since the developments proceed essentially as for the single component lattice gas, we shall not present the detailed analysis[14] and will focus on the main results.

7.7.2 The hydrodynamic limit

In the large wavelength limit, one can identify four hydrodynamic modes in the two-species (non-thermal) lattice gas: the shear mode and the acoustic modes which are independent of color related properties, and one mode which describes color diffusion only. To first order in k, the non-zero imaginary part of the eigenvalues yields the frequency shift of the spectral peaks corresponding to the propagating (acoustic) modes, and to second order in k, the real parts of the eigenvalues yield the dissipation coefficients which determine the line-widths of the spectral components. These properties are reflected in the dynamic structure

[14] The major technical difference is that, instead of the *thermal* scalar product (7.27), here one has the *colored* scalar product $\langle A|B\rangle = \sum_{i=1}^{b} A(\mathbf{c}_i)\,\kappa_i\,B(\mathbf{c}_i)$, where the weight κ_i depends on density and concentration. We refer the reader to the original literature (Hanon and Boon, 1997) for explicit computations.

factor of the 'colored' lattice gas which reads (Hanon and Boon, 1997):[15]

$$\rho^{\text{red}}\, S^{\text{red}}(\mathbf{k},\omega) = \frac{b}{2}\frac{\kappa_r\kappa_b}{\kappa_r+\kappa_b}\frac{2Dk^2}{\omega^2+(Dk^2)^2}$$
$$+\frac{b}{2}\frac{\kappa_r^2}{\kappa_r+\kappa_b}\sum_{\pm}\left(\frac{\Gamma k^2}{(\omega\pm c_s k)^2+(\Gamma k^2)^2}+\frac{\Gamma k}{c_s}\frac{c_s k\pm\omega}{(\omega\pm c_s k)^2+(\Gamma k^2)^2}\right). \tag{7.73}$$

Here κ_r (red) and κ_b (blue) are given by:

$$\kappa_r = d\,\chi_r(1-d\,\chi_r), \qquad \kappa_b = d\,(1-\chi_r)[1-d\,(1-\chi_r)]\,, \tag{7.74}$$

c_s is the speed of sound (here $c_s = \sqrt{3/7}$), Γ is the sound damping coefficient (see (7.52), but without the second term since the colored LGA is non-thermal), and D is the diffusion coefficient. We recognize in (7.73), the typical structure of the hydrodynamic structure factor (compare with (7.58)): at fixed value of k, the spectrum consists of a Brillouin-doublet centered around $\pm kc_s$, and of a central peak characterizing color diffusion. It is straightforward to verify that for $\chi_r = 1$, (7.73) reduces to the usual Landau–Placzek expression of the hydrodynamic spectrum for the non-thermal single-component fluid.

Now if we consider the fluctuations of the observable defined by:

$$\rho_{\text{diff}} = \kappa_b\rho^{\text{red}} - \kappa_r\rho^{\text{blue}}\,, \tag{7.75}$$

its corresponding spectral density can be computed straightforwardly along the lines of the evaluation of the power spectrum $S^{\text{red}}(\mathbf{k},\omega)$, and we obtain the interesting result that the spectrum is then a single Lorentzian characterizing color diffusion alone:

$$\rho_{\text{diff}}\, S_{\text{diff}}(\mathbf{k},\omega) = \frac{b}{2}\frac{\kappa_r\kappa_b}{(\kappa_r+\kappa_b)}\frac{2Dk^2}{\omega^2+(Dk^2)^2}\,. \tag{7.76}$$

Notice that purely diffusive behavior of color is related to an observable which is neither the concentration of one of the components nor the straight difference between the concentrations of the two species.

7.7.3 The power spectrum

The results obtained in the hydrodynamic limit are in accordance with the Landau–Placzek theory for continuous fluids (Boon and Yip, 1980). Now the hydrodynamic theory breaks down at short wavelengths, but the Boltzmann theory should remain valid down to k values where the purely kinetic domain starts. In order to characterize the wavelength domain of the various regimes, we consider the quantities $k\ell_f$, where $\ell_f(\sim 1/\rho)$ is the mean free path, and

[15] As for the single component lattice gas, the shear mode (or transverse momentum mode) decouples from the density fluctuations modes (longitudinal modes), and so does not show up in the dynamic structure factor.

$d = \rho/\rho_{max}$, the reduced density (which is also the average density per channel). Accordingly the hydrodynamic regime is defined by $k\ell_f \ll 1$, the generalized hydrodynamic regime (Boltzmann regime) by $k\ell_f < 1$, and the kinetic regime by $k\ell_f \gtrsim 1$.

We shall discuss the power spectra obtained from automaton simulation data and will compare the results to the analytical Landau–Placzek expressions and to the predictions of the lattice Boltzmann theory. For the latter, one uses the eigenvalue spectrum of the propagator Γ which can be evaluated numerically over the complete k-domain, so extending the computation of the power spectrum to the region of k values where analytical evaluation can no longer be performed. Figure 7.9 shows a typical eigenvalue spectrum as computed numerically, where the 14 modes of the CFHP model can be distinguished.

We observe that in the range $k < 0.4$, corresponding to wavelengths $\lambda > 15\ell_0$ (with ℓ_0 the lattice unit length), the four slow modes are well separated from

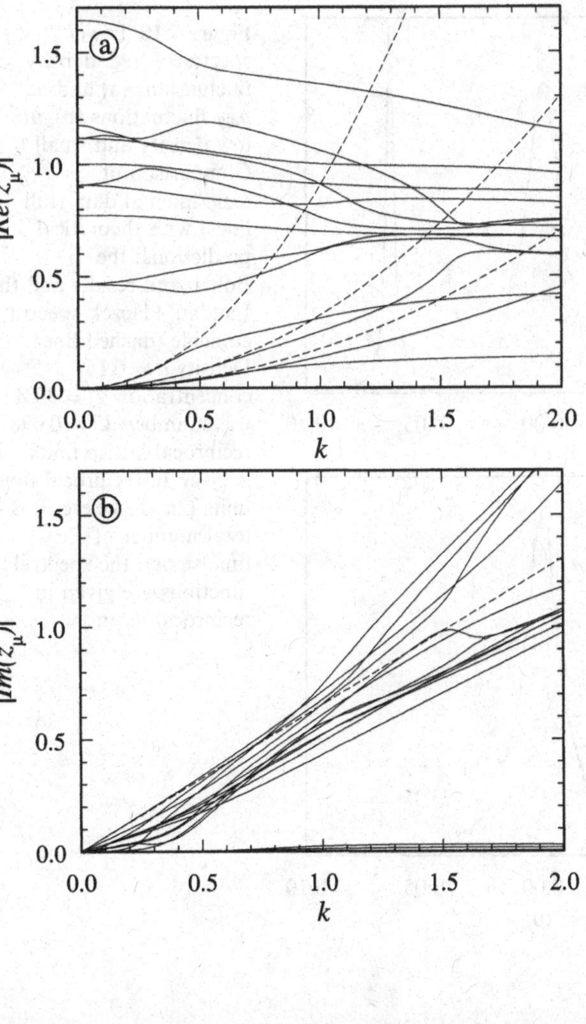

Figure 7.9 Eigenvalue spectrum of the CFHP model propagator: real parts (a) and imaginary parts (b). Full lines are the result of Boltzmann computation; the hydrodynamic limit is shown by dotted lines. The reduced density and concentration are $d = 0.15$ and $\chi_r = 30\%$ respectively. The wavenumber $k = |\mathbf{k}|$ is given in reciprocal lattice units $\times\, 2\pi$.

the kinetic modes, and their behavior is correctly given by the hydrodynamic expressions. We therefore expect that in this domain, the Landau–Placzek theory should provide a correct description of the dynamic structure factor. This is indeed confirmed by the comparison between the results obtained from simulation data and theoretical predictions as shown in Figure 7.10, where we also notice that the hydrodynamic spectrum is indistinguishable from the Boltzmann spectrum.

From the hydrodynamic spectrum one can directly measure the diffusion coefficient $D(d, \chi_r)$; the expression (7.76) of $S_{\rm diff}(\mathbf{k}, \omega)$ is a single Lorentzian with a half-width $\Delta\omega = Dk^2$, so that plotting $\Delta\omega$ as a function of k^2, as in Figure 7.11(a), a linear least-squares fit yields a slope whose value gives a direct measure of the diffusion coefficient. This measure is in agreement with the predictions of the lattice Boltzmann theory up to reduced densities $d \sim 0.25$ as Figure 7.11(b) illustrates. Deviations for larger values of the density are not

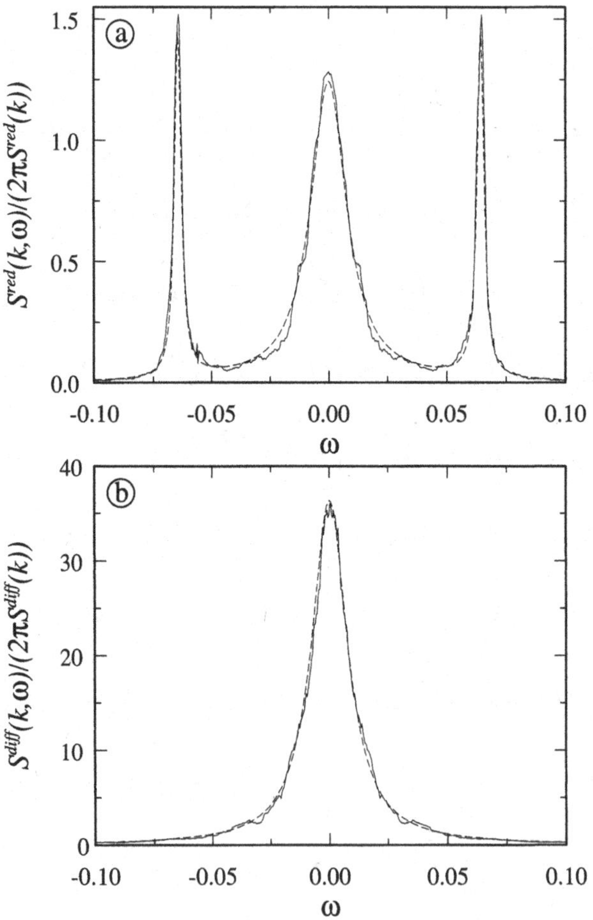

Figure 7.10 Power spectra of 'red' density fluctuations (a) and of $\rho_{\rm diff}$ fluctuations (b), at low density and small k. Comparison of experimental data (full lines) with theoretical predictions: the Boltzmann results and the Landau–Placzek spectra coincide (dashed lines). Density $d = 0.15$; concentration $\chi_r = 30\%$; wavenumber $|\mathbf{k}| = 0.098$ reciprocal lattice units; ω is given in reciprocal time units ($2\pi/T$, where T is total number of time-steps); the spectral functions are given in reciprocal ω units.

surprising since the Boltzmann theory does not take into account correlations which are known to become important at moderate and high densities.

As k increases from 0.4 to 1.4, there is still a distinct scale separation between slow and fast modes (see Figure 7.9(a)), but the eigenvalues of the slow modes depart significantly from the hydrodynamic prediction, indicating the breakdown of the local response hypothesis: the transport coefficients become k-dependent. As a result, the hydrodynamic theory no longer describes the power spectrum correctly,[16] but the complete Boltzmann spectrum is in rather good agreement with the simulation results, as Figure 7.12(a) shows. At higher densities there is a discrepancy between the Boltzmann spectral density and the experimental power spectrum (see Figure 7.12(b)) which reflects the failure of the Boltzmann

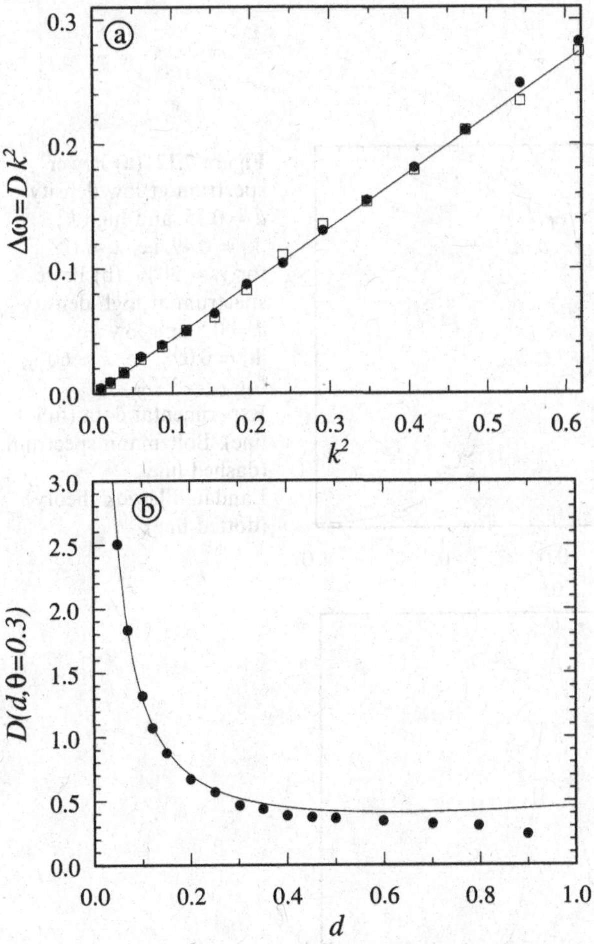

Figure 7.11 Microscopic measure of the diffusion coefficient: (a) Spectral line-width as a function of k^2: $\Delta\omega = D(d, \chi_r)k^2$, for $d = 0.3$ and $\chi_r = 0.3$ (squares) and $\chi_r = 0.5$ (black dots); least-squares fit (full line). (b) Measured values of $D(d, \chi_r = 0.3)$ (black dots) and theoretical prediction from Boltzmann theory (full line).

16 Note that even at rather short wavelength ($\lambda \sim 10\ell_0$) the experimental data can still be reasonably fit with a Landau–Placzek spectral function if the transport coefficients are parameterized, indicating that a hydrodynamic type description holds qualitatively down to quite short wavelengths, provided the transport coefficients are renormalized.

theory to evaluate correctly the transport coefficients in the high density region where correlations are important.

With this discussion we close our investigation of the hydrodynamic fluctuations in the lattice gas (with one and two species). We shall return briefly to the two-component case in Chapter 10 when we discuss two-fluid simulations. But we now turn to the macroscopic dynamics of the LGA; after showing in this chapter that the lattice gas exhibits spontaneous fluctuations which are compatible with those observed in real fluids, we go back to the important question: are the equations describing the macroscopic behavior of the lattice gas compatible with the Navier–Stokes equations of classical fluid dynamics? This question was addressed in Chapter 5 and will now be revisited with a different complementary approach.

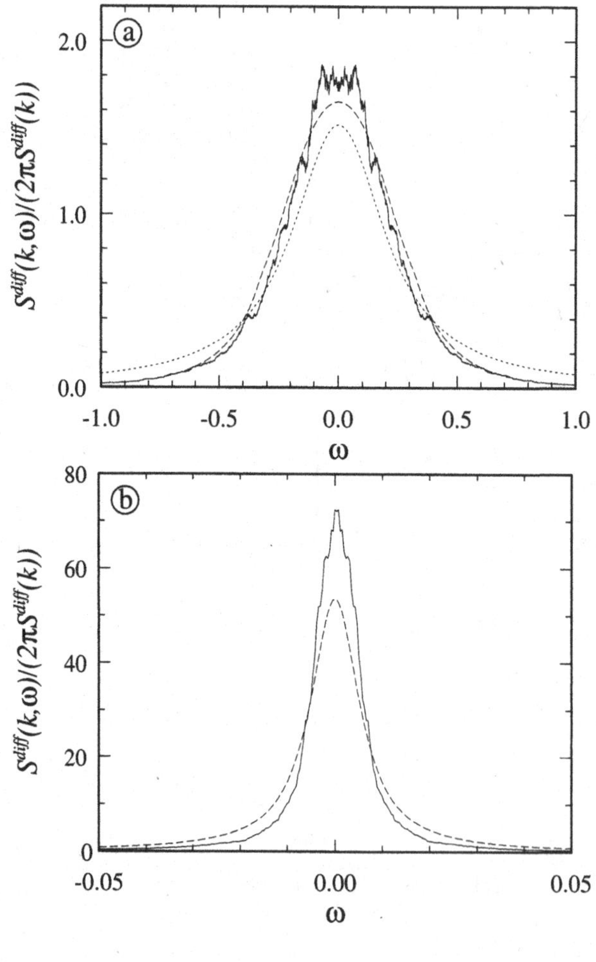

Figure 7.12 (a) Power spectrum at low density $d = 0.15$, and high k: $|\mathbf{k}| = 0.49$, i.e. $\lambda \approx 12\ell_0$, for $\chi_r = 30\%$. (b) Power spectrum at high density $d = 0.5$ and low k: $|\mathbf{k}| = 0.098$, i.e. $\lambda \approx 60\ell_0$, for $\chi_r = 30\%$. Experimental data (full line), Boltzmann spectrum (dashed line), Landau–Placzek theory (dotted line).

Chapter 8

Macrodynamics: projectors approach

It is one of the main objectives of statistical mechanics to provide a microscopic content to the phenomenologically established macroscopic properties and behavior of systems with many degrees of freedom. Although it is not necessary to have a complete knowledge of the details of the microscopic interactions to describe macroscopic phenomena in fluid systems, these phenomena emerge as a consequence of the basic dynamical processes. However, to establish rigorously the connection between the phenomenology and the underlying microscopic processes amounts to solving the many-body problem. Even for systems with oversimplified microscopic dynamics, such as lattice gas automata, this is an impossible task: approximations are unavoidable.

In this chapter we derive the equations governing the macroscopic dynamics of LGAs satisfying the semi-detailed balance condition; we shall start from the microscopic dynamics of the automaton, and use the lattice Boltzmann approximation (Suárez and Boon, 1997a,b).The main objective is to obtain the non-linear hydrodynamic equations, where the Euler and dissipative contributions are expressed in terms of the microscopic evolution rules of the automaton, and whose validity is not restricted to regions close to equilibrium,[1] so that they

[1] Most of the derivations of the LGA hydrodynamic equations given in the literature (Frisch *et al.*, 1986, 1987; Ernst, 1990a; Rothman and Zaleski, 1994) include non-linearities in the Euler part of the equations, but the dissipative term is calculated by linearizing around global equilibrium, thus limiting the validity of the results to a regime in which linear response and simple fluctuation-dissipation relations hold. In the development given by Wolfram (1986), the first few non-linear corrections are included, but this derivation remains at a phenomenological level, as the non-linear coefficients are assumed to be correctly obtained by direct Taylor expansion.

can be used to analyze phenomena taking place in systems arbitrarily far from equilibrium, for instance in thermal LGAs under large temperature gradient.

In order to derive the hydrodynamic equations, we make use of the Boltzmann hypothesis (see Section 4.4.2) that particles entering a collision are uncorrelated. The average outcome of the collision is assumed to be a non-equilibrium distribution which is close to the local equilibrium one. A local equilibrium distribution has the same analytical form as an equilibrium one, but with space- and time-dependent thermodynamic potentials (temperature, chemical potential, etc.). The non-equilibrium configuration can then be split into a local equilibrium component, which depends only on the local values of the set of densities of conserved quantities of the system, and a non-local equilibrium contribution, a 'projection' onto the space orthogonal to the subspace spanned by the constants of motion. This procedure is intimately related to the projection operator methods developed to derive the hydrodynamic equations from the microscopic dynamics in continuous fluids (Zwanzig, 1960; Mori, 1960; Procaccia, Ronis and Oppenheim, 1979).[2] We shall effectively use a simple projector that neglects the coupling of the linear hydrodynamic variables (mass, momentum, energy densities) to bilinear and higher order modes. With the treatment used here – and perhaps at the cost of a somewhat formal analysis – we shall obtain a general formulation of the lattice gas hydrodynamics valid for non-equilibrium states arbitrarily far from global homogeneous equilibrium. We shall also make contact with the linearized equations and show that the dissipative coefficients are given in terms of time-correlation functions in the form of the Green–Kubo expression of the transport coefficients. The present analysis is within the logical continuation of Chapters 6 and 7; in Chapter 5 we give an alternative, less formal, derivation of the lattice gas hydrodynamic equations using the Chapman–Enskog method. Notice that, whatever the method – be it for discrete systems or continuous fluids – there is always a common, unavoidable major step in the analysis: the multiple scales hypothesis, by which one anticipates that there is a natural space- and time-separation between hydrodynamic phenomena and microscopic processes.

8.1 Preliminaries

Before proceeding to the derivation of the hydrodynamic equations, it is useful to give some definitions and summarize some material from earlier chapters.[3]

[2] For lattice gas automata, projection operator techniques have been employed to derive the *linearized* hydrodynamic equations (see Chapter 6) which describe the time evolution of small perturbations around a global equilibrium state (Zanetti, 1989; Ernst and Dufty, 1990; Ernst, 1990).

[3] When, for the sake of clarity in the course of the development, results are given without derivation, the reader is explicitly referred to the appendix for algebraic details.

The observables in the automaton are defined as averages over the *stochastic dynamics* and over an ensemble of initial conditions. The averaging is denoted by angular brackets, $\langle \ldots \rangle_{\text{NE}}$, where the subindex stresses the fact that the ensemble is a non-equilibrium one. The collection of local quantities preserved under collision are labeled by the Greek index φ running from $\varphi = 1$ to $\varphi = N_{\text{hydro}}$, where N_{hydro} denotes the number of hydrodynamic variables. This set should contain the quantities conserved in the dynamics of the physical system modeled by the automaton. The index $\varphi = 1$ corresponds to the number of particles, which is conserved by the collisions.

Before the collision step, the configuration of a node $\mathbf{r}_\star \in \mathcal{L}$ is given by:[4]

$$\{n(\mathbf{r},t)\} \equiv \{n_i(\mathbf{r},t)\}_{i=1}^{b}, \tag{8.1}$$

and after collision by:

$$\{n'(\mathbf{r},t)\} \equiv \{n_i'(\mathbf{r},t)\}_{i=1}^{b}. \tag{8.2}$$

The average of these quantities over the stochastic process and the non-equilibrium ensemble is given by:

$$f_i(\mathbf{r},t) = \langle n_i(\mathbf{r},t)\rangle_{\text{NE}}, \tag{8.3}$$

and:

$$f_i'(\mathbf{r},t) = \langle n_i'(\mathbf{r},t)\rangle_{\text{NE}}, \tag{8.4}$$

respectively. The hydrodynamic variables, that is the densities of conserved quantities, are collected in a vector:

$$\underline{A}(\mathbf{r},t) \equiv \{A_\varphi(\mathbf{r},t)\}_{\varphi=1}^{N_{\text{hydro}}} = \sum_{i=1}^{b} n_i'(\mathbf{r},t)\,\underline{A}_i = \sum_{i=1}^{b} n_i(\mathbf{r},t)\,\underline{A}_i. \tag{8.5}$$

The last equality follows from the fact that the value of the hydrodynamic variables is preserved by the collision step. The hydrodynamic variables include some or all of the following: particle number, momentum, and, in thermal automata, kinetic energy per node. For example, in non-thermal automata whose collision rules preserve the number of particles and the momentum per node, the hydrodynamic variables are the number density, $\rho(\mathbf{r},t)$, and the momentum density, $\rho\mathbf{u}(\mathbf{r},t)$, which can be collected in a two-dimensional vector:

$$\underline{A}(\mathbf{r},t) = \begin{pmatrix} \rho(\mathbf{r},t) \\ \rho\mathbf{u}(\mathbf{r},t) \end{pmatrix} = \sum_{i=1}^{b} n_i(\mathbf{r},t)\,\underline{A}_i = \sum_{i=1}^{b} n_i(\mathbf{r},t) \begin{pmatrix} 1 \\ \mathbf{c}_i \end{pmatrix}, \tag{8.6}$$

[4] For simplicity, we shall omit the subscript $\star$ in the notation of the discrete variables $\mathbf{r}_\star$ and $t_\star$.

with the vector of collisional invariants:[5]

$$\underline{A}_i = \begin{pmatrix} 1 \\ \mathbf{c}_i \end{pmatrix}. \tag{8.7}$$

The average of the hydrodynamic variables is:

$$\langle \underline{A}(\mathbf{r}, t) \rangle_{\mathrm{NE}} = \sum_{i=1}^{b} f_i'(\mathbf{r}, t)\, \underline{A}_i = \sum_{i=1}^{b} f_i(\mathbf{r}, t)\, \underline{A}_i. \tag{8.8}$$

The microscopic prescription of the collision step reads:

$$n_i'(\mathbf{r}, t) = \sum_{\{s'\}} s_i' \xi_{\{s\}\{s'\}}, \tag{8.9}$$

where $\xi_{\{s\}\{s'\}}$ is equal to 1 when, starting from a configuration $\{s\}$, the collision produces configuration $\{s'\}$, and is 0 otherwise. For stochastic automata, ξ is a random Boolean variable. In this case the average post-collisional configuration, starting from a given pre-collisional configuration $\{n(\mathbf{r}, t)\}$ is:

$$\langle n_i'(\mathbf{r}, t) \rangle_{\{n(\mathbf{r},t)\}} = \sum_{\{s\},\{s'\}} s_i'\, A(s \to s')\, \delta\left(\{n(\mathbf{r}, t)\}, \{s\}\right), \tag{8.10}$$

where $A(s \to s')$ is the matrix whose entries are the probabilities of having a configuration $\{s'\}$ as the outcome of a collision starting from a configuration $\{s\}$. The quantity $\delta\left(\{n\}, \{s\}\right)$ is equal to one if the configurations $\{s\}$ and $\{n\}$ are equal, and zero if they are different. A representation of this Kronecker delta in configuration space is (see Section 2.2.4):

$$\delta\left(\{n\}, \{s\}\right) = \prod_{k=1}^{b} n_k^{s_k} (1 - n_k)^{(1-s_k)}. \tag{8.11}$$

With the semi-detailed balance condition (see Section 2.3.1):

$$\sum_{\{s\}} A(s \to s') = 1, \tag{8.12}$$

an arbitrary distribution of the Fermi–Dirac form for the single particle distribution (see Section 4.4):

$$f_i = \frac{1}{1 + \exp\left\{\underline{\phi} * \underline{A}_i\right\}}, \tag{8.13}$$

without correlations between fluctuations in different channels, is invariant under the collision step. The quantities $\underline{\phi}$ are the thermodynamic fields conjugate to

[5] We ignore effects associated with the presence of spurious global invariants (see Brito, Ernst and Kirkpatrick, 1991).

the conserved quantities. The scalar product denoted by $*$ is a contraction of the indices labeling the conserved quantities:

$$\underline{\phi} * \underline{A}_i \equiv \sum_{\varphi=1}^{N_{\text{hydro}}} \phi_\varphi A_{\varphi i}. \tag{8.14}$$

Furthermore, the H-theorem guarantees the existence of a stable global equilibrium state characterized by a probability distribution which is a product of Fermi–Dirac distributions for every channel on every node of the lattice (see Chapter 4).

We define a local equilibrium ensemble, for which the average occupation of channel i is:

$$f_i^{\text{LE}}(\mathbf{r},t) = \frac{1}{1+\exp\{\underline{\phi}(\mathbf{r},t) * \underline{A}_i\}}, \tag{8.15}$$

such that:

$$\begin{aligned}
\langle \underline{A}(\mathbf{r},t)\rangle_{\text{LE}} &\equiv \sum_{i=1}^{b} f_i^{\text{LE}}(\mathbf{r},t)\,\underline{A}_i \\
&= \sum_{i=1}^{b} \frac{1}{1+\exp\{\underline{\phi}(\mathbf{r},t) * \underline{A}_i\}}\underline{A}_i \\
&= \langle \underline{A}(\mathbf{r},t)\rangle_{\text{NE}}.
\end{aligned} \tag{8.16}$$

Equation (8.16) defines the thermodynamic potentials implicitly in terms of the local values of the densities of conserved quantities: $\underline{\phi}(\mathbf{r},t) = \underline{\phi}\left(\langle \underline{A}(\mathbf{r},t)\rangle\right)$. For conserved quantities, since the local equilibrium ensemble has been defined in such a way that $\langle \underline{A}(\mathbf{r},t)\rangle_{\text{LE}} = \langle \underline{A}(\mathbf{r},t)\rangle_{\text{NE}}$, we may omit the label indicating the ensemble over which the average is taken.

The lattice Boltzmann equation (see Section 4.2.2):

$$\begin{aligned}
f_i(\mathbf{r}+\mathbf{c}_i, t+1) &= f_i'(\mathbf{r},t) \\
&= \sum_{\{s\}\{s'\}} s_i' \,\langle \xi \rangle_{\{s\}\to\{\sigma\}} \prod_{k=1}^{b} (f_k(\mathbf{r},t))^{s_k}\,(1-f_k(\mathbf{r},t))^{(1-s_k)},
\end{aligned} \tag{8.17}$$

constitutes the starting point for the derivation of the macroscopic equations that describe the evolution of the hydrodynamic fields. Multiplying Equation (8.17) by $\underline{A}_i$, summing over the index i and using the displacement operator defined by:

$$f_i(\mathbf{r}+\mathbf{c}_i, t) = \exp\{\mathbf{c}_i \cdot \nabla_{\mathbf{r}}\}\; f_i(\mathbf{r},t), \tag{8.18}$$

where the scalar product denoted by $\cdot$ is a contraction of the spatial indices, we obtain:

$$\langle \underline{A}(\mathbf{r},t+1)\rangle = \sum_{i=1}^{b} \exp\{-\mathbf{c}_i \cdot \nabla_{\mathbf{r}}\}\; f_i'(\mathbf{r},t)\,\underline{A}_i. \tag{8.19}$$

Making use of the equality:

$$\langle \underline{A}(\mathbf{r},t)\rangle = \sum_{i=1}^{b} f_i'(\mathbf{r},t)\, \underline{A}_i, \tag{8.20}$$

we can rewrite (8.19) as:

$$\langle \underline{A}(\mathbf{r},t+1)\rangle - \langle \underline{A}(\mathbf{r},t)\rangle = \sum_{i=1}^{b} \left(\exp\{-\mathbf{c}_i \cdot \nabla_{\mathbf{r}}\} - 1\right) f_i'(\mathbf{r},t)\, \underline{A}_i. \tag{8.21}$$

Equation (8.21) has a suitable form for the subsequent development.

8.2 Multiple scales analysis

The microscopic evolution equations (8.21) are finite difference equations containing the full dynamics of the automaton able to describe its behavior at all scales. The microscopic scales are given by the automaton lattice spacing and, for time, by the lapse between two consecutive propagation steps. There is another characteristic microscopic time τ_{LE}, of the order of a few collisions, which corresponds to the relaxation to local equilibrium. In order to derive hydrodynamic-type equations from the evolution equations (8.21), we introduce the parameter ϵ, which represents the ratio between characteristic microscopic and hydrodynamic length scales. This ratio is assumed to be small, a fact which allows us to carry out a multiple scale analysis (see also Chapter 5). Thus, we anticipate the scales of physical interest in the solution of (8.21), and introduce the following variables: a space variable $\mathbf{r}_1 = \epsilon\, \mathbf{r}$, and two time variables, $t_1 = \epsilon\ t$, which is of order ϵ^0 in the regime where Euler equations are valid, and $t_2 = \epsilon^2 t$, which is of order ϵ^0 in the dissipative regime. We further assume that deviations from local equilibrium at each node are small. This corresponds to the Chapman–Enskog picture of a non-equilibrium state (Chapman and Cowling, 1970), where the deviations from local equilibrium are on the order of the gradients of the conserved quantities, i.e. of order ϵ. The substitutions (introduced in Chapter 5):

$$\begin{aligned} \nabla_{\mathbf{r}} &\to \epsilon\, \nabla_{\mathbf{r}_1} \\ \partial_t &\to \epsilon \partial_{t_1} + \epsilon^2 \partial_{t_2}, \end{aligned} \tag{8.22}$$

in Equation (8.21) lead to the hierarchy of equations labeled by the different powers of ϵ:

$$\epsilon^1 : \quad \partial_{t_1} \langle \underline{A}(\mathbf{r},t)\rangle + \nabla_{\mathbf{r}_1} \cdot \langle \underline{\mathbf{J}}(\mathbf{r},t)\rangle_{\text{LE}} = 0, \tag{8.23}$$

$$\begin{aligned} \epsilon^2 : \quad \left(\partial_{t_2} + \frac{1}{2}\partial_{t_1}^2\right) \langle \underline{A}(\mathbf{r},t)\rangle &= \frac{1}{2} \nabla_{\mathbf{r}_1} \nabla_{\mathbf{r}_1} : \left(\langle \mathbf{c}\ \underline{\mathbf{J}}(\mathbf{r},t)\rangle_{\text{LE}}\right) \\ &+ \frac{1}{\epsilon^2} \sum_{i=1}^{b} \left(\exp\{-\mathbf{c}_i \cdot \nabla_{\mathbf{r}}\} - 1\right) \underline{A}_i\ \delta f_i'(\mathbf{r},t), \end{aligned} \tag{8.24}$$

with:

$$\delta f_i'(\mathbf{r},t) = f_i'(\mathbf{r},t) - f_i^{\mathrm{LE}}(\mathbf{r},t), \tag{8.25}$$

and the definition of the currents associated to the conserved fields:

$$\underline{\mathbf{J}}_i = \mathbf{c}_i\, \underline{A}_i, \tag{8.26}$$

$$\langle \underline{\mathbf{J}}(\mathbf{r},t)\rangle_{\mathrm{LE}} = \sum_{i=1}^{b} \underline{\mathbf{J}}_i\, f_i^{\mathrm{LE}}(\mathbf{r},t) = \sum_{i=1}^{b} \mathbf{c}_i\, \underline{A}_i\, f_i^{\mathrm{LE}}(\mathbf{r},t), \tag{8.27}$$

and:

$$\langle \mathbf{c}\,\mathbf{J}(\mathbf{r},t)\rangle_{\mathrm{LE}} = \sum_{i=1}^{b} \mathbf{c}_i\, \underline{\mathbf{J}}_i\, f_i^{\mathrm{LE}}(\mathbf{r},t) = \sum_{i=1}^{b} \mathbf{c}_i\mathbf{c}_i\, \underline{A}_i\, f_i^{\mathrm{LE}}(\mathbf{r},t). \tag{8.28}$$

Equations (8.23) are the Euler equations describing the streaming part of the dynamics, and Equations (8.24) contain the dissipative contributions.

We note that we can rewrite the second term on the l.h.s. of Equation (8.24) by making use of the first order equation (8.23):

$$\begin{aligned}
\partial_{t_1}^2 \langle \underline{A}(\mathbf{r},t)\rangle &= -\nabla_{\mathbf{r}_1}\cdot\left(\partial_{t_1}\langle \underline{\mathbf{J}}(\mathbf{r},t)\rangle_{\mathrm{LE}}\right)\\
&= -\nabla_{\mathbf{r}_1}\cdot\left[\left(\partial_{t_1}\langle \underline{A}(\mathbf{r},t)\rangle\right) * \left(\frac{\partial}{\partial\langle \underline{A}(\mathbf{r},t)\rangle}\langle \underline{\mathbf{J}}(\mathbf{r},t)\rangle_{\mathrm{LE}}\right)\right]\\
&= \nabla_{\mathbf{r}_1}\cdot\left[\left(\nabla_{\mathbf{r}_1}\cdot\langle \underline{\mathbf{J}}(\mathbf{r},t)\rangle_{\mathrm{LE}}\right) * \left(\frac{\partial}{\partial\langle \underline{A}(\mathbf{r},t)\rangle}\langle \underline{\mathbf{J}}(\mathbf{r},t)\rangle_{\mathrm{LE}}\right)\right]\\
&= \nabla_{\mathbf{r}_1}\cdot\left[\left(\nabla_{\mathbf{r}_1}\cdot\langle \underline{\mathbf{J}}(\mathbf{r},t)\rangle_{\mathrm{LE}}\right) * \langle \underline{A}|\underline{A}\rangle_{\mathrm{LE}}^{-1}(\mathbf{r},t) * \langle \underline{A}|\underline{\mathbf{J}}\rangle_{\mathrm{LE}}(\mathbf{r},t)\right],
\end{aligned} \tag{8.29}$$

with the definitions:

$$\left\langle B^{(1)}|B^{(2)}\right\rangle_{\mathrm{LE}}(\mathbf{r},t) = \sum_{i=1}^{b} \kappa_i^{\mathrm{LE}}(\mathbf{r},t)\, B_i^{(1)}\, B_i^{(2)}, \tag{8.30}$$

where (see Section 4.7):

$$\kappa_i^{\mathrm{LE}}(\mathbf{r},t) = f_i^{\mathrm{LE}}(\mathbf{r},t)\left(1 - f_i^{\mathrm{LE}}(\mathbf{r},t)\right), \tag{8.31}$$

and:

$$\left[\langle \underline{A}|\underline{A}\rangle_{\mathrm{LE}}(\mathbf{r},t)\right]^{-1} * \langle \underline{A}|\underline{A}\rangle_{\mathrm{LE}}(\mathbf{r},t) = \langle \underline{A}|\underline{A}\rangle_{\mathrm{LE}}(\mathbf{r},t) * \left[\langle \underline{A}|\underline{A}\rangle_{\mathrm{LE}}(\mathbf{r},t)\right]^{-1} = \underline{\underline{1}}. \tag{8.32}$$

The scalar product defined in (8.30) is in fact a cumulant average (see Appendix A.11):

$$\begin{aligned}
\left\langle B^{(1)}|B^{(2)}\right\rangle_{\mathrm{LE}}(\mathbf{r},t) &= \left\langle\!\left\langle B^{(1)}B^{(2)}\right\rangle\!\right\rangle_{\mathrm{LE}}(\mathbf{r},t)\\
&\equiv \left\langle B^{(1)}B^{(2)}\right\rangle_{\mathrm{LE}}(\mathbf{r},t) - \left\langle B^{(1)}(\mathbf{r},t)\right\rangle_{\mathrm{LE}}\left\langle B^{(2)}(\mathbf{r},t)\right\rangle_{\mathrm{LE}}.
\end{aligned} \tag{8.33}$$

The last equality in (8.29) is a consequence of the relation:

$$\frac{\partial}{\partial \langle \underline{A}(\mathbf{r},t)\rangle} \langle \mathbf{J}(\mathbf{r},t)\rangle_{\text{LE}} = \langle \underline{A}|\underline{A}\rangle_{\text{LE}}^{-1} (\mathbf{r},t) * \langle \underline{A}|\mathbf{J}\rangle_{\text{LE}} (\mathbf{r},t), \tag{8.34}$$

which is proved in Appendix A.11.

8.3 The hydrodynamic equations

Our starting point is the lattice Boltzmann equation (8.17), linearized around local equilibrium. The deviations of the average occupation of channel i at node $\mathbf{r}_\star \in \mathscr{L}$ at time t, from its local equilibrium value are:

$$\delta f_i(\mathbf{r},t) = f_i(\mathbf{r},t) - f_i^{\text{LE}}(\mathbf{r},t), \tag{8.35}$$

$$\delta f_i'(\mathbf{r},t) = f_i'(\mathbf{r},t) - f_i^{\text{LE}}(\mathbf{r},t), \tag{8.36}$$

before and after collision, respectively. We carry out an expansion of Equation (8.17) to obtain:

$$\delta f_i'(\mathbf{r},t) = \sum_j \mathscr{L}_{ij}^{\text{LE}}(\mathbf{r},t)\, \delta f_j(\mathbf{r},t) + \mathscr{O}(\epsilon^2), \tag{8.37}$$

with:

$$\begin{aligned} \mathscr{L}_{ij}^{\text{LE}}(\mathbf{r},t) &= \sum_{\{s\}\{s'\}} s_i' \,\frac{s_j - f_j^{\text{LE}}(\mathbf{r},t)}{\kappa_j^{\text{LE}}(\mathbf{r},t)}\, A(s\to s') \\ &\quad \times \prod_{k=1}^{b} \left(f_k^{\text{LE}}(\mathbf{r},t)\right)^{s_k} \left(1 - f_k^{\text{LE}}(\mathbf{r},t)\right)^{(1-s_k)}. \end{aligned} \tag{8.38}$$

We have assumed that δf_i, and consequently $\delta f_i'$, is small (of order ϵ); this assumption will be discussed below.

The matrix $\mathscr{L}^{\text{LE}}(\mathbf{r},t)$ has the form (see Appendix A.12):

$$\mathscr{L}_{ij}^{\text{LE}}(\mathbf{r},t) = \kappa_i^{\text{LE}}(\mathbf{r},t)\underline{A}_i * \langle \underline{A}|\underline{A}\rangle_{\text{LE}}^{-1} (\mathbf{r},t) * \underline{A}_j + \tilde{\mathscr{L}}_{ij}^{\text{LE}}(\mathbf{r},t). \tag{8.39}$$

The matrix $\tilde{\mathscr{L}}^{\text{LE}}(\mathbf{r},t)$ has the same set of eigenvectors as $\mathscr{L}^{\text{LE}}(\mathbf{r},t)$. The eigenvalues of $\tilde{\mathscr{L}}^{\text{LE}}(\mathbf{r},t)$ corresponding to the set of collisional invariants are zero. The remaining eigenvalues (i.e. the *kinetic* eigenvalues) coincide with those of $\mathscr{L}^{\text{LE}}(\mathbf{r},t)$ which have an absolute value strictly smaller than one. We also notice that the first term of the r.h.s. of (8.39) has the form of a projection operator onto the set of constants of motion (see Appendix A.13).

Given that:

$$\sum_{j=1}^{b} \underline{A}_j\, \delta f_j(\mathbf{r},t) = 0, \tag{8.40}$$

we can replace $\mathscr{L}^{\mathrm{LE}}(\mathbf{r},t)$ by $\tilde{\mathscr{L}}^{\mathrm{LE}}(\mathbf{r},t)$ in Equation (8.37):

$$\delta f'_i(\mathbf{r},t) = \sum_j \tilde{\mathscr{L}}^{\mathrm{LE}}_{ij}(\mathbf{r},t)\,\delta f_j(\mathbf{r},t), \tag{8.41}$$

and perform the following manipulations:

$$\begin{aligned}
\delta f'_i(\mathbf{r},t) &= \sum_{j=1}^{b} \tilde{\mathscr{L}}^{\mathrm{LE}}_{ij}(\mathbf{r},t)\left[f_j(\mathbf{r},t) - f^{\mathrm{LE}}_j(\mathbf{r},t)\right] \\
&= \sum_{j=1}^{b} \tilde{\mathscr{L}}^{\mathrm{LE}}_{ij}(\mathbf{r},t)\left[f'_j(\mathbf{r}-\mathbf{c}_j,t-1) - f^{\mathrm{LE}}_j(\mathbf{r},t)\right] \\
&= \sum_{j=1}^{b} \tilde{\mathscr{L}}^{\mathrm{LE}}_{ij}(\mathbf{r},t)\left[\delta f'_j(\mathbf{r}-\mathbf{c}_j,t-1) + f^{\mathrm{LE}}_j(\mathbf{r}-\mathbf{c}_j,t-1) - f^{\mathrm{LE}}_j(\mathbf{r},t)\right] \\
&= \sum_{j=1}^{b} \tilde{\mathscr{L}}^{\mathrm{LE}}_{ij}(\mathbf{r},t)\left[\delta f'_j(\mathbf{r}-\mathbf{c}_j,t-1) + \left(e^{-\mathbf{c}_j\cdot\nabla}\, e^{-\partial_t} - 1\right) f^{\mathrm{LE}}_j(\mathbf{r},t)\right],
\end{aligned} \tag{8.42}$$

where $e^{-\partial_t}$ is the time-displacement operator. We rewrite Equation (8.42) as:

$$\delta f'_i(\mathbf{r},t) = \sum_{j=1}^{b} \mathscr{S}^{\mathrm{LE}}_{ij}(\mathbf{r},t)\, f^{\mathrm{LE}}_j(\mathbf{r},t) + \sum_{j=1}^{b} \mathscr{T}^{\mathrm{LE}}_{ij}(\mathbf{r},t)\,\delta f'_j(\mathbf{r},t-1), \tag{8.43}$$

where the matrix *operators* $\mathscr{T}^{\mathrm{LE}}$ and $\mathscr{S}^{\mathrm{LE}}$ are given by the expressions:

$$\mathscr{T}^{\mathrm{LE}}_{ij}(\mathbf{r},t) = \tilde{\mathscr{L}}^{\mathrm{LE}}_{ij}(\mathbf{r},t)\ \exp\{-\mathbf{c}_j\cdot\nabla_{\mathbf{r}}\}, \tag{8.44}$$

$$\mathscr{S}^{\mathrm{LE}}_{ij}(\mathbf{r},t) = \tilde{\mathscr{L}}^{\mathrm{LE}}_{ij}(\mathbf{r},t)\ \left(\exp\{-\mathbf{c}_j\cdot\nabla_{\mathbf{r}}\}\exp(-\partial_t) - 1\right). \tag{8.45}$$

Equation (8.43) is iterated to yield:

$$\begin{aligned}
\delta f'_i(\mathbf{r},t) = {} & \\
& \sum_{j,l}\sum_{\tau=0}^{t-1}\left\{{}^{\mathcal{O}}\prod_{\tau'=0}^{\tau-1}\mathscr{T}^{\mathrm{LE}}(\mathbf{r},t-\tau')\right\}_{ij}\ \mathscr{S}^{\mathrm{LE}}_{jl}(\mathbf{r},t-\tau)\, f^{\mathrm{LE}}_l(\mathbf{r},t-\tau) \\
& + \sum_{j=1}^{b}\left\{{}^{\mathcal{O}}\prod_{\tau=0}^{t-1}\mathscr{T}^{\mathrm{LE}}(\mathbf{r},t-\tau)\right\}_{ij}\ \delta f'_j(\mathbf{r},0),
\end{aligned} \tag{8.46}$$

where ${}^{\mathcal{O}}\prod$ is an anti-chronologically time-ordered product, with the convention:

$$\left\{{}^{\mathcal{O}}\prod_{\tau=0}^{-1}\mathscr{T}^{\mathrm{LE}}(\mathbf{r},t-\tau)\right\}_{ij} = \delta_{ij}. \tag{8.47}$$

The requirement that $\delta f'_i(\mathbf{r},t)$ be of order ϵ, and the vanishing of the 'random force' term depending on the initial conditions, $\delta f'_j(\mathbf{r},0)$, are fulfilled by choosing an initial local-equilibrium ensemble in which the values of the thermodynamic variables vary significantly only over a hydrodynamic length scale. In the case that this were not so, there should be an initial transient regime where the hydrodynamic equations are not valid. The duration of this regime is on the order of τ_{LE}, the time necessary to relax to local equilibrium.

We notice that:

$$\left\{ {}^{\mathcal{O}}\prod_{\tau'=0}^{\tau-1} \mathcal{T}^{\text{LE}}(\mathbf{r}, t-\tau') \right\}_{ik} \longrightarrow 0, \quad \tau \gg \tau_{\text{LE}}, \tag{8.48}$$

which is a consequence of the fact that all the eigenvalues of $\tilde{\mathcal{L}}_{ij}^{\text{LE}}(\mathbf{r}, t)$ are strictly smaller than one in absolute value. So, given that the kernel in Equation (8.46) is non-zero only on a microscopic time scale $\tau_{\text{LE}} \sim \mathcal{O}(\epsilon^0)$, and that during this time the local equilibrium distribution does not change significantly (since it depends only on slow variables, whose variation is on a time scale $\mathcal{O}(\epsilon^{-1})$), Equation (8.46) can be approximated by:

$$\begin{aligned} \delta f_i'(\mathbf{r}, t) &= -\sum_{\tau=1}^{t}\sum_{j=1}^{b} \left\{ \left[\tilde{\mathcal{L}}^{\text{LE}}(\mathbf{r}, t)\right]^{\tau} \right\}_{ij} \left(\mathbf{c}_j \cdot \nabla_{\mathbf{r}} + \partial_t\right) f_j^{\text{LE}}(\mathbf{r}, t) \\ &\quad + \mathcal{O}(\epsilon^2 \tau_{\text{LE}}). \end{aligned} \tag{8.49}$$

Within the same level of approximation we can further replace the upper limit in the summation over τ by infinity. Hence, the last term in (8.24) becomes:

$$\begin{aligned} &\sum_{i=1}^{b} \left(\exp\{-\mathbf{c}_i \cdot \nabla_{\mathbf{r}}\} - 1\right) \delta f_i'(\mathbf{r}, t)\, \underline{A}_i \\ &= \epsilon^2 \nabla_{\mathbf{r}_1} \cdot \left[\sum_{i,j} \mathbf{c}_i \underline{A}_i \sum_{\tau=1}^{\infty} \left\{ \left[\tilde{\mathcal{L}}^{\text{LE}}(\mathbf{r}, t)\right]^{\tau} \right\}_{ij} \left(\mathbf{c}_j \cdot \nabla_{\mathbf{r}_1} + \partial_{t_1}\right) f_j^{\text{LE}}(\mathbf{r}, t) \right] \\ &\quad + \mathcal{O}(\epsilon^3) \\ &= \epsilon^2 \nabla_{\mathbf{r}_1} \cdot \left[\sum_{i,j} \mathbf{c}_i \underline{A}_i \sum_{\tau=1}^{\infty} \left\{ \left[\tilde{\mathcal{L}}^{\text{LE}}(\mathbf{r}, t)\right]^{\tau} \right\}_{ij} \left(\mathbf{c}_j \cdot \nabla_{\mathbf{r}_1} f_j^{\text{LE}}(\mathbf{r}, t)\right) \right. \\ &\quad \left. + \left\{ \frac{\partial}{\partial \langle \underline{A}(\mathbf{r}, t)\rangle} f_i^{\text{LE}}(\mathbf{r}, t) \right\} * \partial_{t_1} \langle \underline{A}(\mathbf{r}, t)\rangle \right) \Bigg] + \mathcal{O}(\epsilon^3) \\ &= \epsilon^2\, \nabla_{\mathbf{r}_1} \cdot \left[\sum_{i,j,l} \underline{\mathbf{J}}_i \sum_{\tau=1}^{\infty} \left\{ \left[\tilde{\mathcal{L}}^{\text{LE}}(\mathbf{r}, t)\right]^{\tau} \right\}_{ij} \right. \\ &\quad \left. \times \left(\delta_{jl} - \kappa_j^{\text{LE}}(\mathbf{r}, t)\, \underline{A}_j * \langle \underline{A} | \underline{A} \rangle_{LE}^{-1}(\mathbf{r}, t) * \underline{A}_l\right) \mathbf{c}_l \cdot \nabla_{\mathbf{r}_1} f_l^{\text{LE}}(\mathbf{r}, t) \right] \\ &\quad + \mathcal{O}(\epsilon^3), \end{aligned} \tag{8.50}$$

where we have used Equation (8.23) and the relation:

$$\frac{\partial}{\partial \langle \underline{A}(\mathbf{r}, t)\rangle} f_i^{\text{LE}}(\mathbf{r}, t) = \kappa_i^{\text{LE}}(\mathbf{r}, t) \underline{A}_i * \langle \underline{A} | \underline{A} \rangle_{LE}^{-1}(\mathbf{r}, t), \tag{8.51}$$

proved in Appendix A.12 (see Equation (A.58) with $n = 1$). Notice that the quantity on the r.h.s. of (8.50):

$$\mathcal{Q}^{\mathrm{LE}}_{jl} \equiv \left(\delta_{jl} - \kappa^{\mathrm{LE}}_j(\mathbf{r},t)\, \underline{A}_j * \langle \underline{A}|\underline{A}\rangle^{-1}_{\mathrm{LE}}(\mathbf{r},t) * \underline{A}_l\right) \tag{8.52}$$

is a projection operator onto the set orthogonal to the constants of motion, as discussed in Appendix A.13.

Recombining Equations (8.23) and (8.24), using (8.22) and with the results (8.29) and (8.50), we rewrite the equations of motion to order ϵ^2 in terms of the variables $\mathbf{r}$, t, which are now taken as continuous variables:

$$\partial_t \langle \underline{A}(\mathbf{r},t)\rangle + \nabla_{\mathbf{r}} \cdot \langle \underline{\mathbf{J}}(\mathbf{r},t)\rangle_{\mathrm{LE}} = \nabla_{\mathbf{r}} \cdot \left[\sum_{i,j,l} \underline{\mathbf{J}}_i \sum_{\tau=0}^{\infty}{}' \left\{ \left[\mathcal{L}^{\mathrm{LE}}(\mathbf{r},t)\right]^{\tau} \right\}_{ij} \mathcal{Q}^{\mathrm{LE}}_{jl}(\mathbf{r},t)\, \mathbf{c}_l \cdot \nabla_{\mathbf{r}} f^{\mathrm{LE}}_l(\mathbf{r},t) \right]. \tag{8.53}$$

The prime in the summation indicates that the first term of the sum ($\tau = 0$) is multiplied by 1/2. This factor appears as a consequence of the discreteness of time in the automaton. We have also replaced the matrix $\tilde{\mathcal{L}}^{\mathrm{LE}}(\mathbf{r},t)$ by $\mathcal{L}^{\mathrm{LE}}(\mathbf{r},t)$, since the quantity upon which it acts contains no projections onto the conserved quantities.

Equation (8.53) is the most general form[6] of the hydrodynamic equations in the lattice Boltzmann approximation; the expression of the dissipative term (on the r.h.s. of Equation (8.53)) is valid even far from global equilibrium. So these equations are the fundamental equations describing the macroscopic dynamics of a lattice gas automaton in the hydrodynamic regime; in particular, they can be used to explore non-equilibrium phenomena such as far from equilibrium steady states and hydrodynamic instabilities in lattice gas automata.

8.4 Linear response and Green–Kubo coefficients

Now we shall show that linearization of Equation (8.53) yields the usual linear response results, including the Green–Kubo formulae for the transport coefficients. We first recover the equations of linear hydrodynamics (see Chapters 6 and 7) by carrying out an expansion of (8.53) around the global equilibrium state characterized by the values of the hydrodynamic variables:

$$\langle \underline{A} \rangle_{(\mathrm{eq})} = \sum_{i=1}^{b} \underline{A}_i\, f_i^{(\mathrm{eq})}. \tag{8.54}$$

[6] Notice one restriction: the validity of Equation (8.53) is limited by the fact that we have neglected the coupling of the evolution of the single-particle distribution to n-particle distributions.

The deviation from equilibrium is given by:

$$\langle \delta \underline{A}(\mathbf{r},t)\rangle = \langle \underline{A}(\mathbf{r},t)\rangle - \langle \underline{A}\rangle_{(\mathrm{eq})}. \tag{8.55}$$

Making use of Equation (A.58) (see Section A.11), we expand the single particle distribution $f_i^{\mathrm{LE}}(\mathbf{r},t)$ around equilibrium:

$$f_i^{\mathrm{LE}}(\mathbf{r},t) = f_i^{(\mathrm{eq})} + \kappa_i^{(\mathrm{eq})}\underline{A}_i * \langle \underline{A}|\underline{A}\rangle_{(\mathrm{eq})}^{-1} * \langle \delta\underline{A}(\mathbf{r},t)\rangle + \cdots, \tag{8.56}$$

and we use this expression to linearize the hydrodynamic equations (8.53). In the Euler term we have:

$$\langle \underline{\mathbf{J}}(\mathbf{r},t)\rangle_{\mathrm{LE}} \approx \langle \underline{\mathbf{J}}|\underline{A}\rangle_{(\mathrm{eq})} * \langle \underline{A}|\underline{A}\rangle_{(\mathrm{eq})}^{-1} * \langle \delta\underline{A}(\mathbf{r},t)\rangle, \tag{8.57}$$

and in the dissipative term:

$$\begin{aligned}
&\sum_{i,j,l} \underline{\mathbf{J}}_i \left\{\left[\mathscr{L}^{\mathrm{LE}}(\mathbf{r},t)\right]^{\tau}\right\}_{ij} \left(\delta_{jl} - \kappa_j^{\mathrm{LE}}(\mathbf{r},t)\,\underline{A}_j * \langle \underline{A}|\underline{A}\rangle_{LE}^{-1}(\mathbf{r},t) * \underline{A}_l\right) \\
&\qquad \times \mathbf{c}_l \cdot \nabla_{\mathbf{r}} f_l^{\mathrm{LE}}(\mathbf{r},t) \\
&\quad \approx \sum_{i,j,l} \underline{\mathbf{J}}_i \left\{\left[\mathscr{L}^{\mathrm{eq}}\right]^{\tau}\right\}_{ij} \left(\delta_{jl} - \kappa_j^{(\mathrm{eq})}\,\underline{A}_j * \langle \underline{A}|\underline{A}\rangle_{(\mathrm{eq})}^{-1} * \underline{A}_l\right) \\
&\qquad \times \mathbf{c}_l \cdot \nabla_{\mathbf{r}} \left(\kappa_l^{(\mathrm{eq})}\underline{A}_l * \langle \underline{A}|\underline{A}\rangle_{(\mathrm{eq})}^{-1} * \langle \delta\underline{A}(\mathbf{r},t)\rangle\right) \\
&\quad = \nabla_{\mathbf{r}} \cdot \sum_{i,j} \underline{\mathbf{J}}_i \left\{\left[\mathscr{L}^{\mathrm{eq}}\right]^{\tau}\right\}_{ij} \kappa_j^{(\mathrm{eq})} \\
&\qquad \times \left(\underline{\mathbf{J}}_j - \underline{A}_j * \langle \underline{A}|\underline{A}\rangle_{(\mathrm{eq})}^{-1} * \langle \underline{A}|\underline{\mathbf{J}}\rangle_{(\mathrm{eq})}\right) * \langle \underline{A}|\underline{A}\rangle_{(\mathrm{eq})}^{-1} * \langle \delta\underline{A}(\mathbf{r},t)\rangle \\
&\quad = \nabla_{\mathbf{r}} \cdot \sum_{i,j} \underline{\mathbf{J}}_i \left\{\left[\mathscr{L}^{\mathrm{eq}}\right]^{\tau}\right\}_{ij} \kappa_j^{(\mathrm{eq})}\, \hat{\underline{\mathbf{J}}}_j * \langle \underline{A}|\underline{A}\rangle_{(\mathrm{eq})}^{-1} * \langle \delta\underline{A}(\mathbf{r},t)\rangle,
\end{aligned}$$

where the subtracted current is defined as (see also Section 6.3.2):

$$\hat{\underline{\mathbf{J}}}_j = \underline{\mathbf{J}}_j - \underline{A}_j * \langle \underline{A}|\underline{A}\rangle_{(\mathrm{eq})}^{-1} * \langle \underline{A}|\underline{\mathbf{J}}\rangle_{(\mathrm{eq})} \equiv (\mathscr{Q}\,\underline{\mathbf{J}})_j\,. \tag{8.58}$$

In the second equality of Equation (8.58), we have used the definition of the equilibrium projectors (Zanetti, 1989; Ernst, 1990a; Ernst and Dufty, 1990) $(\mathscr{P} + \mathscr{Q} = 1)$ acting on an arbitrary vector $\{B_j\}_{j=1}^{b}$:

$$(\mathscr{P} B)_j = \underline{A}_j * \langle \underline{A}|\underline{A}\rangle_{(\mathrm{eq})}^{-1} * \langle \underline{A}|B\rangle_{(\mathrm{eq})}, \tag{8.59}$$

$$(\mathscr{Q} B)_j = B_j - \underline{A}_j * \langle \underline{A}|\underline{A}\rangle_{(\mathrm{eq})}^{-1} * \langle \underline{A}|B\rangle_{(\mathrm{eq})} \equiv \hat{B}_j. \tag{8.60}$$

The projected quantities have the following properties (see Appendix A.13):

$$\langle \underline{A} \mid \mathscr{P} B\rangle_{(\mathrm{eq})} = \langle \underline{A} \mid B\rangle_{(\mathrm{eq})}, \tag{8.61}$$

$$\langle \underline{A} \mid \mathscr{Q} B\rangle_{(\mathrm{eq})} \doteq \left\langle \underline{A} \mid \hat{B}\right\rangle_{(\mathrm{eq})} = 0. \tag{8.62}$$

In particular, for the subtracted current:

$$\left\langle \underline{A} \mid \hat{\underline{\mathbf{J}}} \right\rangle_{(\mathrm{eq})} = \langle \underline{A} \mid \mathcal{Q}\underline{\mathbf{J}} \rangle_{(\mathrm{eq})} = 0, \tag{8.63}$$

which means that $\hat{\underline{\mathbf{J}}}$ has no projections onto the constants of motion.

Using the definition:

$$\kappa_i^{(\mathrm{eq})}\hat{\underline{\mathbf{J}}}_i(\tau) = \sum_{j=1}^{b} \{[\mathcal{L}^{\mathrm{eq}}]^{\tau}\}_{ij}\ \kappa_j^{(\mathrm{eq})}\ \hat{\underline{\mathbf{J}}}_j, \tag{8.64}$$

we can write:

$$\sum_{i,j} \underline{\mathbf{J}}_i\ \{[\mathcal{L}^{\mathrm{eq}}]^{\tau}\}_{ij}\ \kappa_j^{(\mathrm{eq})}\ \hat{\underline{\mathbf{J}}}_j = \left\langle \hat{\underline{\mathbf{J}}} \mid \hat{\underline{\mathbf{J}}}(\tau) \right\rangle_{(\mathrm{eq})}, \tag{8.65}$$

where the replacement of $\underline{\mathbf{J}}_i$ by $\hat{\underline{\mathbf{J}}}_i$ is possible given that $\hat{\underline{\mathbf{J}}}_j$ has no projection onto the set of constants of motion.

Using these results in (8.53), we obtain the linearized equations:

$$\begin{aligned}\partial_t \langle \delta \underline{A}(\mathbf{r},t) \rangle + \nabla_{\mathbf{r}} \cdot \left[\langle \underline{\mathbf{J}} | \underline{A} \rangle_{(\mathrm{eq})} * \langle \underline{A} | \underline{A} \rangle_{(\mathrm{eq})}^{-1} * \langle \delta \underline{A}(\mathbf{r},t) \rangle \right] \\ = \nabla_{\mathbf{r}}\nabla_{\mathbf{r}} : \left[\underline{\underline{\Lambda}}^{(\mathrm{eq})} * \langle \delta \underline{A}(\mathbf{r},t) \rangle \right],\end{aligned} \tag{8.66}$$

where the matrix of linear transport coefficients is given by the expression:

$$\underline{\underline{\Lambda}}^{(\mathrm{eq})} = \left[\sum_{\tau=0}^{\infty}{}' \left\langle \hat{\underline{\mathbf{J}}} \mid \hat{\underline{\mathbf{J}}}(\tau) \right\rangle_{(\mathrm{eq})} \right] * \langle \underline{A} | \underline{A} \rangle_{(\mathrm{eq})}^{-1}. \tag{8.67}$$

If the decay of the correlation function that appears on the r.h.s. in (8.67) were slow on the microscopic time scale, one would be able to use the Euler–Maclaurin sum formula (Bender and Orszag, 1978) to replace the summation over time by a time integral. However, for lattice gas automata the decay of $\left\langle \hat{\underline{\mathbf{J}}} \mid \hat{\underline{\mathbf{J}}}(\tau) \right\rangle_{(\mathrm{eq})}$ takes place in times of the order of a few automaton time steps, and the discrete sum must be retained.

Equation (8.67) shows that $\underline{\underline{\Lambda}}^{(\mathrm{eq})}$ is proportional to the sum over time (an integral in the continuous time limit) of the equilibrium time correlation function (see details in Appendix A.14):

$$\begin{aligned}\sum_{\tau=0}^{\infty}{}' \left\langle \hat{\underline{\mathbf{J}}} \mid \hat{\underline{\mathbf{J}}}(\tau) \right\rangle_{(\mathrm{eq})} &= \sum_{\tau=0}^{\infty}{}' \left[\sum_{i,j} \hat{\underline{\mathbf{J}}}_i\ \hat{\underline{\mathbf{J}}}_j\ \{[\mathcal{L}^{\mathrm{eq}}]^{\tau}\}_{ij}\ \kappa_j^{(\mathrm{eq})} \right] \\ &= \sum_{\tau=0}^{\infty}{}' \left[\frac{1}{V} \sum_{\mathbf{r}\mathbf{r}'} \sum_{i,j} \left\langle \left(\hat{\underline{\mathbf{J}}}_i \delta n_i(\mathbf{r},\tau) \right) \left(\hat{\underline{\mathbf{J}}}_j \delta n_j(\mathbf{r}',0) \right) \right\rangle_{(\mathrm{eq})}^{(\mathrm{B})} \right],\end{aligned} \tag{8.68}$$

where $V = \mathcal{N}$ is the number of nodes of the lattice, and the superscript (B) indicates that the time-correlation function is evaluated in the Boltzmann approximation. The transport coefficients are then well-defined thanks to the fact that:

$$\langle \delta n_i(\mathbf{r}, \tau)\delta n_j(\mathbf{r}, 0)\rangle^{(B)}_{(eq)} \tag{8.69}$$

decays to zero sufficiently fast for the sum in (8.68) to be finite (Schmitz and Dufty, 1990). The result is that the dissipative coefficients exhibit a Green–Kubo form (Rivet, 1987b; Ernst, 1990b) analogous to the continuous theory expression (Kubo, 1958; Zwanzig, 1965).

This concludes the derivation of the non-linear hydrodynamic equations for lattice gas automata satisfying the semi-detailed balance condition, and of their subsequent linearization yielding correlation function expressions for the dissipative coefficients.

8.5 Long-time tails

Before closing this chapter, an important comment is in order. There is a restriction in our derivation of the hydrodynamic equations in the sense that by using the Boltzmann approximation we have neglected all mode-coupling contributions (Ernst, 1990b; Brito and Ernst, 1992). The consequence of mode-coupling effects is that the correlation function:

$$\langle \delta n_i(\mathbf{r}, \tau)\delta n_j(\mathbf{r}, 0)\rangle \tag{8.70}$$

generally exhibits long-time behavior usually in the form of algebraic decay: *long-time tails* $\sim t^{-D/2}$, where D is the space dimension.[7] This implies that (i) in dimensions lower than or equal to 2 the sum in (8.68) diverges, and the hydrodynamic equations are valid only for regimes in which mode-coupling effects are negligible, and (ii) in dimensions 3 and higher, the form of the hydrodynamic equations remains valid, but the transport coefficients are renormalized.

In mode-coupling theory, one starts with the idea that the long-time behavior can be explained on the basis of hydrodynamic arguments. Consider the case of the velocity of a tagged particle; the mode which describes the decay of its velocity correlations, the shear mode, and the mode which describes particle displacements, the diffusion mode, are coupled. The assumption is that eventually the particle velocity will be equal to the fluid velocity, so that the velocity of the particle is expressed in terms of the particle probability density and of the fluid velocity fluctuations. The former obeys the diffusion equation and the latter the linearized Navier–Stokes equation. So the basic assumption combines

[7] The first observation of long-time tails was reported by Alder and Wainwright (1970) who measured the time decay of the velocity autocorrelation function for hard disks and hard spheres, by molecular dynamics simulations.

the solutions of the two equations, and the result for the normalized velocity autocorrelation function $\psi(t)$ reads (Ernst, 1991):

$$\psi(t) \simeq \frac{D-1}{D}(1-d)\frac{v_0}{bd}\left[4\pi(v+D_s)t\right]^{-D/2}, \tag{8.71}$$

where bd/v_0 is the number density per elementary unit volume of the lattice (e.g. $v_0 = \sqrt{3}/2$ in the triangular lattice), and D_s denotes the self-diffusion coefficient. This result is, in fact, the same as in continuous fluid theory (Ernst, Hauge and van Leeuwen, 1971), apart from the factor $(1-d)$ which comes from the exclusion principle.

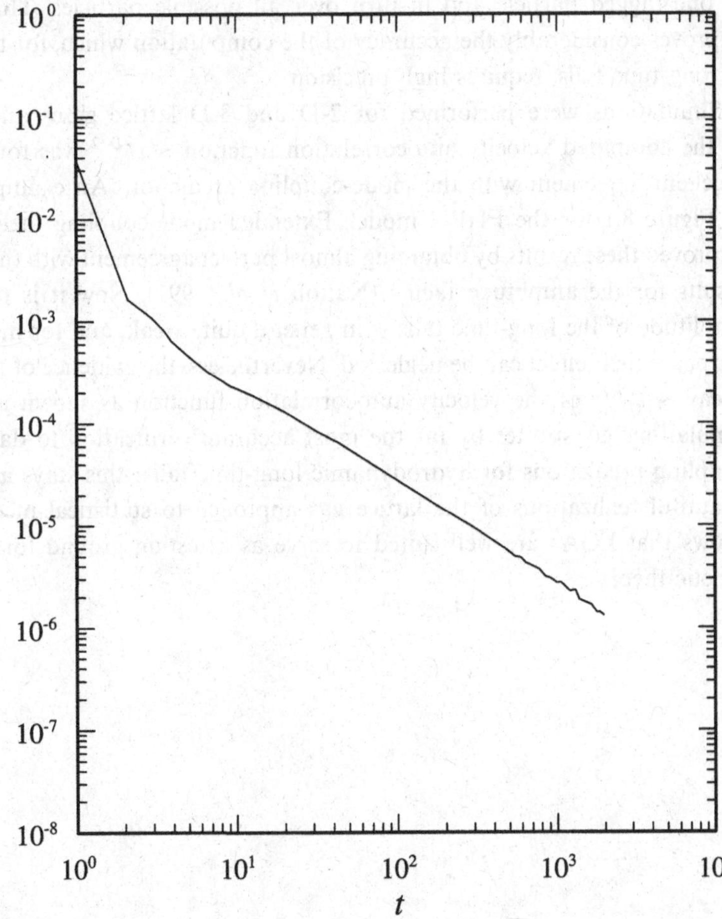

Figure 8.1 Normalized velocity autocorrelation function $\psi(t)$ of a tagged particle computed with the moment-propagation method in the FHP-3 lattice gas: the long-time behavior $\sim t^{-D/2}$ (here $D = 2$) is observed over more than two decades; reduced density $d = 0.8$; error $\sim 10^{-8}$. (van der Hoef and Frenkel, 1990).

In order to compute the velocity autocorrelation function of a particle in the lattice gas where all particles are indistinguishable, 'the' particle must be labeled differently so that one can follow its dynamics. A very efficient procedure was developed by Frenkel *et al.* (Frenkel and Ernst, 1989; van der Hoef and Frenkel, 1990, 1991): the 'moment-propagation' method, which exploits the fact that in a collision a particle loses its identity. So the collision rules for the tagged particle are stochastic, and the stochastic rules can be chosen such that any outgoing particle is equally likely to be the tagged particle. Consequently there are many different possible paths for the particle, each with a certain probability that can be evaluated as the product of the scattering probabilities. Then the velocity autocorrelation function can be computed by averaging over all possible paths of one tagged particle and in turn over all possible particles. This procedure improves considerably the accuracy of the computation which, for the detection of long-time tails, requires high precision.

Simulations were performed for 2-D and 3-D lattice gases and the decay of the computed velocity autocorrelation function $\sim t^{-D/2}$ was found to be in excellent agreement with the mode-coupling prediction. An example is shown in Figure 8.1 for the FHP-3 model. Extended mode-coupling theory has even improved these results by obtaining almost perfect agreement with the simulation results for the amplitude factor (Naitoh *et al.*, 1991). Now it is true that the amplitude of the long-time tails is in general quite weak, and for most practical purposes their effect can be neglected. Nevertheless, the evidence of the algebraic decay $\sim t^{-D/2}$ of the velocity autocorrelation function as shown in lattice gas simulations constitutes by far the most accurate verification to date of mode-coupling predictions for hydrodynamic long-time tails: this stays as one of the beautiful realizations of the lattice gas approach to statistical mechanics, and shows that LGAs are well suited to serve as a testing ground for concepts in kinetic theory.

Chapter 9

Hydrodynamic regimes

In Chapters 5 and 8, we derived the macrodynamic equations for single species thermal and non-thermal lattice gases. These macrodynamic equations govern the dynamics of the lattice gas in the hydrodynamic limit, i.e. they describe the long-time evolution of large-scale fluctuations of the macroscopic variables.[1] 'Long-time' and 'large-scale' should be understood as long and large compared to the microscopic natural time and space scales, the automaton time step and the lattice mesh, respectively.

Now we will show, at least for non-thermal models, that there exist regimes for which the macrodynamical equations of the lattice gas become identical to the standard hydrodynamic equations describing real fluids. We restrict our analysis to non-thermal models for two reasons. First, because the concepts are more easily and efficiently put to work (the case of thermal models is much more complicated); and second, because most of lattice gas simulations of hydrodynamic phenomena – in particular those reported in Chapter 10 – have been performed with non-thermal models.

We start from the macrodynamical equations for non-thermal models (see Section 5.7.3):

$$\partial_t \rho + \partial_\beta (\rho u_\beta) = 0, \tag{9.1}$$

[1] The macroscopic variables are the ensemble-averages of the conserved quantities (see Section 4.5).

and

$$\partial_t j_\alpha + \partial_\beta\Big(g j_\alpha u_\beta\Big) + \partial_\alpha P =$$

$$\partial_\beta\Bigg(\nu(\rho)\Big(\partial_\alpha j_\beta + \partial_\beta j_\alpha - \frac{2}{D}\partial_\gamma j_\gamma \delta_{\alpha\beta}\Big) + \nu'(\rho)\Big(\partial_\gamma j_\gamma \delta_{\alpha\beta}\Big)\Bigg), \tag{9.2}$$

$$+\mathcal{O}(\epsilon^3\eta) + \mathcal{O}(\epsilon^2\eta^2) + \mathcal{O}(\epsilon\eta^3),$$

where $g(\rho)$ is the non-Galilean factor:

$$g(\rho) = \frac{b-2\rho}{b-\rho}\frac{b}{D(D+2)}\Big(\frac{4\xi_4}{\xi_2^2} + \frac{D^2}{b}\Big), \tag{9.3}$$

and P is the hydrostatic pressure:

$$P(\rho, \mathbf{j}^2) = \frac{\xi_2}{b}\rho - \frac{\Lambda}{D}\frac{g(\rho)}{\rho}\mathbf{j}^2, \tag{9.4}$$

with

$$\Lambda = \frac{D^2\xi_2^2 - D2\xi_4 b}{D^2\xi_2^2 + 4\xi_4 b}. \tag{9.5}$$

The geometrical coefficients ξ_2 and ξ_4 are defined in Section 4.5.2, and their values are given for various models in Table 4.1.

The expansion parameter ϵ in (9.2) is the space-scale separation parameter; the hydrodynamic limit corresponds to $\epsilon \ll 1$, the condition under which the macrodynamical equations were derived. The expansion parameter η represents the order of magnitude of the macroscopic momentum flux $\mathbf{j}$, that is, the (small) 'deviation' from zero-velocity basic equilibrium (see Section 4.5).

9.1 The acoustic limit

The hydrodynamic equations for continuous fluids, such as the Navier–Stokes equations, are known to exhibit acoustic solutions, that is, solutions for which the fluid velocity and the density fluctuations are small enough so that the non-linear terms become negligible. The problem was considered in Chapter 6; we reiterate the reasoning here in a slightly different presentation for non-thermal fluids.

Consider the density fluctuation $\delta\rho$ defined by:

$$\rho(t, \mathbf{r}) = \rho_0 + \delta\rho(t, \mathbf{r}),$$

where ρ_0 is the constant and uniform mean density, and the velocity fluctuation

$\mathbf{u}(t, \mathbf{r})$ (with zero mean velocity). When $\rho(t, \mathbf{r})$ and $\mathbf{u}(t, \mathbf{r})$ are small, we can ignore the non-linear terms (in $\delta\rho$ and $\mathbf{u}$) in Equations (9.1) and (9.2) to obtain:

$$\begin{cases} \partial_t \dfrac{\delta\rho}{\rho_0} + \partial_\beta u_\beta = 0, \\ \partial_t u_\alpha + \dfrac{\xi_2}{b} \partial_\alpha \dfrac{\delta\rho}{\rho_0} = v \partial_\beta \partial_\beta u_\alpha + \left(v' + \dfrac{D-2}{D} v\right) \partial_\alpha \partial_\beta u_\beta, \end{cases} \tag{9.6}$$

where v and v' stand implicitly for $v(\rho_0)$ and $v'(\rho_0)$ respectively. Using the standard 'nabla' notation, the preceding set of equations can be rewritten as:

$$\begin{cases} \partial_t \dfrac{\delta\rho}{\rho_0} + \nabla \cdot \mathbf{u} = 0, \\ \partial_t \mathbf{u} + \dfrac{\xi_2}{b} \nabla(\dfrac{\delta\rho}{\rho_0}) = v \Delta \mathbf{u} + \left(v' + \dfrac{D-2}{D} v\right) \nabla(\nabla \cdot \mathbf{u}), \end{cases} \tag{9.7}$$

where we clearly recognize the standard equations of viscously damped acoustics.

Here a comment is in order about Galilean invariance. The acoustic equations (9.6) or (9.7) are Galilean invariant while the full lattice gas macrodynamical equations are not (because of the non-Galilean factor $g(\rho)$). In the acoustic limit, Galilean invariance is recovered because the non-linear term $\partial_\beta(g(\rho) u_\beta j_\alpha)$ is absent in (9.6) and (9.7).

Proceeding one step further in simplification, we consider the inviscid flow, i.e. we consider short (yet macroscopic) time scales where viscous damping is negligible. The dynamics is then governed by the standard Eulerian acoustic equations:

$$\begin{cases} \partial_t \dfrac{\delta\rho}{\rho_0} + \nabla \cdot \mathbf{u} = 0, \\ \partial_t \mathbf{u} + \dfrac{\xi_2}{b} \nabla(\dfrac{\delta\rho}{\rho_0}) = 0. \end{cases} \tag{9.8}$$

This set of equations has sound wave solutions[2] of the following form:

$$\begin{cases} \mathbf{u}(t, \mathbf{r}) = u_1' \exp\left(ikc_s t - i\mathbf{k} \cdot \mathbf{r}\right) \hat{\mathbf{k}}, \\ \dfrac{\delta\rho(t, \mathbf{r})}{\rho_0} = \dfrac{\hat{\mathbf{k}} \cdot \mathbf{u}(t, \mathbf{r})}{c_s}, \end{cases} \tag{9.9}$$

where u_1' is an arbitrary (small) velocity amplitude, $\hat{\mathbf{k}} \equiv \mathbf{k}/k$ is the longitudinal unit vector (parallel to the wave vector $\mathbf{k}$), and c_s is the sound speed of the lattice gas which depends only on geometric parameters:

$$c_s = \sqrt{\frac{\xi_2}{b}}. \tag{9.10}$$

[2] These sound waves are called 'longitudinal waves' because the fluid velocity $\mathbf{u}$ is everywhere parallel to the wave vector.

As in classical hydrodynamics, we define the 'Mach number' Ma as the ratio of the fluid velocity to the sound speed:

$$Ma \equiv \frac{u'}{c_s}. \tag{9.11}$$

From (9.9), we observe that the relative density fluctuation $\delta\rho/\rho_0$ is of the same order of magnitude as the Mach number.

On longer time scales, the viscous terms become relevant, and Equations (9.7) exhibit solutions describing sound wave propagation with an amplitude decaying exponentially slowly in time:

$$\begin{cases} \mathbf{u}(t,\mathbf{r}) = u_1' \exp\left(ik\sqrt{c_s{}^2 - \alpha^2 k^2}\, t - i\mathbf{k}\cdot\mathbf{r}\right) \mathrm{e}^{-\alpha k^2 t}\, \hat{\mathbf{k}}, \\ \dfrac{\delta\rho(t,\mathbf{r})}{\rho_0} = \dfrac{\hat{\mathbf{k}}\cdot\mathbf{u}(t,\mathbf{r})}{c_s}\, \mathrm{e}^{-i\phi}. \end{cases} \tag{9.12}$$

Here the damping coefficient α is a combination of the kinematic viscosity and bulk viscosity coefficients:

$$\alpha = \frac{D-1}{D}\nu + \frac{1}{2}\nu',$$

and the phase shift ϕ is given by:

$$\begin{cases} \sin\phi = \dfrac{\alpha k}{c_s}, \\ \cos\phi \geq 0. \end{cases}$$

In Chapter 10, we explain carefully how post-processed numerical simulation data of damped sound waves provide an efficient way to 'measure' the damping coefficient α, and the sound speed c_s in a lattice gas.

In addition to longitudinal sound waves, there exists a second family of solutions: 'shear waves' or 'transverse waves'. They are characterized by a uniform density field, and a velocity field orthogonal to the wave vector:

$$\begin{cases} \mathbf{u}(t,\mathbf{r}) = u_1' \exp(-i\mathbf{k}\cdot\mathbf{r})\, e^{-\nu k^2 t}\, \hat{\mathbf{k}}_\perp, \\ \dfrac{\delta\rho(t,\mathbf{r})}{\rho_0} = 0. \end{cases} \tag{9.13}$$

These are not propagating waves (there is no phase rotation); their amplitude decays exponentially over long time scales. The shear wave solution is not of direct interest for lattice gas simulations of fluid dynamics, but appears to be very useful to obtain an independent measure of the kinematic viscosity coefficient ν. From the damping coefficient α of sound waves and the damping coefficient ν of shear waves, we can then compute the shear and bulk viscosities separately (see Chapter 10).

9.2 The incompressible limit

For continuous 'real' fluids, a hydrodynamic regime is said to be 'incompressible' when the density of any material element of the fluid stays constant ($= \rho_0$) during its motion. In other words, the derivative $D\rho/Dt$ of the density of a fluid element following the motion of the fluid must vanish:

$$\partial_t \rho + \mathbf{u} \cdot \nabla(\rho) = 0.$$

The continuity equation and the momentum equation for real fluids then become:

$$\begin{cases} \nabla \cdot \mathbf{u} = 0, \\ \partial_t \mathbf{u} + \mathbf{u} \cdot \nabla(\mathbf{u}) = -\nabla(\dfrac{P}{\rho_0}) + \nu \Delta \mathbf{u}. \end{cases} \tag{9.14}$$

The first equation is the 'incompressibility condition', and the second is the 'incompressible Navier–Stokes' equation (see for example Batchelor (1967), page 174).

For 'real' fluids, the incompressible Navier–Stokes equation can be seen as a low Mach number approximation of the fully compressible Navier–Stokes equations.[3] The main feature of incompressible regimes is that the density can be taken as approximatively equal to a constant ρ_0, except in the pressure term which must fluctuate so that the two equations in (9.14) be compatible. This amounts to saying that pressure fluctuations are slaved to the fluid velocity field so that the divergence of $\mathbf{u}$ remains zero.

The same kind of limit can be taken for the lattice gas; the macrodynamical equations (9.1) and (9.2) then reduce to:

$$\begin{cases} \partial_\beta u_\beta = 0, \\ \partial_t u_\alpha + g(\rho_0)\partial_\beta \left(u_\alpha u_\beta\right) + \dfrac{1}{\rho_0}\partial_\alpha P = \nu(\rho_0)\partial_\beta\partial_\beta u_\alpha, \end{cases}$$

or, equivalently:

$$\begin{cases} \nabla \cdot \mathbf{u} = 0, \\ \partial_t \mathbf{u} + g(\rho_0)\mathbf{u} \cdot \nabla(\mathbf{u}) = -\nabla(\dfrac{P}{\rho_0}) + \nu(\rho_0)\Delta \mathbf{u}. \end{cases} \tag{9.15}$$

These equations are not identical to the continuous fluid incompressible Navier–Stokes equations (9.15) because of the non-Galilean factor $g(\rho_0)$ which is constant but (in general) not equal to one. We now show that there exists a simple rescaling by which we recover the standard incompressible Navier–Stokes equations.

[3] There are some tricky points in the Mach number expansion (see e.g. Majda, 1984).

Let us denote by u_0 and by ℓ_0 the typical fluid velocity and the typical space scale of the problem (expressed in lattice units) respectively. For example, consider the flow around a fixed solid obstacle in a wind tunnel-like configuration: the typical velocity u_0 would be the up-stream mean velocity in the simulated wind tunnel, and the typical length ℓ_0 would be some transverse size of the obstacle. With a length scale and a velocity scale, we can build the typical time scale:

$$t_0 = \frac{\ell_0}{g(\rho_0)u_0}$$

This time scale is equivalent to the 'circulation time' commonly used in continuous fluid dynamics. The presence of the non-Galilean factor $g(\rho_0)$ in this definition is necessary to render the rescaled equations of motion Galilean-invariant.

With space, time and velocity scales, we can define the following dimensionless variables:

$$T \equiv \frac{t}{t_0}, \qquad \mathbf{R} \equiv \frac{\mathbf{r}}{\ell_0}, \qquad \mathbf{U} \equiv \frac{\mathbf{u}}{u_0}, \qquad \Pi \equiv P\frac{1}{g(\rho_0)\rho_0 u_0^2}. \tag{9.16}$$

In terms of these rescaled variables and fields, the incompressible macrodynamical equations (9.15) are rewritten as:

$$\begin{cases} \nabla \cdot \mathbf{U} = 0, \\ \partial_T \mathbf{U} + \mathbf{U} \cdot \nabla(\mathbf{U}) = -\nabla(\Pi) + \dfrac{1}{Re}\Delta\mathbf{U}, \end{cases} \tag{9.17}$$

where the operators ∇ and Δ are taken with respect to the rescaled space variable $\mathbf{R}$. The parameter:

$$Re = \frac{u_0 \ell_0 g(\rho_0)}{\nu(\rho_0)}. \tag{9.18}$$

will be called the 'Reynolds number', because it plays, in lattice gases, the same role as the Reynolds number for real fluids.

The equations (9.17) are identical to the dimensionless form of the 'real world' incompressible Navier–Stokes equations (9.14). They involve only one dimensionless control parameter: the Reynolds number. The rescaling (9.16) is sufficient to transform the incompressible macrodynamical equations of non-thermal lattice gases into the classical incompressible Navier–Stokes equations for real Newtonian fluids, which are Galilean-invariant.

9.3 Comments

9.3.1 Invariances

At the microscopic level, a lattice gas is neither fully rotation-invariant, nor fully translation-invariant, neither is it Galilean-invariant. However, these invariances, which are verified for real fluids, are recovered for certain hydrodynamic regimes. The rotation invariance is recovered at the level of the macrodynamical equations, provided the lattice gas satisfies fourth order crystallographic isotropy (see Chapters 2 and 5). The translation invariance is also recovered at the level of the macrodynamical equations, at least in the hydrodynamic limit (weak gradients, or, equivalently, large-scale distances). The Galilean invariance is only recovered for particular hydrodynamic regimes: the acoustic regime and the incompressible regime. In the incompressible regime, the Galilean invariance is recovered in a less trivial way than for acoustic regimes where the non-Galilean factor disappears simply because the non-linear term is absent; in the incompressible regime, its effect is eliminated by a proper 'time stretching' (rescaling).

9.3.2 Four-dimensional models

We should pay special attention to four-dimensional (4-D) models such as FCHC models. These models can be used, without modification, to simulate three-dimensional incompressible fluid dynamics, with passive scalar advection-diffusion. Indeed, let us call u_α, $\alpha = 1,\ldots,3$ the first three components of the 4-D fluid velocity vector $\mathbf{u}$, and u_4 its fourth component. Now, consider a piece of 4-D lattice that is extremely thin in the fourth direction (say, one lattice unit length wide), with periodic conditions in this fourth direction, so that the macroscopic fields have vanishing dependence on the fourth space-coordinate. Then, the 4-D incompressible macrodynamical equations (9.15) degenerate into 3-D incompressible macrodynamical equations for ρ and u_α, $\alpha = 1,\ldots,3$, plus an advection-diffusion equation for the fourth component u_4:

$$\begin{cases} \partial_t \rho + \nabla \cdot \mathbf{u} = 0, \\ \partial_t \mathbf{u} + g(\rho_0)\mathbf{u} \cdot \nabla(\mathbf{u}) = -\nabla \left(\dfrac{P}{\rho_0} \right) + \nu(\rho_0)\Delta \mathbf{U}, \\ \partial_t u_4 + g(\rho_0)\mathbf{u} \cdot \nabla(u_4) = \nu(\rho_0)\Delta u_4, \end{cases} \tag{9.19}$$

where the divergence, gradient and Laplacian are three-dimensional operators. The rescaling (9.16), applied to (9.19) yields the 'real-world' three-dimensional incompressible Navier–Stokes equations with a passively advected diffusing scalar quantity, which can be viewed, for example, as a passive tracer (like a dye) concentration with no effect on the fluid density.

9.3.3 Lattice gases to simulate real fluid dynamics

We have shown that, within certain limits, lattice gases can exhibit large-scale (hydrodynamic) behavior identical to the dynamics of real continuous Newtonian fluids. Therefore, lattice gas automata can be used, and have been used as an alternative to standard computational fluid dynamics techniques to simulate fluid motions. Chapter 10 is devoted to various aspects of lattice gas simulations.

Chapter 10

Lattice gas simulations

One of our main objectives has been to show that single-species non-thermal lattice gases can exhibit large-scale collective behavior governed by the same continuous, isotropic and Galilean-invariant equations as real Newtonian fluids. This is true despite the intrinsically Boolean, spatially discrete, anisotropic and non-Galilean invariant structure of lattice gases. Moreover, in the past 10 years, further lattice gas models have been designed to incorporate more complicated physical features such as reactive processes, magneto-hydrodynamic phenomena or surface tension (see Section 11.4 in Chapter 11).

On one hand, there has been considerable effort in basic research to understand the subtleties of the statistical mechanics of lattice gases and on the other hand intense work has been accomplished to take advantage of the similarities between lattice gases and real fluids in order to simulate fluid motions with simple and easily implemented lattice gas algorithms. Indeed, because of their fully Boolean cellular automaton structure, lattice gases are excellent candidates for efficient implementations on both dedicated and general purpose computers with serial, vectorial, parallel or even massively parallel architecture. In addition, various physical effects can be added at low cost. For example, the presence in a flow of a rigid fixed obstacle is extremely easy to take into account: it just requires replacing the standard collision rule by a bounce-back rule (see Section 2.4.1) on all nodes covered by the obstacle. Modifying the shape or the position of the obstacle is almost immediate, and no mesh modification is necessary.

Besides technicalities, the lattice gas methods for fluid dynamics offer another fundamental advantage: the simulation of flows where bifurcations are suspected. A bifurcation occurs when a given regime loses its stability in favor of another regime, as soon as a dimensionless control parameter exceeds a critical value. The new regime originates from small perturbations which grow instead of being damped. This growth saturates because of non-linear effects, and leads to the new regime. These small perturbations operate some kind of 'natural selection' by eliminating unstable regimes. In natural fluid systems, thermal noise plays this role. In standard (non-microscopic) computational fluid dynamics methods, the unstable regime can survive artificially, and an externally imposed disturbance is sometimes necessary to trigger or at least to fasten the onset of the new regime. Since the lattice gas has fundamentally a microscopic structure and possesses spontaneous microscopic fluctuations (see Chapter 7), the macroscopic variables are naturally noisy: they are averages of naturally fluctuating microscopic quantities. Thus no external perturbation is needed to trigger bifurcations. Examples of lattice gas simulations displaying bifurcations are described in Section 10.5.

Since this chapter is dedicated to the application of the lattice gas method to simulate fluid dynamics, we start with a rapid overview of lattice gas implementation strategies and of frequently encountered programming difficulties. We then give some notions on how to perform physically relevant numerical simulations of fluid flows. Lattice gas numerical experiments are then described to illustrate these concepts.

10.1 Lattice gas algorithms on dedicated machines

The possibility of building dedicated machines, that is, electronic devices specially designed to run lattice gas algorithms was considered as early as 1984. Several machines were constructed and successfully used, like the RAP family of D. d'Humières and A. Cloucqueur (ENS, Paris), and the Cellular Automaton Machines (CAM) of T. Toffoli and N. Margolus (MIT).

We briefly describe one of the earliest of these dedicated machines: the RAP-1 prototype (October 1985).[1] The acronym 'RAP' stands for 'Réseau d'Automates Programmables' (Programmable Automata Array). The RAP-1 machine is a desk-top, relatively versatile, dedicated hardware, which can handle two-dimensional lattice gas models, with up to 16 channels per node. It is not, strictly speaking, a computer, since it contains no microprocessor. It is essentially built with:

- 16 arrays of video random access memory (VRAM) chips to store and

[1] For articles on dedicated machines, see Section 11.7.1 of the 'Guide for further reading' (Chapter 11).

manipulate the binary digits encoding the instantaneous occupation level (0 or 1) of the 16 channels on the 512 × 256 nodes of the lattice.

- two arrays of static random access memory (SRAM) chips to store the look-up table needed for the collision phase, and the look-up table for the color-encoding of the occupation levels to be displayed.
- Shift registers to perform the propagation phase.
- Logical functions on programmable array logics (PAL) chips, for various control and data manipulation tasks.
- An interface board for communication with an external personal computer.
- A video board for display on an external monitor of the instantaneous state of the lattice nodes, 50 times per second.
- A power supply.

The RAP-1 machine can update the Boolean configuration of 512 × 256 nodes with 16 channels each, at a rate of 50 times per second, so that the instantaneous state of the whole lattice can be displayed at the European standard video frequency. Figure 10.1 shows the RAP-1 facility, with its external personal computer and its video display monitor.

The automaton behavior of the RAP-1 machine can be programmed from the external personal computer, through a set of three tables:

Figure 10.1 The RAP-1 facility. On the left is the external personal computer required to program and run the RAP machine. On the right is the RAP-1 machine itself with its display monitor.

- A propagation table that drives the shift registers machinery so as to produce the correct information streaming during the propagation phase.
- A collision table, accessed as a look-up table, to perform the desired local information reorganization node by node during the collision phase.
- A color coding table, which tells the video board how to display, by color blending, the Boolean state of each node.

The simplest one-dimensional and two-dimensional lattice gas models (HPP, FHPs, Conway's game of life, colored FHP, etc.) fit easily within the machine capacity. Note that the look-up table strategy for the collision phase does not imply that the machine can handle only deterministic models. Indeed, one or more of the 16 channels per node can be set at random, and used to select among two or more sub-tables within the same collision table. This trick is used to implement FHP models for which binary collisions involve a random choice between two output states (see Chapter 3, Section 3.2).

Several fluid dynamics experiments were performed on the RAP-1 machine, to provide early demonstrations of the capabilities of the lattice gas method for fluid dynamics simulations. Figure 1(b) in the preface is a good example which shows the flow behind a fixed solid flat plate in a wind tunnel-type geometry; the streamlines are obtained after post-processing the raw data produced by RAP-1 running the FHP-3 lattice gas model. Further results obtained with RAP-1 are given in Section 10.5.

10.2 Lattice gas algorithms on general purpose computers

Lattice gas algorithms have been implemented, since 1985, on a large variety of computers, ranging from personal computers to powerful massively parallel supercomputers. Implementations were done on machines running various operating systems, with programs written in various programming languages. Obviously it would be out of place to describe all existing implementations; we simply want to shed some light on the main programming problems and strategies in general, independently of the language used, of the operating system, and, as far as possible, of the particular hardware architecture.

10.2.1 Channel-wise vs. node-wise storage

'How to store the Boolean field in the memory of the machine?' is the first problem that the lattice gas code designer has to face. The random access memory of a general purpose computer is organized in words of typically 32 or 64 bits, and the basic operations recognized by most processors manipulate words rather than individual bits. Consider a lattice gas model with b channels per node, residing on a piece of lattice with $\mathcal{N}$ nodes. The storage problem is the following: how to dispatch the $b \times \mathcal{N}$ bits necessary to encode the instantaneous Boolean configuration into the available memory words, so as to minimize memory wasting and to optimize the update speed of the automaton. Two strategies are commonly used; each strategy has advantages and drawbacks, that we discuss hereafter. To be simple, suppose we want to implement a lattice gas model such as HPP (see Section 3.1) which requires 4 channels per node, on a (non-realistic) computer with 4 bits per memory word. The essentials of the discussion hold for any number of channels per node, and for computers with any number of bits per word. We denote by $\mathcal{N}$ the total number of lattice nodes that the lattice gas simulation code must manage.

The first strategy is the 'channel-wise storage': each 4-bit word stores the *same channel* of 4 different nodes. Thus, 4 arrays of memory words are necessary to store the whole Boolean field, one for each channel, and each array must contain $\mathcal{N}/4$ words. Figure 10.2(a) illustrates this storage mode. If the number of node $\mathcal{N}$ is not an integer multiple of the number of bits per memory word, then the number of words per array must be $\mathcal{N}/4 + 1$, if $\mathcal{N}/4$ denotes the integer division. The last word of each array is then only partially used. This strategy has minor memory wasting.

The second strategy is the 'node-wise storage', which, in some way, is 'orthogonal' to the the channel-wise storage strategy. Each 4-bit word stores the 4 different channels of the *same node*. For a lattice with $\mathcal{N}$ nodes, $\mathcal{N}$ words

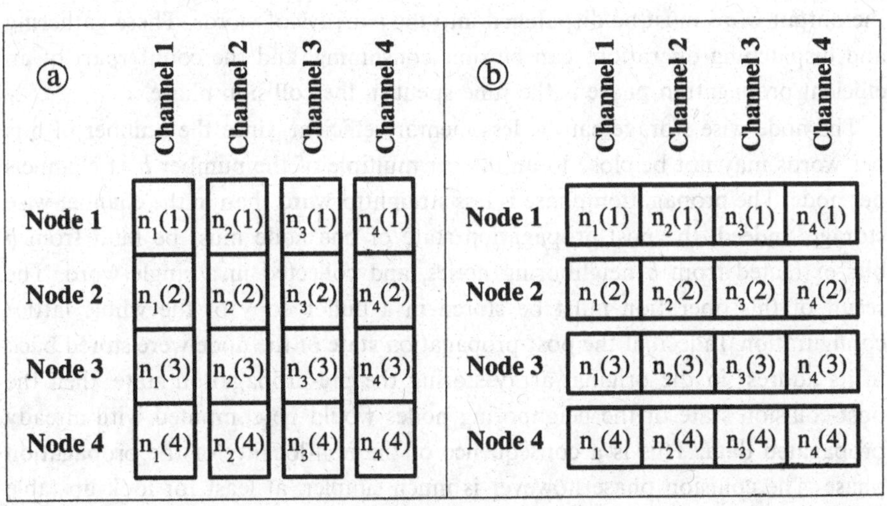

Figure 10.2 The two main storage modes of a Boolean lattice gas in the memory words of a computer. The number of bits per memory words is taken to be 4 for simplicity. The solid rectangular boxes symbolize the 4-bit memory words. (a) Bit-wise storage; (b) node-wise storage.

are necessary to store the whole Boolean field. This storage mode is illustrated in Figure 10.2(b). If the number of bits per memory word is smaller than the number of channels per node, then two or more words may be necessary to store the Boolean state of one single node. If, on the contrary, the number of bits per memory word is larger than b, then the Boolean state of two or more nodes can be stored in one single memory word. There may be some waste of memory if the number of bits per memory word is not close to an integer multiple of b. However, the wasted bits can be used to store auxiliary information (obstacle, body force, etc.).

Neither of these two storage modes can be preferred *a priori*; the right choice depends on several factors. The channel-wise storage wastes very little memory. The propagation phase is conceptually simple, since it reduces to systematic data shifts within each of the b arrays of words. If the collision rule is simple enough to be coded with a non-prohibitive number of logical operations, then the collision phase is also very efficient, since the logical operations (AND, OR, NOT ...) act on all the bits of their data words at the same time, so that a single 'turn off the loop' performs the collision on 64 nodes on a computer with 64-bit words for example. However, when the collision is too complex to be efficiently coded with combined logical operations,[2] one must resort to a look-up table strategy which operates as follows: the b bits of a single node must be picked from b different memory words, and gathered into a single auxiliary word encoding the pre-collision state of the node under consideration. This word is then sent as an address to the collision look-up table, which delivers the post-collision state. Then, the converse operation must be performed: the b bits of the output word must be dispatched into the b arrays of words. These gathering and dispatching operations can be time-consuming, and the counterpart of an efficient propagation phase is the time spent in the collision phase.

The node-wise storage can be less memory-efficient, since the number of bits per words may not be close to an integer multiple of the number b of channels per node. The propagation phase is less straightforward than in the channel-wise storage. Indeed, the post-propagation state of one node must be built from b bits extracted from b neighboring nodes, and collected in a single word. The result of this operation must be stored in a buffer copy of the whole lattice configuration. Indeed, if the post-propagation state of the node were stored back in its address in the original array storing the pre-propagation state, then the post-collision state of the neighboring nodes would be computed with already propagated data. This is a consequence of the non-locality of the propagation phase. The collision phase however is much simpler, at least for look-up table strategies, since each word (or fraction of word) of the buffer copy can directly serve as an address for the look-up table. The results furnished by the look-up

[2] This is the case for most models, except the simplest HPP and FHP models.

table are then stored back in the original array, so that they are ready for the next propagation phase.

To sum up the discussion, the channel-wise storage is in general preferable for models with simple collisions and/or if the available amount of memory is the bottle-neck of the computer to be used; the channel-wise storage offers better efficiency for models with complicated collision schemes, and/or on a computer where the real bottle-neck is the CPU-time.

10.2.2 Collision strategies

Several strategies have been designed to handle the difficult problem of the collisions. In the previous section, we mentioned the implementations with logical operators and with a look-up table. Here we give some further details for these strategies, and describe other algorithms that have been used in some cases.

10.2.2.1 Collision algorithms with logical operators

When the number of channels per node is relatively small, and if the collision rule is sufficiently simple, it may be efficient to design a combination of logical operators that transforms the pre-collision state into the post-collision state. With this combination, one can handle random Boolean variables for non-deterministic models. The solution is equivalent to a transcription with logical operators of the polynomial collision term of the microdynamic equation (see for example Equation (3.6) for the FHP-1 model). This transcription simply amounts to changing the multiplications and the additions by AND and OR operations, respectively. (The OR substitution is safely possible because in the microdynamic equations only one term in each sum can be non-zero.) Except for simple models such as HPP, FHP-1 and FHP-2, this method is prohibitively slow.

10.2.2.2 Collision algorithms with look-up tables

This solution is memory-consuming but rather fast, especially on machines where indirect addressing is efficient. The principle is simple: take the pre-collision state as an address in a table, and the information stored at this address gives the post-collision state. The table must have 2^b entries for a model with b channels per node. This rapidly becomes cumbersome for models such as FCHC which require up to 24 or even 26 channels per node. On a 64-bit machine, a look-up table for FCHC-8 requires 512 megabytes of random access memory, just to hold the table. Of course, a single table makes the model deterministic. This is often incompatible with the condition of G-invariance required for correct macrodynamics. Three solutions exist: (i) use several tables with a random swap between each table, (ii) keep a single table designed to incorporate some kind of 'frozen randomness' to reduce the resulting lack of G-invariance to an acceptable level, (iii) use a multi-entry table (see next section). The first solution is often prohibitively memory-consuming. the second solution works relatively well, especially for models such as FCHC for which the huge number of possible states introduces some natural shuffling. The third solution works well for models such as GBL, for which only a few states lead to random choices with only a few possibilities. The next section is devoted to this kind of collision algorithms with multiple entries.

10.2.2.3 Collision algorithms with multi-entry look-up tables

Very frequently, only a small fraction of the 2^b possible Boolean states leads to several post-collision states, but the number of these possible post-collision states may be too large for a randomization strategy based on a random choice between several *complete* look-up tables. The multi-entry look-up table method solves this problem as follows: a first table stores, in a compact format, a starting address and an ending address for each of the 2^b possible Boolean states. These starting and ending addresses, obtained for a given (pre-collision) state, define a zone in a second table, where all possible post-collision states are stored. It is then easy to select randomly an address between the starting and ending addresses, and to pick the post-collision state stored at this address. This mechanism can be improved to allow for non-equally probable post-collision states.

10.2.2.4 Collision algorithms using symmetries

It is possible but time consuming to use the lattice symmetries to reduce the size of the collision tables. The principle is simple: the possible Boolean states are sorted into families, gathering states which can be transformed one into the other by a lattice isometry. For each family, one single 'family leader' is selected, and its possible output states are stored in a table.

10.2.3 Obstacles

When the node-wise storage mode is selected, there exists a simple and efficient way to implement bounce-back boundary conditions (see Section 2.4.1). Suppose that the b microscopic velocity vectors $\mathbf{c}_i$ are labeled such that $\mathbf{c}_{i+b/2} = -\mathbf{c}_i$. If the Boolean states are stored node-wise in computer words, so that the most significant bit of each word stores the occupation level of channel 1, and so on in sequence up to channel b, then the bounce-back collision is simply a permutation of the $b/2$ most significant bits with the next $b/2$ following bits. This requires only two masking operations, two shifts, and an 'OR' operation.

In addition, it is useless to apply this bounce-back rule in the bulk of an obstacle, provided all obstacle nodes are initialized with zero particle density. It is sufficient to apply the bounce-back rule to the 'surface' of the obstacle only, that is, to the obstacle nodes that can be occupied because of particle propagation. Indeed, the bounce-back rule sends back (to where it came from) any particle arriving at the surface, so preventing any fluid particle 'contaminating' the bulk of the obstacle. The thickness of the 'surface' layer depends on the set of microscopic velocity vectors: if some of these vectors connect a node to its second or third nearest neighbors, then the layer must be two or three times thicker.

10.3 Essential features of a lattice gas simulation code

In the previous section, we addressed the main problems and choices involved in designing the hard core of a lattice gas simulation code: the automaton updating machinery. However, such a code cannot be reduced, even conceptually, to its hard core. Many auxiliary software tools are needed to properly operate the code and to extract qualitative physically relevant data. In addition to pure programmer skill, these auxiliary tools sometimes require deep physical insight.

These tools can be roughly classified into three main categories: initialization tools, raw physical data extraction, and post-processing.

10.3.1 Initialization

Like any computational fluid dynamics method, the lattice gas method requires one to decide what should be the initial macroscopic fields (uniform, harmonic, etc.). Furthermore, for lattice gas methods, there is an additional task before starting the time-evolution: the initial *microscopic* Boolean field must be set so as to correspond to the desired initial macroscopic fields. This involves a random initialization of the $n_i(\mathbf{r}_\star)$ with 0s or 1s, with the minimal constraint that the resulting averaged macroscopic fields be actually *the* desired initial macroscopic fields. Often there is an additional constraint: the random initialization of the $n_i(\mathbf{r}_\star)$ must be realized with probabilities that correspond to (local) equilibrium distributions, such as those discussed in Chapter 4. This eliminates (or shortens) the transient regime during which local equilibrium would have to settle. Therefore the average populations $f_i(\mathbf{r}_\star)$ corresponding to the desired initial macroscopic fields must be computed according to the relations obtained in Section 4.5. Then, at each node and for each channel, a pseudo-random number generator must be used to assign to $n_i(\mathbf{r}_\star)$ the value 0 or 1, with a probability $f_i(\mathbf{r}_\star)$. A convenient way to do this is to use a random number generator delivering values φ with a uniform distribution between 0.0 and 1.0 (most built-in random number generators do so); the integer part of $\varphi + f_i(\mathbf{r}_\star)$, is then either 0 or 1, with respective probabilities $f_i(\mathbf{r}_\star)$ and $1 - f_i(\mathbf{r}_\star)$.

If the initial macroscopic fields are uniform, the probabilities $f_i(\mathbf{r}_\star)$ do not depend on $\mathbf{r}_\star$. Then they can be pre-computed once, for all the nodes (instead of once for each node), leading to considerable speed-up.

Note also that if CPU time is a crucial issue – especially for non-homogeneous initial fields that require computation of the probabilities $f_i(\mathbf{r}_\star)$ for each node – a simplified procedure can be used to obtain the $f_i(\mathbf{r}_\star)$, usually with no dramatic physical consequences: one can use approximation formulae for the probabilities $f_i(\mathbf{r}_\star)$, such as those obtained in Sections 4.5.1 and 4.5.2, but taking into account the first order correction terms only. This simplifies the computation considerably.

10.3.2 Raw physical data extraction

The lattice gas method for fluid dynamics simulations uses a fictitious microscopic word. Like in the 'real world', macroscopic relevant quantities are recovered through some averaging procedure, and the resulting averaged quantities are noisy (like in the 'real world' !), with a 'signal to noise ratio' that increases roughly as the square root of the number of samples used for averaging. But this residual noise is much larger in the lattice gas context than for real fluids because, in general, the number of samples over which the averages can be realistically computed is much smaller (by several orders of magnitude) than the number of particles in the smallest real macroscopic volume. Three kinds of averaging can be performed, separately or in combination, in lattice gases: ensemble-averaging, space-averaging, and time-averaging. The choice of the right method is non-trivial and may involve subtle physical considerations.

10.3.2.1 Ensemble-averaging

Ensemble-averaging requires the simulation to be run with many different realizations of the Boolean fields, initialized independently with the *same* macroscopic fields, that is, with the same probability distribution. The macroscopic fields are then computed at any time, by summation over all these microscopically different replicas. This strategy is particularly well suited to parallel computing, since all the replicas evolve independently, without any need for information exchanges. This noise-reduction technique has however a serious drawback for fluid flows when symmetry-breaking is expected: different replicas may break the symmetry differently, and the averaging over these replicas may spuriously destroy the broken symmetry. An example is discussed in Section 10.5.3.

10.3.2.2 Space- and time-averaging

By space- and/or time-averaging one computes the mean value of the macroscopic fields by averaging over small regions ('cells') of space or space-time. The larger the cells, the smaller the residual noise, and the poorer the effective space- and/or time-resolution. So there should be a compromise between the loss in resolution and the loss in accuracy due to the residual noise. This averaging procedure can also be viewed as a 'poor man's' filtering process, where the microscopic data are convoluted with a square function that is equal to one in a small space or space-time domain, and zero everywhere else. This interpretation gives a natural generalization of the space- and/or time-averaging: the square function can be replaced by any smooth function whose Fourier transform decays smoothly at high space- and/or time-frequencies. However, the compromise between noise reduction and effective resolution still remains: setting the cut-off frequency too low will efficiently reduce the noise, but may reject physically interesting small structures; conversely, setting the cut-off frequency too high will produce prohibitively noisy fields.

10.3.3 Post-processing

Like standard floating point computational fluid dynamics methods, lattice gas simulations often require serious processing for drawing physically significant results from raw data, i.e. the averaged macroscopic fields. This may involve physical aspects such as computing macroscopic fields different from those delivered naturally by the lattice gas simulation. For example, one may wish to compute the pressure field and the vorticity field from the density and velocity fields. This phase may also involve graphical aspects, such as choosing the right way to display the results to highlight a particular phenomenon, which is of crucial importance, especially for three-dimensional simulations. Since these problems are common to all computational fluid dynamics methods, we shall not embark on a detailed discussion here. We shall focus attention on post-processing problems involving specific aspects of the lattice gas method, such as drag and lift coefficients measurements, and Strouhal number evaluations.

10.3.3.1 Drag and lift coefficients measurements

A viscous fluid moving around a fixed solid obstacle always exerts a force on the object. This force can have longitudinal and transverse components with respect to the mean flow direction. The longitudinal component, expressed in a suitable dimensionless form, is the 'drag coefficient':

$$C_{\mathrm{D}} = \frac{F_{\parallel}}{\frac{1}{2}\rho u^2 S}, \tag{10.1}$$

where ρ is the mass density, u is the mean flow velocity, and S is the area of transverse section of the obstacle, that is, the area of projection of the obstacle onto a plane orthogonal to the mean flow velocity.

The transverse component of the force, expressed in the same way, yields the 'lift coefficient':

$$C_L = \frac{F_\perp}{\frac{1}{2}\rho u^2 S}. \tag{10.2}$$

These quantities are of obvious interest, and can be rather accurately extracted from a lattice gas simulation (see Rivet, 1993). First, one has to compute the momentum variation $\Delta\mathbf{p}$ during one time step, due to the collisions on the obstacle. This must be done at each time step during the simulation, and the values must be saved to be subsequently post-processed. Then one has to compute the area of transverse section of the obstacle, in natural lattice units, and the average mass density *per unit volume*, in natural lattice units. The volume of the unit lattice cell must be taken into account if it differs from 1 (in natural lattice units). The drag and lift coefficients are obtained from:

$$C_D = \frac{\Delta P_\parallel}{g(\rho)\frac{1}{2}\rho u^2 S}, \tag{10.3}$$

and

$$C_L = \frac{\Delta P_\perp}{g(\rho)\frac{1}{2}\rho u^2 S}. \tag{10.4}$$

The factor $g(\rho)$ appears to express the momentum variation per natural lattice time unit (time step) as a force, that is, a momentum variation per natural physical time unit (see Section 9.2). It was found that the values of the drag coefficients computed for spheres and cylinders with the lattice gas method fall within a small percentage of experimental values.

10.3.3.2 Strouhal number evaluation

When the fluid velocity around an obstacle increases so that the Reynolds number exceeds a critical value, the initially laminar stationary flow can produce an oscillating wake, such as the Bénard–von Kármán vortex street (see Section 10.5.3). The frequency f_w of the oscillation (in dimensionless form) is the 'Strouhal number':

$$St = \frac{f_w L}{u}, \tag{10.5}$$

where L is the typical size of the obstacle, and u is the mean flow velocity.

The Strouhal number – and how it varies as a function of the Reynolds number – may be a precious indicator of what happens in the wake (König, Eisenlohr and Eckelmann, 1990). This is the case for example in the wake of a cylinder, where the three-dimensional structure of the wake may be complex. The Strouhal number can be computed by 'lattice gas hot-wire anemometry', i.e. by implementing the numerical analogue of a 'hot-wire velocity probe', a laboratory device that delivers the local fluid velocity as a function of time (see

Bonetti and Boon, 1989). In the lattice gas simulation, the 'numerical' hot-wire probe is an averaging cell, over which the averaged velocity is computed *at each time step* during the simulation, and stored for further post-processing. The resulting (discrete) signal is very noisy, but a low frequency mode clearly dominates. The next step is to get the best possible estimate of the frequency of this dominant mode, which is done with standard signal processing algorithms such as modulation and filtering. The frequency so obtained yields the Strouhal number through the formula:

$$St = \frac{f_w L}{g(\rho)u}, \tag{10.6}$$

where L and u are expressed in natural lattice units. The lattice gas-specific non-Galilean factor $g(\rho)$ appears as a scaling factor between the physical time unit and the natural lattice time unit (see Section 9.2). Strouhal numbers obtained from LGA simulations of the wake behind a cylinder are also found to be in agreement (within a small percentage) with experimental values.

10.4 Measurement of basic lattice gas properties

In order to perform reliable quantitative lattice gas simulations, it is essential to have correct evaluations of the non-Galilean factor $g(\rho)$, of the sound speed c_s, and of the kinematic and bulk viscosity coefficients (ν and ν'). The theoretical viscosity coefficients are obtained within the Boltzmann approximation (see Section 5.7.4), and thus may yield inaccurate values. Therefore it is important to obtain an 'experimental measure' of these coefficients by numerical simulations. For the non-Galilean factor and the sound speed, the macrodynamic theory (see Chapter 5) provides more accurate values (here the Boltzmann approximation is not necessary) at least for models satisfying the semi-detailed balance condition. For models without semi-detailed balance, we must use the Boltzmann approximation from the beginning, even to evaluate $g(\rho)$ and $c_s(\rho)$ (c_s may depend on the density when semi-detailed balance is not verified). So it is important to use reliable methods to 'measure' numerically the values of these coefficients for several densities. This calibration operation is a prerequisite that should be carried once for all for any lattice gas model that is going to be used for hydrodynamic simulations. A convenient and efficient method for this calibration is to 'measure' the lattice gas quantities from sound or shear wave simulations (see Section 9.1 on acoustic regimes).

10.4.1 Measuring $g(\rho)$ and $\nu(\rho)$

Consider a lattice gas simulation running on a lattice domain with periodic boundary conditions along each direction. The initial state of the lattice gas is prepared with the following values of macroscopic variables:

$$\begin{cases} \rho(0,\mathbf{r}) = b \times d\,, \\ u_x(0,\mathbf{r}) = u_{1x}\,, \\ u_y(0,\mathbf{r}) = u_{1y}\cos(k\,x)\,. \end{cases}$$

The size of the domain in the x direction must be an integer multiple of wavelength $2\pi/k$, and both u_{1x} and u_{1y} should be sufficiently small so that the non-linear terms in the LGA hydrodynamic equations become unimportant (see Section 9.1). These conditions correspond to a uniform flow in the x direction, superimposed to a sinusoidal shear flow in the y direction, with a wave vector in the x direction. The same arguments as in Section 9.1 lead to the conclusion that the shear wave propagates in the x direction with a phase velocity $g(\rho)u_{1x}$ and decays exponentially as $\exp(-\nu(\rho)k^2 t)$.

Obtaining $g(\rho)$ and $\nu(\rho)$ is then easy: at regularly spaced time intervals, the macroscopic velocity field is computed, then a spatial Fourier transform is applied. The phase and the logarithm of the amplitude of the mode corresponding to k are extracted, and their time-variations are adjusted to linear functions (by a standard least squares fit method): the slopes yield respectively the non-Galilean factor $g(\rho)$, and the kinematic viscosity $\nu(\rho)$. The root mean square difference between the simulation values and the best fitting straight line can serve as an estimate of the measurement accuracy.

10.4.2 Measuring $c_s(\rho)$ and $\nu'(\rho)$

The same method applied to sound (compression) waves provides a measurement of the sound damping coefficient. As discussed in Section 9.1, the amplitude of sound waves decays as $\exp(-\Gamma k^2 t)$, where

$$\Gamma = \frac{D-1}{D}\nu + \frac{1}{2}\nu'\,.$$

The damping coefficient Γ is obtained from a least squares fit of the logarithm of the amplitude as a function of time. Knowing $\nu(\rho)$ from shear wave experiments, and Γ from sound wave experiments, the value of the bulk viscosity $\nu'(\rho)$ follows straightforwardly.

The sound velocity cannot be accurately extracted by fitting the phase of the Fourier mode to a linear function. Indeed, the phase speed in the viscous case is *not* c_s but $\sqrt{c_s{}^2 - \Gamma^2 k^2}$ (see Equation (9.12)). One convenient way to obtain c_s

is to compute the Fourier transform of both velocity and density fluctuations, and to select the mode corresponding to k. The ratio of their moduli gives the sound speed c_s.

10.4.3 An example: the FCHC-3 model

As an illustration of the calibration procedure, we consider the FCHC-3 lattice gas. Remember that the FCHC-3 model is a four-dimensional FCHC model with no rest particle ($b = 24$), with semi-detailed balance and self-duality (see Section 3.7.3 on the FCHC-3 model). The collision rule involves a look-up table

Table 10.1 Non-Galilean factor $g(\rho)$ and sound speed $c_s(\rho)$ for the FCHC-3 lattice gas model. d is the density per channel, $\rho = d \times b$ is the density per node. g_{th} is the theoretical value of $g(\rho)$ (see Equation (9.3)), g_{exp} is the value obtained by shear wave numerical simulations, g_{err} is the root mean square estimated accuracy of the simulation. The same subscripts are used for the sound speed $c_s(\rho)$ (see Equation (9.10) for the theoretical value).

d	ρ	g_{th}	g_{exp}	g_{err}	c_{sth}	c_{sexp}	c_{serr}
0.32	7.680	0.3529	0.3531	0.0040	0.7071	0.7097	0.0040
0.34	8.160	0.3232	0.3230	0.0040	0.7071	0.7079	0.0040
0.36	8.640	0.2916	0.2906	0.0040	0.7071	0.7068	0.0040
0.38	9.120	0.2580	0.2571	0.0040	0.7071	0.7079	0.0040
0.40	9.600	0.2222	0.2215	0.0040	0.7071	0.7073	0.0040
0.42	10.080	0.1839	0.1827	0.0040	0.7071	0.7065	0.0040
0.44	10.560	0.1428	0.1416	0.0040	0.7071	0.7046	0.0040
0.46	11.040	0.0987	0.0980	0.0040	0.7071	0.7053	0.0040
0.48	11.520	0.0512	0.0509	0.0040	0.7071	0.7055	0.0040
0.50	12.000	0.0000	0.0005	0.0040	0.7071	0.7067	0.0040

Table 10.2 Kinematic and bulk viscosity coefficients $\nu(\rho)$ and $\nu'(\rho)$ for the FCHC-3 lattice gas model. The convention for the meaning of the subscripts is the same as in Table 10.1.

d	ρ	ν_{th}	ν_{exp}	ν_{err}	ν'_{th}	ν'_{exp}	ν'_{err}
0.32	7.680	0.0351	0.0392	0.0007	0.0000	0.0014	0.0015
0.34	8.160	0.0321	0.0359	0.0007	0.0000	0.0001	0.0011
0.36	8.640	0.0296	0.0329	0.0006	0.0000	0.0021	0.0013
0.38	9.120	0.0274	0.0301	0.0006	0.0000	0.0022	0.0013
0.40	9.600	0.0256	0.0280	0.0006	0.0000	0.0010	0.0011
0.42	10.080	0.0242	0.0262	0.0006	0.0000	0.0014	0.0015
0.44	10.560	0.0230	0.0245	0.0006	0.0000	0.0006	0.0009
0.46	11.040	0.0222	0.0234	0.0005	0.0000	0.0007	0.0008
0.48	11.520	0.0217	0.0226	0.0005	0.0000	0.0010	0.0008
0.50	12.000	0.0215	0.0225	0.0005	0.0000	0.0003	0.0005

strategy with one single table. So the FCHC-3 lattice gas is only approximately G-invariant. Simulations are performed over a domain of $256 \times 64 \times 64$ lattice nodes (see Dubrulle *et al.*, 1990). Tables 10.1 and 10.2 give the theoretical values, the measured values and the estimated accuracies, for $g(\rho)$ and $c_s(\rho)$, and for the coefficients $\nu(\rho)$ and $\nu'(\rho)$ of the FCHC-3 lattice gas model.

10.5 Examples of lattice gas simulations

We now turn to more concrete applications as we present numerical experiments performed by lattice gas methods for some classical fluid dynamical phenomena.[3]

10.5.1 The Kelvin–Helmholtz instability

Consider two fluid layers initially separated by a flat interface, and moving with respect to each other with parallel velocities. When the velocity difference is sufficiently large, the flat interface becomes unstable, and a sinusoidal disturbance of the interface grows and eventually saturates because of the non-linearities. This instability produces an array of identical vortices rolling 'between' the two fluid layers, a phenomenon known as the 'Kelvin–Helmholtz instability' (see Drazin and Reid, 1981). This type of flow was simulated in two dimensions on the RAP-1 machine using the FHP-3 lattice gas model, and variants incorporating two different species with diffusion, or reaction-diffusion. The advantage is that

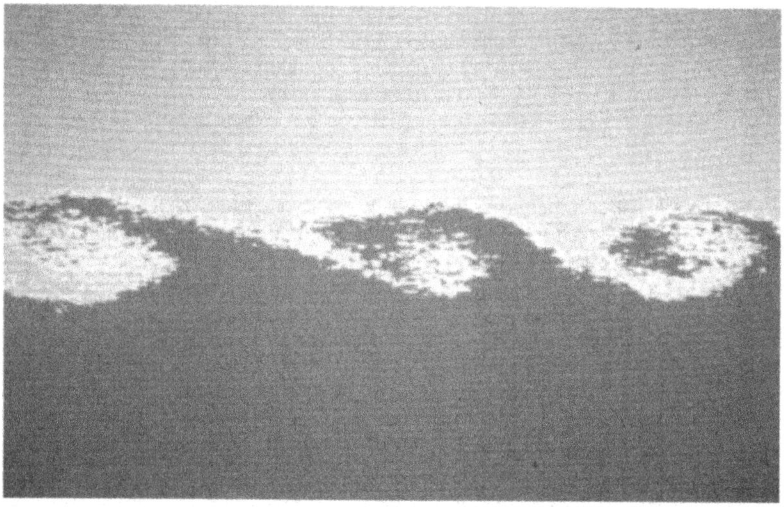

Figure 10.3 Kelvin–Helmholtz instability simulated on RAP-1, with a diffusive two-species version of the FHP-3 model. Snapshot after 1200 automaton steps.

[3] For further illustrations of lattice gas simulations, see Section 11.7 in Chapter 11.

the second species may serve as a tracer for better visualization of the vortices. Figure 10.3 shows the simulation of the Kelvin–Helmholtz instability after 1200 time steps. The upper layer flows to the left, and is more concentrated in one of the two species, whereas the lower layer flows to the right and is more concentrated in the other species. The gray levels label the concentration differences. The two layers are initially separated by a flat interface, and after 1000 time steps the Kelvin–Helmholtz instability is clearly visible. The patterns in Figure 10.3 can be compared (qualitatively) to those in Figure 10.4 showing (real) clouds in two atmospheric layers undergoing a Kelvin–Helmholtz instability.

Figure 10.5 illustrates a simulation with a slightly different collision rule where the two species are subject to a reactive process according to a majority rule: if there are more particles of species A on a node, then the particles of the less abundant species B are transformed into particles of species A. Consequently, the gradients are sharper, and the dispersive effect of particle diffusion is reduced; so one can run the simulation for longer times (4000 time steps) and avoid fuzzy contours.

10.5.2 Particle aggregation

With a simple modification of the FHP-3 rules, one can study qualitatively the fractal aggregation of solid particles. A randomly chosen rest particle plays the

Figure 10.4 Photograph of real clouds between two atmospheric layers with differential motion: the Kelvin–Helmholtz instability occurring between these layers is made visible by the clouds. (Drazin and Reid, 1981.)

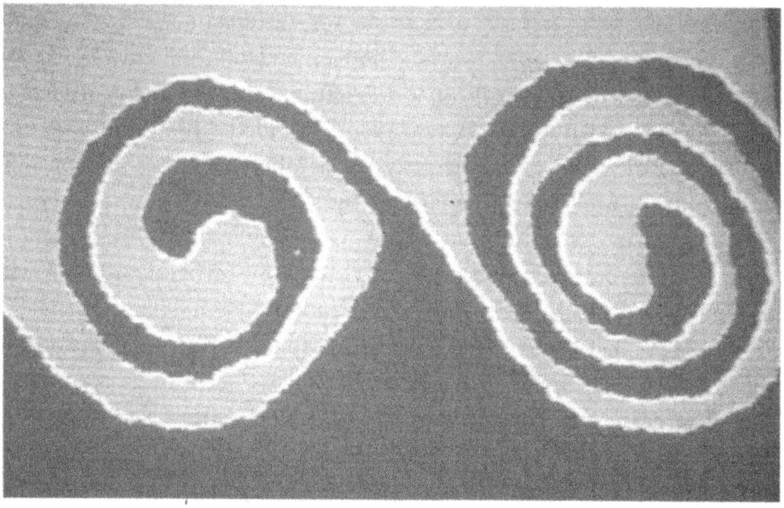

Figure 10.5 Snapshot (after 4000 automaton steps) of the Kelvin–Helmholtz instability simulated on RAP-1, with a two-species reaction-diffusion version of the FHP-3 model. The reactive process is used to reduce the dispersive effect of diffusion, in order to obtain sharper contours. Note that this lattice gas model differs from the model used in the simulation illustrated in Figure 10.3.

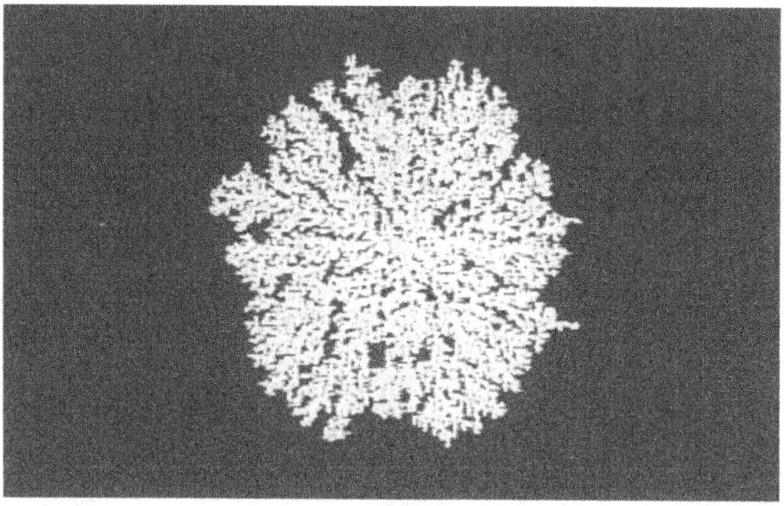

Figure 10.6 Fractal aggregation of particles starting with a 'seed' located at the center of the figure, simulated with a FHP-3 lattice gas model with 'sticky' rest particles.

role of the initial seed: it is 'sticky' in the sense that any moving particle arriving at one of the neighboring sites (next to the seed) stops and stays, and becomes sticky too. In the course of time, an aggregate of fixed sticky particles forms and grows with an apparently fractal structure. Figure 10.6 showing such a fractal aggregate was obtained after just a few seconds of simulation on RAP-1.

10.5.3 Two-dimensional flow past an obstacle

The FCHC lattice gas models described in Chapter 3 are basically four-dimensional models which were initially designed to simulate three-dimensional incompressible flow with, in addition, the possibility of tracking a passive scalar in the flow (see Section 9.3.2). With one more dimensional reduction (as described in Section 9.3.2), four-dimensional models can also be used – in a degenerate way – to simulate two-dimensional hydrodynamics with *two* passive scalars.

Having recourse to four-dimensional models to simulate three-dimensional hydrodynamics is quite logical since there exists no simple really three-dimensional lattice gas model with the required symmetry giving correct isotropy in the hydrodynamic equations. But it may seem somewhat odd to use four-dimensional models for *two-dimensional* hydrodynamics, since simpler really two-dimensional hydrodynamically correct lattice gases exist. In fact there is a good reason for this apparently paradoxical idea. Each model has an intrinsic kinematic viscosity coefficient that depends on the details of the collision rules. This coefficient cannot be made indefinitely small, at least for models with semi-detailed balance (see Section 2.3.1). This limitation imposes an upper bound to the range of achievable Reynolds numbers, with a given amount of computer memory. Because FCHC models have much lower viscosity than usual two-dimensional lattice gases, using FCHC for two-dimensional hydrodynamics allows us to reach higher Reynolds numbers, and thereby to broaden the range of achievable hydrodynamic regimes, with a more efficient use of computer resources. The price to pay is the complexity of the model.

Figures 10.7 and 10.8 illustrate the capability of FCHC models to simulate two-dimensional hydrodynamics. The FCHC-8 lattice gas was implemented on a four-processor CRAY-2 computer. The simulation domain is a rectangle of 4096×1024 lattice nodes. The upper and lower boundaries are solid walls, with bounce-back rules. A wind-tunnel type flow is driven by momentum injection at the inlet (left boundary) and momentum removal at the outlet (right boundary) so as to maintain a roughly constant overall left to right fluid velocity with modulus ~ 0.3 in lattice units. To the left (at 100 nodes of the inlet) and at mid-height, there is a rectangular solid obstacle spanning over 32×100 nodes. The resulting Mach number and Reynolds number are $Ma = 0.47$, and $Re = 564$, respectively. Although this value of the Mach number may seem relatively high to be compatible with incompressibility, the density fluctuations were checked to stay below 3.5% of the mean value $d = 0.38$.

The simulation domain is divided into 32×32 square cells, over which the fluid velocity and density are space-averaged (with no time-averaging). No ensemble-averaging is performed because the fluid flow is likely to break spontaneously the 'top–bottom' symmetry. Statistically, the independent realizations over which the ensemble-average would be computed may break this symmetry differently, and consequently, ensemble-averaging would 'erase' the symmetry-breaking.

Figure 10.7 shows the macroscopic velocity field 40 000 time steps after the impulsive start of the flow. This corresponds to 69.2 circulation times (see Chapter 9). The velocity is coded with one arrow per averaging cell, and the mean flow is subtracted in order to emphasize the wake structure. A long regular Bénard–von Kármán vortex street is clearly visible in the wake of the obstacle (for basics about Bénard–von Kármán vortex streets, see Lamb, 1945). The vorticity field is not a direct product of the lattice gas simulation. It must be computed from the velocity field during the post-processing phase. Figure 10.8 is a view of the vorticity field corresponding to the velocity field shown in Figure 10.7. The gray scale is such that positive values of the vorticity (counter-clockwise vortices) appear in white, and negative values (clockwise

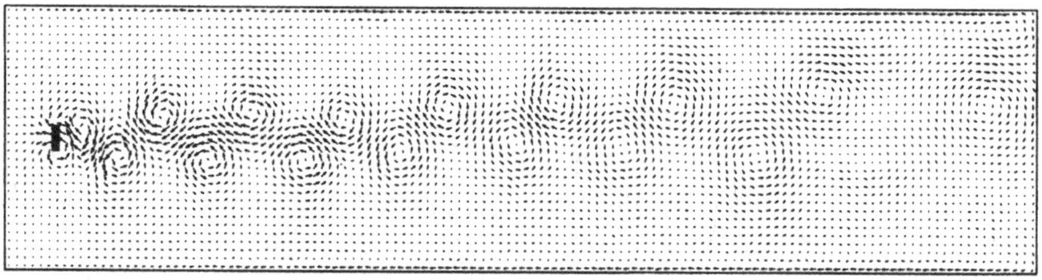

Figure 10.7 Two-dimensional incompressible flow around a rectangular obstacle in wind-tunnel geometry, simulated with the FCHC-8 lattice gas model on a CRAY-2 computer. The fluid velocity field with mean flow subtracted is represented 40 000 time steps (69.2 circulation times) after the impulsive start of the flow.

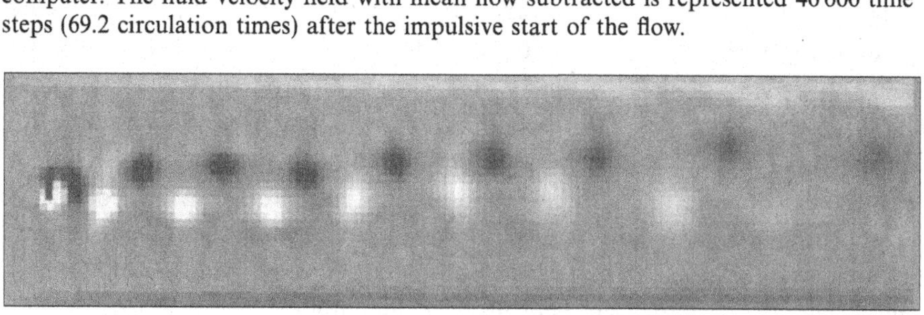

Figure 10.8 Gray-level coding of the vorticity field corresponding to the velocity field in Fig. 10.7.

vortices) appear in black. The square 'pixels' clearly visible in Figure 10.8 are the averaging cells. Purposely, no smoothing was performed in this picture, because it would have hidden the coarse discretization of the physical space, imposed by the necessity of space-averages. As explained in Section 10.3.2, the size of the averaging cells results from a compromise between the accepted noise level, and the desired physical resolution: larger cells would have led to less noisy values of the velocity field, at the expense of a poorer space-resolution; smaller cells would have led to a better resolution, with an increased noise level. With both space- *and* time-averaging, one would have a gain in 'signal-to-noise' ratio for a given space-resolution, but at the cost of a loss in time-resolution.

10.5.4 Three-dimensional flow past an obstacle

The initial vocation of the FCHC family is obviously the simulation of three-dimensional hydrodynamics. Several models of this family have been successfully used to recover subtle, typically three-dimensional fluid dynamics phenomena, such as the loss of translation-invariance in the wake behind a cylinder. The details of the vortex patterns in the wake, and their time-evolution were obtained from the lattice gas simulation, and several features observed in real fluids have been recovered.

Experimental studies of incompressible flows around circular cylinders reveal that the *two-dimensional* wake behind the cylinder undergoes a Hopf bifurcation when the Reynolds number exceeds a critical value close to 47. This bifurcation characterizes a regime transition from a stationary wake with a symmetrical recirculation zone, to the oscillating Bénard–von Kármán vortex street illustrated in Section 10.5.3. The three-dimensional structure of the wake raises complex questions. In the limit of an infinitely long cylinder, is the wake translation-invariant in the direction of the axis of the cylinder? Is there a range of Reynolds numbers where the translation-invariant wake remains stable (a Bénard–von Kármán array of straight vortices parallel to the cylinder)? In that case, what is the critical Reynolds number beyond which the translation-invariant wake becomes unstable, and how does the transition to a more complex regime occur? There are several reasons why laboratory experiments are difficult to carry out and their results hard to exploit. For example, in practice, cylinders are necessarily finite and their ends must be firmly held to avoid vibrations and spurious coupling between fluid oscillations and obstacle vibrations. The question is then: how and how much do the boundary conditions at the ends of the cylinder influence the wake patterns and the relative stability of each regime? For reviews of this interesting problem, see Williamson (1989) for the experimental aspects, and Huerre and Monkewitz (1990) for the theoretical aspects.

We now describe a lattice gas simulation of this flow performed with the

FCHC-8 model. Figure 10.9 illustrates the geometry of the simulation. The solid arrow labeled 'U' shows the direction of the incoming flow. The boundary conditions are periodic in the transverse direction X and in the span-wise direction Y. In the stream-wise direction Z, the inlet and outlet conditions are as described in the two-dimensional simulation of Section 10.5.3. The upstream velocity is 0.19 in lattice units, and the corresponding Mach number and Reynolds number are $Ma = 0.29$ and $Re = 73$, respectively. The Mach number has a low enough value to maintain the density fluctuations within 1.5% of the mean value $d = 0.42$ (for more details, see Rivet, 1993).

Figure 10.10(a) is a snapshot taken 16 000 time steps (74.4 circulation times) after the impulsive start of the flow. It shows a two-dimensional cross-section through the fluid velocity field, in a plane perpendicular to the cylinder axis. A Bénard–von Kármán vortex street is clearly visible in the wake of the cylinder, as in the two-dimensional case. The vortices are evidenced in Figure 10.10(b) with a gray-level encoding of the span-wise (Y) component of the vorticity, in the same plane. The macroscopic velocity field is obtained by space-averaging over $8 \times 8 \times 8$ cubic cells, and the vorticity field is computed *a posteriori* from the macroscopic velocity field, during the post-processing phase; the square pixels are the cross-sections of the cubic space-averaging cells. No smoothing procedure has been applied to the picture for the reason discussed in Section 10.5.3.

Any cross-section by a plane perpendicular to the cylinder would show the same kind of structure: oscillating vortex streets with alternately positive and negative vortices detaching from the obstacle. But are all these oscillating vortex streets phase-synchronized? In other words, are the vortex rolls straight and parallel to the cylinder, as we would naively expect? To answer this question, we make cuts by planes parallel both to the cylinder and to the mean velocity direction. Figures 10.11(a), 10.11(b) and 10.11(c) are three snapshots taken respectively 10 000, 16 000 and 24 000 time steps (46.4, 74.4 and 111.7 circulation times) after the impulsive start of the flow. The pictures give a finite-distance perspective view of the simulation domain, seen from a direction roughly at

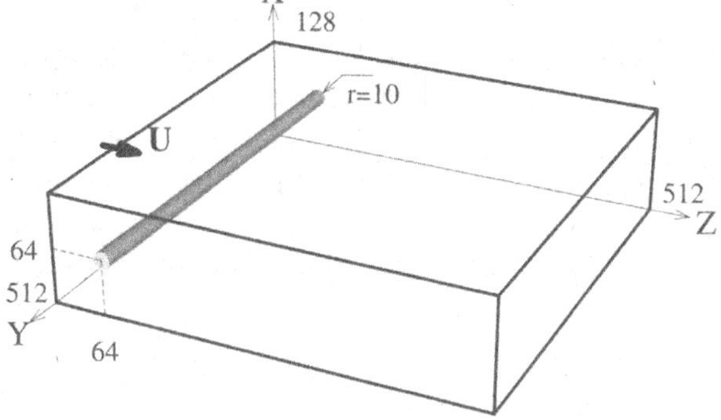

Figure 10.9 Geometry of the simulation box for three-dimensional incompressible flow around a cylinder (all dimensions in lattice nodes).

55° from the X axis. They show a two-dimensional section in a plane tangent to the cylinder and parallel to the stream-wise (Z) direction. In this plane, the velocity field is represented so as to make the span-wise structure of the vortex rolls clearly visible; the transverse (X) component of the macroscopic velocity

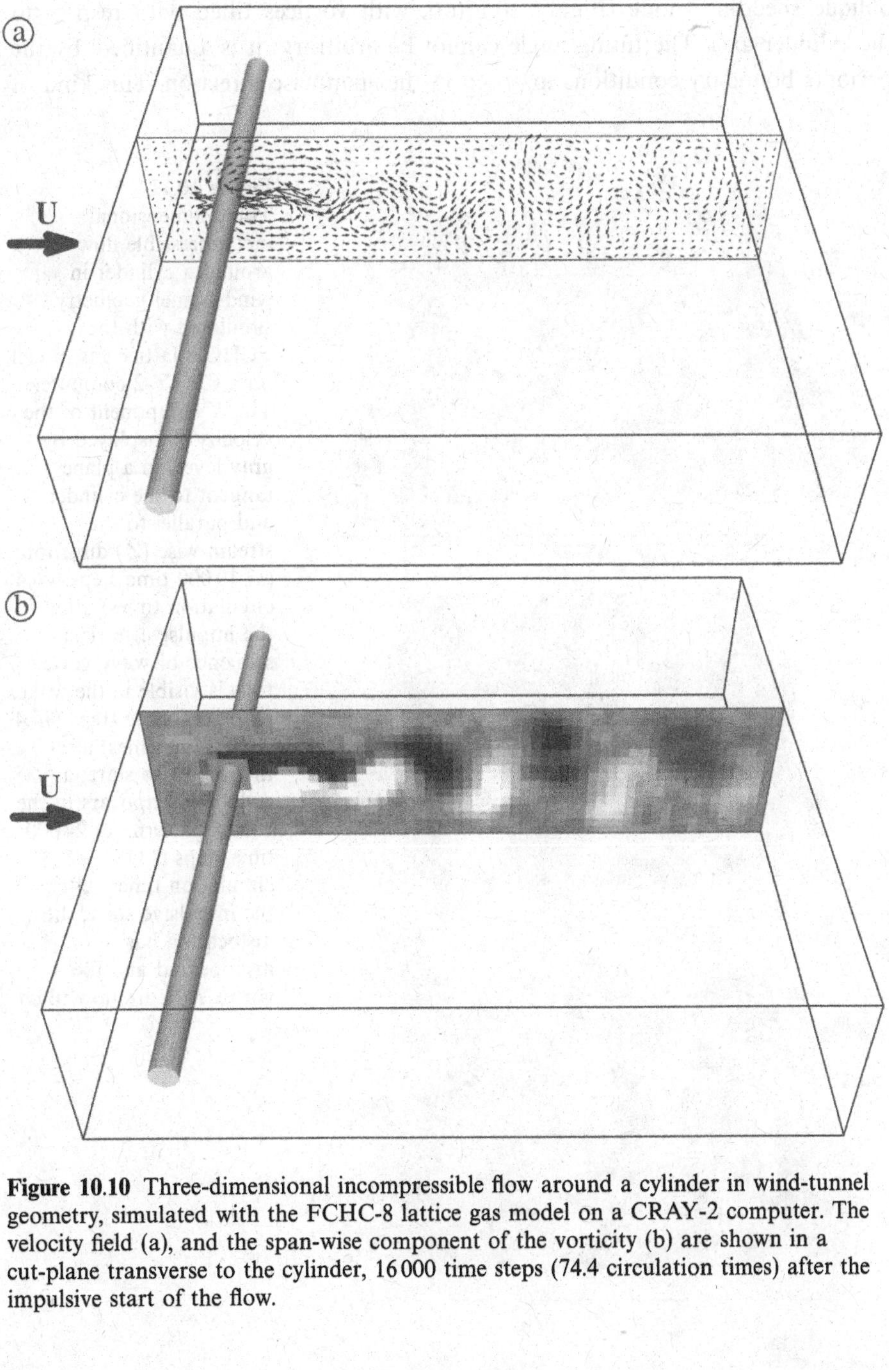

Figure 10.10 Three-dimensional incompressible flow around a cylinder in wind-tunnel geometry, simulated with the FCHC-8 lattice gas model on a CRAY-2 computer. The velocity field (a), and the span-wise component of the vorticity (b) are shown in a cut-plane transverse to the cylinder, 16 000 time steps (74.4 circulation times) after the impulsive start of the flow.

field is represented by a gray-scale coding. The alternating dark and bright stripes are the signatures of the vortex rolls. Without symmetry breaking, they would be straight and parallel to the cylinder, which is obviously not the case: they first appear wavy but roughly parallel to the cylinder (Figure 10.11(a)); then a dislocation occurs (Figure 10.11(b)) in the vortex pattern, leading to an oblique shedding mode (Figure 10.11(c)), with vortices tilted with respect to the cylinder axis. The tilting angle cannot be arbitrary: it is 'quantified' by the periodic boundary conditions imposed in the span-wise direction. This kind of

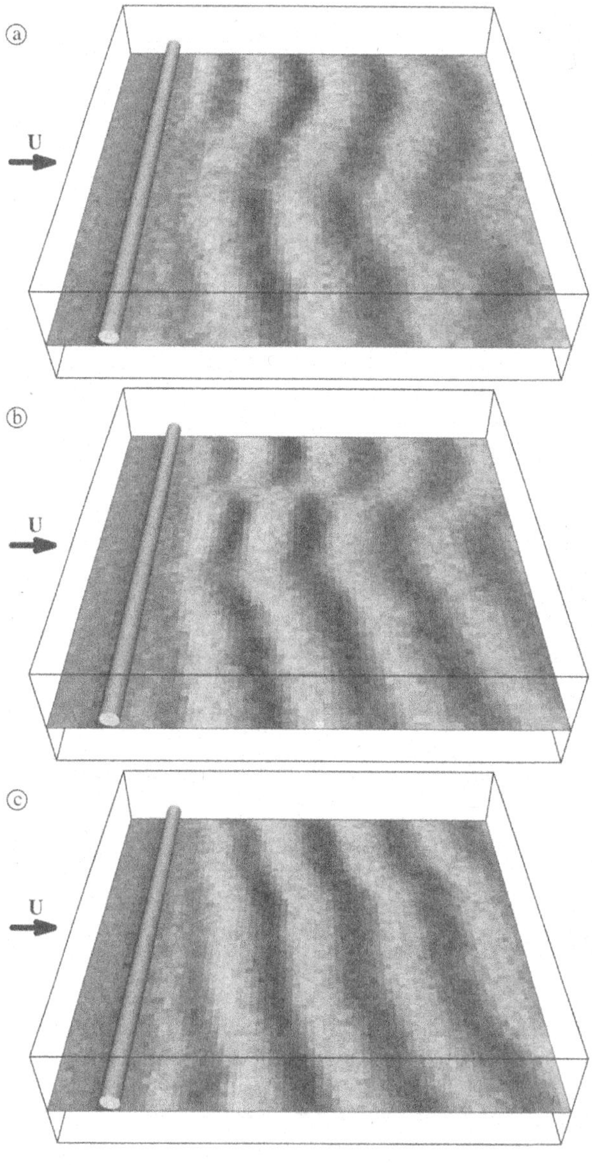

Figure 10.11
Three-dimensional incompressible flow around a cylinder in wind-tunnel geometry, simulated with the FCHC-8 lattice gas model on a CRAY-2 computer. The X component of the velocity is displayed by gray levels in a plane tangent to the cylinder and parallel to the stream-wise (Z) direction. (a) 10 000 time steps (46.4 circulation times) after the impulsive start; a sequence of wavy vortex rolls is visible in the wake. (b) 16 000 time steps (74.4 circulation times) after the impulsive start; a dislocation appears in the vortex pattern. (c) 24 000 time steps (111.7 circulation times) after the impulsive start; the dislocation has disappeared and the vortex rolls are now tilted.

oblique vortex shedding is frequently reported in the literature on experimental and theoretical work on cylinder wakes.

10.5.5 Two-dimensional flow of a two-phase fluid in a porous medium

Fluid flows with moving interfaces, especially in complex geometries (such as a porous medium), are particularly challenging problems in computational and theoretical fluid dynamics (see for instance Homsy, 1987). They are also clearly important for basic and applied research in geology and soil mechanics.

The problem of two-dimensional flow of a two-phase fluid in a porous medium has been efficiently and successfully addressed by lattice gas methods (see for example Lutsko, Boon and Somers, 1992). The model is a variant of the Rothman and Keller model (see Rothman and Keller, 1988) implemented on a network of 400 Transputers. The Rothman and Keller model is an extension of the two-species FHP model, modified to include surface tension. Incorporating surface tension effects *a priori* requires non-local interactions between particles (see Rothman and Zaleski, 1997). In the implementation used by Lutsko, Boon and Somers, the surface tension effect is obtained by transporting the information about the locality of the fluid phases, along the channels during the propagation phase, so that the collision phase remains local.

Here the simulation runs on a lattice domain of 2000×800 nodes, where the porous medium effect is simulated by fixed scattering nodes distributed randomly over the simulation domain (here 1% of the lattice nodes are occupied by fixed scatterers). The viscosity ratio between the two fluids is ~ 10, and the surface tension coefficient is 0.28. The noise reduction of the data uses space-averaging over 10×10 cells, and time-averaging over 2500 time steps. Figure 10.12 shows the destabilization of the initially flat interface when the low-viscosity (black) fluid is pushed through the medium filled with the highly viscous (grey) fluid: viscous fingering occurs spontaneously, like in the Saffman–Taylor instability.

In natural situations and in laboratory experiments, it is difficult to access the porous medium in order to observe the destabilization of the interface and to analyze its structure whose details are therefore not well known. In the numerical experiments described above, Lutsko, Boon and Somers exploited the lattice gas simulation data to show that the destabilized interface between the two fluids exhibits a multi-fractal structure (Lutsko, Boon and Somers, 1992), a result which should be of interest for further experimental investigations.

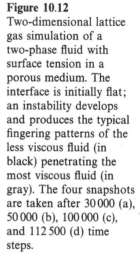

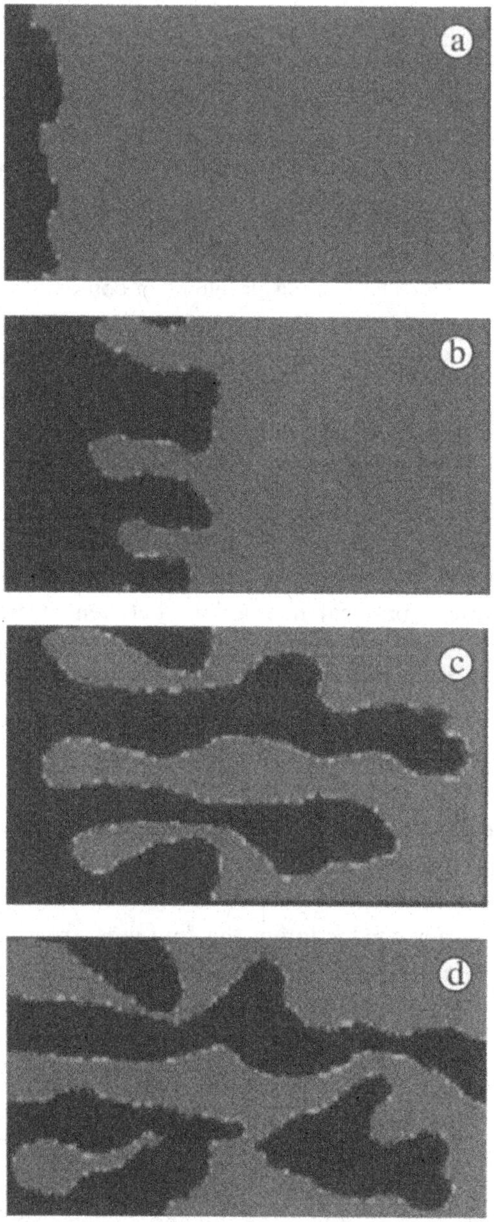

Figure 10.12
Two-dimensional lattice gas simulation of a two-phase fluid with surface tension in a porous medium. The interface is initially flat; an instability develops and produces the typical fingering patterns of the less viscous fluid (in black) penetrating the most viscous fluid (in gray). The four snapshots are taken after 30 000 (a), 50 000 (b), 100 000 (c), and 112 500 (d) time steps.

Chapter 11

Guide for further reading

Since 1985, lattice gas automata have become a widely and actively explored field. Academic groups and industrial laboratories around the world have invested considerable effort in their research activities to drive the subject in various new and promising directions. As a result, the literature on lattice gases and related topics has grown so rapidly and has become so voluminous that an exhaustive list and a detailed review of all relevant lattice gas publications would practically make a book by itself.[1]

Here we select research areas where lattice gases have played an important role, and, for each of them, we quote articles considered as representative, historically or presently. Our review is by no means complete and our choices are certainly selective; unavoidably we haven't done justice to those whose work escaped our selection. Nevertheless we have attempted to cover a range of works expanding over various aspects of the subject in the available literature. Our goal will be reached if the reader finds this chapter a helpful tool for exploring the subject beyond the scope of this book.

[1] Systematic explorations of the literature were performed by G.D. Doolen and by D. d'Humières from 1986 to 1995; alphabetical lists of references (often with abstracts) can be found in special issues of *Physica D*, **47**, pp. 299–337, 1991 (for the period 1986–1990), *J. Stat. Phys.*, **68**, pp. 611–669, 1992 (for the period 1990–1992), and *Fields Inst. Comm.*, **6**, pp. 275–346, 1996 (for the period 1992–1995).

11.1 The historical 'roots'

11.1.1 Discrete kinetic theory

In the early sixties, the problem of shock waves in dilute gases was a subject of increasing research effort, for fundamental as well as industrial reasons. Because a shock wave – for example in supersonic flow – may have a typical thickness of the same order as the mean free path of the particles in the dilute gas, the problem cannot be treated with continuous fluid mechanics (as described by the Navier–Stokes equations) which is valid in the 'hydrodynamic limit', that is, when the smallest excited scale (macroscopic scale) is much larger than the mean free path (microscopic scale). A logical approach then is gas kinetic theory, and in particular the Boltzmann equation. Now solving an integro-differential equation such as the Boltzmann equation is far from a trivial matter. To circumvent the problem, 'discrete kinetic models' were proposed; in these models, the dilute gas is modeled as a set of particles with *continuous* position space, but with *discrete* velocity space: the particles are assumed to have only a finite number of velocities. The resulting discrete Boltzmann equations are no longer integro-differential, but partial differential equations instead. This type of discrete kinetic theory can be seen as the 'ancestor' of the lattice gas approach. In fact the idea goes even further back in time: in the mid-nineteenth century, J.C. Maxwell proposed a model gas based on a similar idea. Discrete kinetic theories produced a vast literature (mostly since 1960); the following list gives some references.[2]

- Maxwell, J.C., *Scientific Papers II*, Cambridge University Press, Cambridge (UK), 1890, p. 26.
- Broadwell, J.E., *'Shock structure in a simple discrete velocity gas'*, *Phys. Fluids*, 7(8), p. 1243, 1964.
- Broadwell, J.E., *'Study of rarefied shear flow by the discrete velocity method'*, *J. Fluid Mech.*, **19**(3), p. 401, 1964.
- Gatignol, R., *'Etude de la structure d'une onde de choc pour un gaz à répartition discrète de vitesses'*, *C. R. Acad. Sci. Paris*, **261**, p. 2841, 1965.
- Gatignol, R., *'Théorie Cinétique des Gaz à Répartition Discrète de Vitesse'*, in *Lecture Notes in Physics*, **36**, Springer-Verlag, Berlin, 1975.
- Longo, E., Monaco, R. and Platkowski, T., *'Sound propagation and shock wave structure by the semi-discrete Boltzmann equation'*, *J. Mech. Theo. Appl.*, **7**, pp. 233–243, 1988.
- Cabannes, H., *'On the initial-value problem in discrete kinetic theory'*, *Eur. J. Mech. B*, **10**, pp. 207–224, 1991.

[2] The references are organized by chronological order in each sub-section.

11.1.2 The early days

Discrete kinetic theory describes model gases with discrete velocity space, but where time and position space are continuous. The idea came about, in the early seventies, to further simplify these models by restricting particle displacements to the geometry of a regular lattice where the particles move from node to node at regular time intervals. The concept of the lattice gas was born. The earliest model, residing on a two-dimensional square lattice, was proposed as a simple model for statistical mechanics. But it was not until ten years later that the idea arose to use lattice gases for numerical simulations of fluid dynamics. A quite possible reason for this delay is that computer science and technology were not sufficiently developed for fluid dynamical applications of lattice gases in the early seventies. By 1985 the concept of massively parallel computers had spread in the scientific community, which technically paved the way for lattice gas simulations.

The following short list presents some basic references on the first lattice gas model (HPP), and the first 'isotropically correct' model (FHP) for fluid simulation.[3]

- Hardy, J. and Pomeau, Y., *'Thermodynamics and hydrodynamics for a modeled fluid'*, *J. Math. Phys.*, **13**, pp. 1042–1051, 1972.
- Hardy, J., Pomeau, Y. and de Pazzis, O., *'Time evolution of a two-dimensional model system: Invariant states and time correlation functions'*, *J. Math. Phys.*, **14**, pp. 1746–1759, 1973.
- Hardy, J. and Pomeau, Y., *'Microscopic model for viscous flow in two dimensions'*, *Phys. Rev. A*, **16**, pp. 720–726, 1977.
- Frisch, U., Hasslacher, B. and Pomeau, Y., *'Lattice gas automata for the Navier–Stokes equation'*, *Phys. Rev. Lett.*, **56**, pp. 1505–1508, 1986.

11.1.3 Cellular automata

Lattice gas automata can always be cast in such a way that they constitute a sub-class of cellular automata (see the discussion in the Preface). The reader interested in the general class of cellular automata, as well as in some aspects of basic computer science, should refer to:

- von Neumann, J., *Theory of Self-Reproducing Automata*, University of Illinois Press, 1966.
- Hopcroft, J.E. and Ullman, J.D., *Introduction to Automata Theory, Languages and Computation*, Addison-Wesley, New-York, 1979.

[3] These models are discussed in detail in Chapter 3.

- Wolfram, S., *Theory and Applications of Cellular Automata*, World Scientific, Singapore, 1986.

11.2 Three-dimensional models

The promising results of the lattice gas method for fluid dynamics simulations in two dimensions stimulated the search for three-dimensional models. Unfortunately, the 'miracle' that renders the macroscopic behavior of FHP models isotropic does not happen so naturally in three dimensions. Several solutions were proposed, including the 'FCHC models' that reside on a four-dimensional lattice (see Section 3.7.3). But these models have at least 24 channels per node, and the design of collision rules soon becomes a subtle optimization puzzle. Indeed, the collision rules must satisfy the correct physical properties (conservation laws, G-invariance, possibly semi-detailed balance, etc.), and in addition – for high Reynolds number flow – they should yield the smallest possible value for the kinematic viscosity. Moreover, they must be numerically efficient, and shouldn't of course exceed the computer's technical limitations. These questions are discussed in detail in the following papers where the interested reader will also find elaborations on collision rules strategy.

- d'Humières, D., Lallemand, P. and Frisch, U., *'Lattice gas models for 3-D hydrodynamics'*, *Europhys. Lett.*, **2**, pp. 291–297, 1986.
- Hénon, M., *'Isometric collision rules for the 4-D FCHC lattice gas'*, *Complex Systems*, **1**, pp. 475–494, 1987.
- Hénon, M., *'Optimization of collision rules in the FCHC Lattice Gas and addition of rest particles'*, in the proceedings of the workshop *'Discrete Kinetic Theory, Lattice Gas Dynamics and Foundations of Hydrodynamics'*, Torino (Italy), 20–24 September 1988, ed. R. Monaco, World Scientific, Singapore, 1989.
- Somers, J.A. and Rem, P.C., *'The construction of efficient collision tables for fluid flow computations with cellular automata'*, in *Cellular Automata and Modeling of Complex Physical Systems*, pp. 161–177, ed. P. Manneville, Springer-Verlag, Berlin, 1989.

11.3 Theoretical analyses

Initially, rather than tools for computer simulations, lattice gas models were considered as 'test-benches' for theoretical ideas and methods. So it is not surprising that the literature on lattice gas theory is so rich, since many aspects of the subject have been explored by various approaches. In what follows we offer the reader our selection of references which we have organized in five categories.

11.3.1 General lattice gas theory

This set of articles and theses is merely a sample of the basic literature on lattice gas theory. As there is a close connection between theory and computational issues, most of the items listed below address both aspects. Note that the articles defining lattice gas models, listed in Section 11.1.2, also belong to the present list, since they contain a theoretical analysis of the lattice gas models originally introduced in these papers.

- d'Humières, D. and Lallemand, P., *'Lattice gas automata for fluid mechanics'*, *Physica*, **140A**, pp. 337–347, 1986.
- Wolfram, S., *'Cellular automaton fluids 1: basic theory'*, *J. Stat. Phys.*, **45**, pp. 471–525, 1986.
- Hénon, M., *'Viscosity of a lattice gas'*, *Complex Systems*, **1**, pp. 762–789, 1987.
- Frisch, U., d'Humières, D., Hasslacher, B., Lallemand, P., Pomeau, Y. and Rivet, J.P., *'Lattice gas hydrodynamics in two and three dimensions'*, *Complex Systems*, **1**, pp. 649–707, 1987.
- Rivet, J.P., *'Hydrodynamique par la méthode des gaz sur réseaux'*, Ph.D. Thesis, Université de Nice (France), 1988.
- Noullez, A., *'Automates de gaz sur réseaux: aspects théoriques et simulations'*, Ph.D. Thesis, Université Libre de Bruxelles (Belgium), 1990.

11.3.2 Statistical physics and thermodynamics

As emphasized in the Preface, lattice gases are as much models for hydrodynamics (describing collective fluid motions) as they are model systems for statistical mechanics, and both aspects are still actively explored today. The standard tools of statistical mechanics (mode coupling theory, fluctuation-dissipation methods, projection operator techniques, etc.) are quite naturally applicable to lattice gases (as several chapters show in this book). Fundamental as well as complementary information on the statistical mechanical aspects of LGAs can be found in:

- Hardy, J., de Pazzis, O. and Pomeau. Y., *'Molecular dynamics of a classical lattice gas: Transport properties and time correlation functions'*, *Phys. Rev. A*, **13**, pp. 1949–1961, 1976.
- Rivet, J.P., *'Green–Kubo formalism for lattice gas hydrodynamics and Monte-Carlo evaluation of shear viscosities'*, *Complex Systems*, **1**(4), pp. 838–851, 1987.

- Dufty, J.W. and Ernst, M.H., *'Hydrodynamics modes and Green–Kubo relations for lattice gas cellular automata'*, *J. Phys. Chem.*, **93**, p. 7015, 1989.
- Ernst, M.H., van Velzen, G.A. and Binder, P.M., *'Breakdown of the Boltzmann equation in cellular-automata lattice gases'*, *Phys. Rev. A*, **39**, pp. 4327–4329, 1989.
- Ernst, M.H., *'Statistical mechanics of cellular automata fluids'*, in *Liquids, Freezing and the Glass Transition*, eds. J.P. Hansen, D. Levesque and J. Zinn-Justin, North Holland, Amsterdam, 1991.
- Ernst, M.H. and Dufty, J.W., *'Green–Kubo relations for lattice gas cellular automata'*, *J. Stat. Phys.*, **58**, p. 57, 1990.
- van der Hoef, M.A. and Frenkel, D., *'Long-time tails of the velocity autocorrelation function in 2-D and 3-D lattice gas cellular automata: a test of mode-coupling theory'*, *Phys. Rev. A*, **41**, p. 4277, 1990.
- Ernst, M.H., *'Mode coupling theory and tails in CA-fluids'*, *Physica D*, **47**, p. 198, 1991.
- Naitoh, T., Ernst, M.H., van der Hoef, M.A. and Frenkel, D., *'Extended mode coupling and simulations in cellular automata fluids'*, *Phys Rev. A*, **44**, p. 2484, 1991.
- Bussemaker, H.J. and Ernst, M.H., *'Biased lattice gases with correlated equilibrium states'*, *J. Stat. Phys.*, **68**, pp. 431–455, 1992.
- Brito, R., Bussemaker, H.J. and Ernst, M.H., *'A fluctuation formula for the non-Galilean factor in lattice gas automata'*, *J. Phys. A*, **25**, pp. L949–L954, 1992.
- Grosfils, P., Boon, J.P., Brito, R. and Ernst, M.H., *'Statistical hydrodynamics of lattice gas automata'*, *Phys. Rev. E*, **48**(4), pp. 2655–2668, 1993.
- Boghosian, B.M. and Taylor, W., *Correlations and Renormalization in Lattice Gases*, M.I.T. internal report no. MIT-CTP-2265, April 1994.
- Grosfils, P., *'Hydrodynamique statistique des gaz sur réseaux'*, Ph.D. Thesis, Université Libre de Bruxelles (Belgium), 1994.
- Suárez, A., Boon, J.P. and Grosfils, P., *'Long-range correlations in nonequilibrium systems: Lattice gas automaton approach'*, *Phys. Rev. E*, **54**(2), pp. 1208–1224, 1996.
- Suárez, A. and Boon, J.P., *'Non-linear hydrodynamics of lattice gases with semi-detailed balance'*, *Int. J. Mod. Phys. C*, **8**(4), pp. 653–674, 1997.
- Suárez, A. and Boon, J.P., *'Nonlinear lattice gas hydrodynamics'*, *J. Stat. Phys.*, **87**(5/6), pp. 1123–1130, 1997.

- Hanon, D. and Boon, J.P., *'Diffusion and correlations in lattice-gas automata'*, *Phys. Rev. E*, **56**(6), pp. 6331–6339, 1997.

There have been several attempts to extend lattice gas automata to the quantum domain. Models have been proposed for the Dirac equation and for the Schrödinger equation. Also motivated by the observation that a natural architecture for nanoscale quantum computation is that of a quantum cellular automaton, authors have developed theoretical analyses and algorithmic schemes for quantum lattice gas automata. Here are some references on a subject which, although still in its infancy, looks promising.

- Succi, S. *'Numerical solution of the Schrödinger equation with discrete kinetic theory'*, *Phys. Rev. E*, **53**(2), pp. 1969–1975, 1996.
- Meyer, D. *'From quantum cellular automata to quantum lattice gases'*, *J. Stat. Phys.*, **85**, pp. 551–574, 1996.
- Boghosian, B.M., and Taylor, W. *'A quantum lattice gas model for the many-particle Schrödinger equation'*, *Phys. Rev. E*, **56**, pp. 705–716, 1997.
- Meyer, D. *'Quantum lattice gases and their invariants'*, *Int. J. Mod. Phys. C*, **8**(4), pp. 717–725, 1997.
- Boghosian, B.M., and Taylor, W. *'Simulating quantum mechanics on a quantum computer'* *Physica D*, **120**, pp. 30–42, 1998.
- Meyer, D. *'Quantum mechanics of lattice gas automata'*, *J. Phys. A*, **31**, pp. 2321–2340, 1998.

11.3.3 Violation of semi-detailed balance

All theoretical developments in Chapters 4 to 8 are based on the semi-detailed balance assumption (see Section 2.3.1): the collision rules are 'democratic' with respect to all possible microscopic states, so that if initially all microscopic states are equally probable, they remain so after collision. Under this assumption, the value of the kinematic viscosity coefficient (evaluated within the Boltzmann approximation) has a lower bound that may severely limit the achievable upper value of the Reynolds number (for a given computational effort). This lower bound disappears if the semi-detailed balance constraint is relaxed, but then there is a price to pay on the theoretical side: the Boltzmann hypothesis must be *imposed ab initio*. The reader interested in this aspect is referred to:

- Dubrulle, B., *'Method of computation of the Reynolds number for the two models of lattice gas involving violation of semi-detailed balance'*, *Complex Systems*, **2**, pp. 577–609, 1988.
- Dubrulle, B., Frisch, U., Hénon, M. and Rivet, J.P., *'Low-viscosity lattice gases'*, *J. Stat. Phys.*, **59**, pp. 1187–1226, 1990.

- Ernst, M.H. and Bussemaker, H.J., *'Algebraic spatial correlations in lattice gas automata violating detailed balance'*, *J. Stat. Phys.*, **81**(1/2), pp. 515–536, 1995.
- Bussemaker, H.J., *'Pattern formation and correlations in lattice gas automata'*, Ph.D. Thesis, Universiteit Utrecht (The Netherlands), 1995.
- Tribel, O. and Boon, J.P., *'Entropy and correlations in lattice gas automata without detailed balance'*, *Int. J. Mod. Phys. C*, **8**(4), pp. 641–652, 1997.

11.3.4 Invariants and conservation laws

The following articles treat the problem of identifying and classifying in a systematic way the various kinds of invariants in lattice gases (see Section 2.3.3).

- Pomeau, Y., *'Invariants in cellular automata'*, *J. Phys. A*, **17**, pp. L415–L418, 1984.
- d'Humières, D., Qian, Y.H. and Lallemand, P., *'Finding the linear invariants of lattice gases'*, in the proceedings of the workshop *Computational Physics and Cellular Automata*, Ouro Preto (Brazil), 8–11 August 1989.
- Bernardin, D., *'Global invariants and equilibrium states in lattice gases'*, *J. Stat. Phys.*, **68**(3/4), pp. 457–496, 1992.

11.3.5 Obstacles and Knudsen layers

There is a wide variety of problems in hydrodynamics where flows are studied in the presence of obstacles and boundaries with simple or complicated shapes. It is an advantage of the lattice gas approach to offer convenient implementation of solid bodies with non-trivial geometries, by using 'special collision rules' on the boundary nodes defining the shape of the solid body (see Section 2.4.1). However, a careful analysis based on the Boltzmann equation shows that the location where the macroscopic mean velocity actually vanishes is *not* exactly on the surface of the obstacle, but rather a fraction of a lattice mesh size outside. Furthermore, the value of this shift may depend on the orientation of the surface of the obstacle. This is a delicate matter that must be taken into account for accurate simulations, in particular for obstacles with moderate size (compared to the full size of the automaton). This rather technical point is discussed in:

- Lavallée, P., Boon, J.P. and Noullez, A., *'Boundaries in lattice gas flows'*, *Physica D*, **47**, pp. 233–240, 1991.
- Cornubert, R., d'Humières, D. and Levermore, D., *'A Knudsen layer theory for lattice gases'*, *Physica D*, **47**, pp. 241–259, 1991.

- Ginzbourg, I. and d'Humières, D., *'Local second-order boundary method for lattice Boltzmann models'*, *J. Stat. Phys.*, **84**(5/6), pp. 927–971, 1996.
- Ginzbourg, I. and Adler, P.M., *'Boundary flow conditions analysis for the three-dimensional lattice Boltzmann model'*, *J. Phys. II*, **4**, pp. 191–214, 1994.

11.4 Models with particular features

The success of the lattice gas method for the study of macroscopic motions of athermal fluids stimulated efforts to incorporate further physical features to describe systems such as multi-component fluids, reactive processes, reaction-diffusion systems, surface tension and free interfaces, magnetic fields, as well as acoustic phonons and elastic seismic waves. Examples can be found in the following categories:

11.4.1 Fluid mixtures and colloids

- Burges, C. and Zaleski, S., *'Buoyant mixtures of cellular automaton gases'*, *Complex Systems*, **1**, pp. 31–50, 1987.
- Chen, H. and Matthaeus, W.H., *'Cellular automaton formulation of passive scalar dynamics'*, *Phys. Fluids.*, **30**, pp. 1235–1237, 1987.
- Ladd, A.J.C., Colvin, M.E. and Frenkel, D., *'Application of lattice-gas cellular automata to the Brownian motion of solids in suspension'*, *Phys. Rev. Lett.*, **60**, p. 975, 1988.
- Bernardin, D. and Sero-Guillaume, O.E., *'Lattice gases mixtures models for mass diffusion'*, *Eur. J. Mech. B*, **9**, p. 21, 1990.
- van der Hoef, M.A. and Frenkel, D., *'Tagged particle diffusion in 3-D lattice gas cellular automata'*, *Physica D*, **47**, p. 191, 1991.
- van der Hoef, M.A., *'Simulation study of diffusion in lattice gas fluids and colloids'*, Ph.D. Thesis, FOM-Instituut voor Atoom- en Molecuulfysica, Amsterdam (The Netherlands), December 1992.

11.4.2 Reaction-diffusion systems

- Clavin, P., Lallemand, P., Pomeau, Y. and Searby, G., *'Simulation of free boundaries in flow systems by lattice gas models'*, *J. Fluid Mech.*, **188**, pp. 437–464, 1988.
- Dab, D. and Boon, J.P., *'Cellular automata approach to reaction-diffusion systems'*, pp. 257–273, in *Cellular Automata and Modeling of Complex*

Physical Systems, eds. P. Manneville, N. Boccara, G.Y. Vichniac and R. Bidaux, Springer-Verlag, Berlin, 1989.

- Sero-Guillaume, O.E. and Bernardin, D., *'A lattice gas model for heat transfer and chemical reaction'*, *Eur. J. Mech. B*, **9**(2), pp. 177–196, 1990.
- Dab, D., Lawniczak, A., Boon, J.P. and Kapral, R., *'Cellular automaton model for reactive systems'*, *Phys. Rev. Lett.*, **64**(20), pp. 2462–2465, 1990.
- Boon, J.P., Dab, D., Kapral, R. and Lawniczak, A., *'Lattice gas automata for reactive systems'*, *Phys. Rep.*, **273**, pp. 556–647, 1996.

11.4.3 Immiscible fluids and free interfaces

- Clavin, P., d'Humières, D., Lallemand, P. and Pomeau, Y., *'Automates cellulaires pour les problèmes a frontières libres en hydrodynamique à 2 et 3 dimensions'*, *C. R. Acad. Sci. Paris II*, **303**, pp. 1169–1174, 1986.
- d'Humières, D., Lallemand, P. and Searby, G., *'Dynamics of two-dimensional bubbles by the lattice gas method'*, *Complex Systems*, **1**, pp. 333–350, 1987.
- Clavin, P., Lallemand, P., Pomeau, Y. and Searby, G., *'Simulation of free boundaries in flow systems by lattice gas models'*, *J. Fluid Mech.*, **188**, pp. 437–464, 1988.
- Rothman, D.H. and Keller, J.M., *'Immiscible cellular automaton fluids'*, *J. Stat. Phys.*, **52**, pp. 1119–1127, 1988.
- Rothman, D.H. and Zaleski, S., *'Spinodal decomposition in a lattice-gas automaton'*, *J. Phys. Paris*, **50**, pp. 2161–2174, 1989.
- Somers, J.A. and Rem, P.C., *'Analysis of surface tension in two-phase lattice gases'*, *Physica D*, **47**, pp. 39–46, 1991.
- Gunstensen, A.K. and Rothman, D.H., *'A Galilean-invariant immiscible lattice gas'*, *Physica D*, **47**(1/2), pp. 53–63, 1991.
- Appert, C., Rothman, D.H. and Zaleski, S. *'A liquid-gas model on a lattice'*, *Physica D*, **47**(1/2), pp. 85–96, 1991.
- Rothman, D.H. and Zaleski, S., *'Lattice-gas models of phase separation: interfaces, phase transitions, and multiphase flow'*, *Rev. Mod. Phys.*, **66**(4), pp. 1417–1479, 1994.
- Adler, C., d'Humières, D. and Rothman, D.H., *'Surface tension and interface fluctuations in immiscible lattice gases'*, *J. Phys. I*, **4**, pp. 29–46, 1994.
- Olson, J.F. and Rothman, D.H., *'Three-dimensional immiscible lattice gas: application to sheared phase separation'*, *J. Stat. Phys.*, **81**(1/2), pp. 199–222, 1995.

- Flekkøy, E.G. and Rothman, D.H., *'Fluctuating fluid interfaces'*, *Phys. Rev. Lett.*, **75**(2), pp. 260–263, 1995.

11.4.4 Flow in porous media

- Rothman, D.H., *'Cellular automaton fluids: a model for flow in porous media'*, *Geophysics*, **53**, pp. 509–518, 1988.
- Rothman, D.H., *'Macroscopic laws for immiscible two-phase flow in porous media: results from numerical experiments'*, *J. Geophys. Res.*, **95**, pp. 8663–8674, 1990.
- Kohring, G.A., *'Calculation of the permeability of porous media using hydrodynamic cellular automata'*, *J. Stat. Phys.*, **63**(1/2), pp. 411–418, 1991.
- Chen, S., Diemer, K., Doolen, G., Eggert, K., Fu, C., Gutman, S. and Travis, B.J., *'Lattice gas automata for flow through porous media'*, *Physica D*, **47**(1/2), pp. 72–84, 1991.
- Lutsko, J.L., Boon, J.P. and Somers, J.A., *'Lattice gas automata simulations of viscous fingering in porous media,'*, *Lecture Notes in Physics*, **398**, pp. 124–135, Springer-Verlag (Berlin), 1992.

11.4.5 Thermo-hydrodynamics

- Chopard, B. and Droz, M., *'Cellular automata model for heat conduction in fluid'*, *Phys. Rev. Lett. A*, **126**, pp. 476–480, 1988.
- Luo, L.S., Chen, H., Chen, S., Doolen, G.D. and Lee, Y.C., *'Generalized hydrodynamic transport in lattice-gas automata'*, *Phys. Rev. A*, **43**, pp. 7097–7100, 1991.
- Chen, S., Chen, H., Doolen, G.D., Gutman, S. and Lee, Y.C., *'A lattice gas model for thermodynamics'*, *J. Stat. Phys.*, **62**, p. 1121, 1991.
- Ernst, M.H. and Das, S.P., *'Thermal cellular automata'*, *J. Stat. Phys.*, **66**, pp. 465–483, 1991.
- Grosfils, P., Boon, J.P. and Lallemand, P., *'Spontaneous fluctuation correlations in thermal lattice gas automata'*, *Phys. Rev. Lett.*, **68**, pp. 1077–1080, 1992.
- Das, S.P. and Ernst, M.H., *'Thermal transport properties in a square lattice gas'*, *Physica A*, **187**, pp. '191–209, 1992.

11.4.6 Elastic waves

- Rothman, D.H., *'Modeling seismic P-waves with cellular automata'*, *Geophys. Res. Lett.*, **14**(1), pp. 17–20, 1987.

- Mora, P., *'The lattice Boltzmann phononic lattice solid'*, *J. Stat. Phys.*, **68**(3/4), pp. 591–610, 1992.
- Maillot, B., *'Modèles semi-microscopiques d'ondes élastiques'*, Ph.D. Thesis, Institut de Physique du Globe, Paris (France), June 1994.

11.4.7 Other models

- Hatori, T. and Montgomery, D.C., *'Transport coefficients for magneto-hydrodynamics cellular automata'*, *Complex Systems*, **1**, pp. 735–752, 1987.
- Montgomery, D.C. and Doolen, G.D., *'Two cellular automata for plasma computations'*, *Complex Systems*, **1**, pp. 831–838, 1987.
- Chen, H. and Matthaeus, W.H., *'New cellular automaton model for magnetohydrodynamics'*, *Phys. Rev. Lett.*, **58**, pp. 1845–1848, 1987.
- Markus, M. and Hess, B., *'Isotropic cellular automaton for modeling excitable media'*, *Nature*, **347**, pp. 56–58, 1990.
- Chen, S., Chen, H., Doolen, G.D., Lee, Y.C. and Brand, H., *'Lattice gas models for non-ideal gas fluids'*, *Physica D*, **47**, p. 97, 1991.
- Eloranta, K. *'Cellular automata for contour dynamics'*, *Physica D*, **89**(1/2), pp. 184–203, 1995.

11.5 Lattice Boltzmann method

We have stressed the fact that, because of the Boolean nature of the lattice gas variables combined with the (usually) stochastic collision rules, LGAs possess intrinsic fluctuations, and we have discussed in detail the insight that can be gained from the analysis of these fluctuations and their correlations (see Chapter 7). Now when we are interested in macroscopic observables, we must perform time- and/or space- and/or ensemble-averaging of microscopic variables. The macroscopic quantities so obtained are 'noisy' because averaging over a finite sample, while reducing the microscopic fluctuations, does not eliminate them completely. So we face a trade-off between the size of the averaging samples, the desired space- and time-resolution for the macroscopic variables, and the available size of computer memory. In the early nineties, the idea emerged to implement directly the lattice Boltzmann equation (see Section 4.2.2) rather than the microdynamic equations of the lattice gas (see Section 2.2). Indeed, while the microdynamic equations are for Boolean variables that need averaging, the lattice Boltzmann equation governs the time-evolution of the mean populations f_i, which are 'floating-point' *noiseless* variables, and can be simulated directly to obtain average quantities. So the lattice Boltzmann method can be seen as a finite-difference scheme for fluid dynamics numerical simulations. The price to pay is some loss of robustness: a lattice Boltzmann simulation can 'blow up' when the simulation parameters are chosen too close to their limits (which is not the case for lattice gas simulations).

In most practical cases, it is not the full lattice Boltzmann equation which is used, but rather its linearized version. The linearization is performed around an equilibrium solution, so that the collision operator (in the r.h.s. of Equation (4.9)) reduces to a linear operator (the 'linearized collision operator'), which acts on the first order perturbations to the mean populations.

Basic references to the lattice Boltzmann method as well as several applications are found in:

- Succi, S., Benzi, R. and Higuera, F. *'The lattice Boltzmann equation: A new tool for computational fluid dynamics'*, *Physica D*, **47**(1/2), pp. 219–230, 1991.
- Gunstensen, A.K., Rothman, D.H., Zaleski, S. and Zanetti, G., *'Lattice Boltzmann model of immiscible fluids.'*, *Phys. Rev. A*, **43**(8), pp. 4320–4327, 1991.
- Benzi, R., Succi, S. and Vergassola, M., *'The lattice Boltzmann equation: theory and applications'*, *Phys. Rep.*, **222**(3), pp. 145–197, 1992.
- Alexander, F.J., Chen, H., Chen, S. and Doolen, G.D., *'A lattice Boltzmann model for compressible fluids'*, *Phys. Rev. A*, **46**(4), pp. 1967–1970, 1992.
- Chen, S., Wang, Z., Shan, X. and Doolen, G.D., *'Lattice Boltzmann computational fluid dynamics in three-dimension'*, *J. Stat. Phys.*, **68**(3/4), pp. 379–400, 1992.
- Mora, P., *'The lattice Boltzmann phononic lattice solid'*, *J. Stat. Phys.*, **68**(3/4), pp. 591–610, 1992.
- Somers, J.A., *'Direct simulation of fluid flow with cellular automata and the lattice Boltzmann equation'*, *Appl. Sci. Res.*, **51**, pp. 127–133, 1993.
- Martinez, D.O., Matthaeus, W.H., Chen, S. and Montgomery, D.C., *'Comparison of spectral method and lattice Boltzmann simulations of two-dimensional hydrodynamics'*, *Phys. Fluids*, **6**(3), pp. 1285–1298, 1994.
- Wagner, L., *'Dependence of drag on a Galilean invariance-breaking parameter in lattice Boltzmann flow simulations'*, *Phys. Rev. E*, **49**(3), pp. 2115–2118, 1994.
- Oxaal, U., Flekkøy, E.G. and Feder, J., *'Irreversible dispersion at a stagnation point: Experiments and lattice Boltzmann simulations'*, *Phys Rev Lett.*, **72**(22), pp. 3514–3517, 1994.
- Ginzbourg, I. and d'Humières, D., *'Local second-order boundary method for lattice Boltzmann models'*, *J. Stat. Phys.*, **84**(5/6), pp. 927–971, 1996.

11.6 Lattice Bhatnagar–Gross–Krook model

The linearized Boltzmann collision operator (see Section 11.5) can be simplified by reducing its matrix to a diagonal form with one single non-zero eigenvalue. This method proposed initially in 1954 by Bhatnagar, Gross and Krook amounts to replacing the full spectrum of eigenvalues by one single relaxation time. The method applies straightforwardly to the lattice Boltzmann scheme and is quite useful in particular for tuning, within certain limits, the kinematic viscosity to lower values than can be achieved by the LGA method. The procedure known as the lattice BGK method (LBGK) is discussed from the point of the theory as well as for various applications in:

- Bhatnagar, P., Gross, E.P. and Krook, M.K., *'A model for collision processes in gases'*, *Phys. Rev.*, **94**, pp. 511–525, 1954.
- Qian, Y.H., d'Humières, D. and Lallemand, P., *'Lattice BGK models for Navier–Stokes equation'*, *Europhys. Lett.*, **17**, pp. 479–484, 1992.
- Qian, Y.H. and Orszag, S.A., *'Non-linear correction to Navier–Stokes equation derived from lattice BGK models'*, *Europhys. Lett.*, **21**, pp. 255–259, 1992.
- Succi, S., d'Humières, D., Qian, Y.H. and Orszag, S.A., *'On the small-scale dynamical behavior of lattice BGK and lattice Boltzmann schemes'*, *J. Sci. Comp.*, **8**, pp. 219–230, 1993.
- Qian, Y.H., *'Simulating thermohydrodynamics with lattice BGK models'*, *J. Sci. Comp.*, **8**, pp. 231–242, 1993.
- Flekkøy, E.G., *'Lattice Bhatnagar–Gross–Krook models for miscible fluids'*, *Phys. Rev. E*, **47**(6), pp. 4247–4257, 1993.

11.7 Numerical simulations and implementations

Because of the binary and discrete nature of cellular automata, and because of the global updating of the LGA, lattice gases are very good candidates for easy and efficient implementation on general purpose massively parallel (super)computers, as well as on dedicated computers. The next two sub-sections give references to articles on dedicated hardware implementations, and general purpose computer implementations and simulations.

11.7.1 Implementation on dedicated hardware

- Margolus, N., Toffoli, T. and Vichniac, G., *'Cellular-Automata Supercomputers for Fluid Dynamics Modeling'*, *Phys. Rev. Lett.*, **56**, pp. 1694–1696, 1986.

- Toffoli, T. and Margolus, N., *Cellular Automata Machines: a New Environment for Modeling*, M.I.T. Press, Cambridge (MA), 1986.
- Cloucqueur, A. and d'Humières, D., *'RAP-1, a cellular automaton machine for fluid dynamics'*, *Complex Systems*, **1**, pp. 584–597, 1987.
- Cloucqueur, A. and d'Humières, D., *'RAP, a family of cellular automaton machines for fluid dynamics'*, *Helv. Phys. Acta*, **62**, pp. 525–541, 1989.
- Margolus, N. and Toffoli, T., *'Cellular automata machines'*, pp. 219–249, in *Lattice Gas Methods for Partial Differential Equations*, Addison-Wesley, 1990.
- Adler, C., Boghosian, B.M., Flekkøy, E., Margolus, N. and Rothman, D.H., *'Simulating three-dimensional hydrodynamics on a cellular-automata machine'*, *J. Stat Phys.*, **81**(1/2), pp. 105–128, 1995.

11.7.2 Simulations on general purpose computers

- d'Humières, D., Lallemand, P. and Shimomura, T., *Cellular Automata, a New Tool for Hydrodynamics*, Los Alamos National Laboratory, Report LA-UR-85-4051, 1985.
- d'Humières, D., Pomeau, Y. and Lallemand, P., *'Simulation d'allées de Von Karman 2-D à l'aide d'un gaz sur réseau'*, *C. R. Acad. Sci. Paris II*, **301**, pp. 1391–1394, 1985.
- d'Humières, D. and Lallemand, P., *'Ecoulement d'un gaz sur un réseau dans un canal bidimensionnel: développement du profile de Poiseuille'*, *C. R. Acad. Sci. Paris II*, **302**, pp. 983–988, 1986.
- d'Humières, D. and Lallemand, P., *'Numerical simulations of hydrodynamics with lattice gas automata in two dimensions'*, *Complex Systems*, **1**(4), pp. 598–631, 1987.
- d'Humières, D., Lallemand, P. and Searby, G., *'Numerical experiments on lattice gases mixtures and Galilean invariance'*, *Complex Systems*, **1**(4), pp. 632–647, 1987.
- Rivet, J.P., *'Simulation d'écoulements tri-dimensionels par la méthode des gaz sur réseaux: premiers résultats'*, *C. R. Acad. Sci. Paris II*, **305**, pp. 751–756, 1987.
- Rivet, J.P., Hénon, M., Frisch, U. and d'Humières, D., *'Simulating fully three-dimensional external flow by lattice gas methods'*, *Europhys. Lett.*, **7**, pp. 231–236, 1988.
- Kadanoff, L.P., McNamara, G.R. and Zanetti, G., *'From automata to fluid flow: comparisons of simulations and theory'*, *Phys. Rev. A*, **40**, pp. 4527–4541, 1989.

- Somers, J.A. and Rem, P.C., *'Flow computation with lattice gases'*, *Appl. Sci. Res.*, **48**, pp. 391–435, 1991.
- Hénon, M. and Scholl H., *'Lattice-gas simulation of a nontransverse large-scale instability for a modified Kolmogorov flow'*, *Phys. Rev. A*, **43**, pp. 5365–5366, 1991.
- Somers, J.A. and Rem, P.C., *'Obtaining numerical results from the 3D FCHC-lattice gas'*, *Lecture Notes in Physics*, **398**, pp. 59–78, Springer-Verlag, Berlin, 1992.
- Hénon, M., *'Implementation of the FCHC lattice gas model on the Connection Machine'*, *J. Stat. Phys.*, **68**(3/4), pp. 353–378, 1992.
- Rivet, J.P., *'Spontaneous symmetry breaking in the 3-D wake of a long cylinder simulated by the lattice gas method and drag coefficient measurements'*, *Appl. Sci. Res.*, **51**, pp. 123–126, 1993.

11.8 Books and review articles

We close with a list of books and proceedings volumes published since 1986; we also include general review articles on lattice gases and related topics.

- Wolfram, S. (editor), Collection of contributions to the workshop on *Large Non-Linear Systems*, Santa Fe, (NM, USA), 27–29 October 1986; Published as a special issue of *Complex Systems*, **1**(4), 1987.
- Frisch, U., *'Une nouvelle stratégie pour l'hydrodynamique: les réseaux d'automates'*, *J. Astr. Fr.*, **32**, pp. 17–20, 1988.
- Boon, J.P., *'Physique, complexité et automates'*, *Bull. Soc. Fr. Phys.*, October 1989.
- Monaco, R. (editor), Proceedings of the workshop on *'Discrete Kinetic Theory, Lattice Gas Dynamics and Foundations of Hydrodynamics*, Torino (Italy), 20–24 September 1988, World-Scientific, Singapore, 1989.
- Doolen, G.D. (editor), *Lattice Gas Methods for Partial Differential Equations*, Addison-Wesley, 1990.
- Boon, J.P., *'Lattice gas automata: a new approach to the simulation of complex flows'*, in *Microscopic Simulations of Complex Flows*, ed. M. Mareschal, Plenum Press, New York, 1990.
- Doolen, G.D. (editor), *Lattice Gas Methods for Partial Differential Equations: Theory, Applications and Hardware*, special issue of *Physica D*, **47**(1/2), 1991. (Proceedings of the NATO Advanced Research Workshop on *Lattice Gas Methods for Partial Differential Equations: Theory, Applications and Hardware*, Los Alamos National Laboratory (NM, USA), 6–8 September 1989).

- Chen, S., Doolen, G.D. and Matthaeus, W.H., *'Lattice gas automata for simple and complex fluids'*, *J. Stat. Phys.*, **64**(5/6), pp. 1133–1162, 1991.
- Ernst, M.H., *'Statistical mechanics of cellular automata fluids'*, in the proceedings of the NATO ASI on *Liquids, Freezing and the Glass Transition*, Les Houches (France), July 3–28, 1989, eds. J.P. Hansen, D. Levesque and J. Zinn-Justin, North Holland, Amsterdam, 1991.
- Boon, J.P. (editor), special issue of *J. Stat. Phys.*, **68**(3/4), 1992. (Proceedings of the NATO Advanced Research Workshop on *Lattice Gas Automata Theory, Implementation, and Simulation*, Observatory of Nice (France), 25–28 June 1991).
- Boon, J.P., Frisch, U. and d'Humières, D., *'L'hydrodynamique modélisée sur réseau'*, *La Recherche*, **253**, pp. 390–399, April 1993.
- Rothman, D.H. and Zaleski, S., *Lattice-Gas Cellular Automata*, Cambridge University Press, Cambridge (UK), 1997.
- Chopard, B. and Droz, M., *Cellular Automata Modeling of Physical Systems*, Cambridge University Press, Cambridge (UK), 1998.
- Wolf-Gladrow, D., *An Introduction to Lattice-Gas Cellular Automata and Lattice Boltzmann Models*, Springer-Verlag, Berlin, 1999.
- Succi, S., *The Lattice Boltzmann Equation for Fluid Dynamics and Beyond*, Oxford University Press, (to be published).

Appendix

Mathematical details

A.1 A non-local collision model can be re-coded as a local collision model (see Section 1.3.5)

Proof

Consider a non-local collision model with b channels per node. The collision phase is such that the post-collision state $n'_j(\mathbf{r}_\star)$ of any node labeled by its position vector $\mathbf{r}_\star$ is a function $\mathscr{F}$ of the pre-collision state $n_i(\mathbf{r}_\star - \mathbf{e}_k)$ of B nodes (the node itself plus $B-1$ other nodes), with position vectors $\mathbf{r}_\star - \mathbf{e}_k$:

$$n'_j(\mathbf{r}_\star) = \mathscr{F}_j\Big(n_i(\mathbf{r}_\star - \mathbf{e}_k)_{(i=1,\ldots,b;\,k=1,\ldots,B)}\Big), \quad \text{for } j = 1, \ldots, b. \tag{A.1}$$

In the most general case, the non-local collision process at a given node can also involve the state of the node itself. So, one of the $\mathbf{e}_k$ s (say $\mathbf{e}_1$ for example) must be the null vector. In addition, the number B of 'collision vectors' $\mathbf{e}_k$ is finite, and the $\mathbf{e}_k$ s are globally invariant under the action of the point-group of the lattice.

The propagation phase involves the set of b propagation vectors $\mathbf{c}_i$ which are also globally invariant under the point-group of the lattice. The final (post-propagation) state $n'_i(\mathbf{r}_\star)_{(i=1,\ldots,b)}$ of node $\mathbf{r}_\star$ is given by:

$$n'_i(\mathbf{r}_\star) = n'_i(\mathbf{r}_\star - \mathbf{c}_i), \quad \text{for } i = 1, \ldots, b. \tag{A.2}$$

We define a new model with the same lattice structure, but with $\tilde{b} = b \times B$

channels per node. As before, each node is labeled by its position vector $\mathbf{r}_\star$, but now each channel is labeled by a composite label $\tilde{i} = (i,k)$ with $i = 1,\ldots,b$ and $k = 1,\ldots,B$.

The collision phase of the new model is such that the post-collision state $\tilde{n}'_{(j,l)}(\mathbf{r}_\star)$ of any node $\mathbf{r}_\star$ is given by:

$$\tilde{n}'_{(j,l)}(\mathbf{r}_\star) = \mathscr{F}_j\left(\tilde{n}_{(i,k)}(\mathbf{r}_\star)_{(i=1,\ldots,b;\ k=1,\ldots,B)}\right), \quad \text{for } j = 1,\ldots,b;\ l = 1,\ldots,B. \tag{A.3}$$

The new collision process is now purely local, since the right hand side of (A.3) involves only the state $\tilde{n}_{(i,k)}$ of node $\mathbf{r}_\star$.

The propagation phase of the new model involves the propagation vectors $\tilde{\mathbf{c}}_{(i,k)} = \mathbf{c}_i + \mathbf{e}_k$, with i and k ranging from 1 to b and from 1 to B, respectively. The final (post-propagation) state $\tilde{n}'_{(i,k)}(\mathbf{r}_\star)_{(i=1,\ldots,b;k=1,\ldots,B)}$ of node $\mathbf{r}_\star$ is given as usual by:

$$\tilde{n}'_{(i,k)}(\mathbf{r}_\star) = \tilde{n}'_{(i,k)}(\mathbf{r}_\star - \mathbf{c}_{(i,k)}), \quad \text{for } i = 1,\ldots,b;\ k = 1,\ldots,B. \tag{A.4}$$

It is clear that if the initial state $\tilde{n}_{(i,k)}(\mathbf{r}_\star, t_\star = 0)$ of the new model is set such that $\tilde{n}_{(i,k)}(\mathbf{r}_\star, t_\star = 0) = n_i(\mathbf{r}_\star - \mathbf{e}_k, t_\star = 0)$, then, at any further discrete time $t_\star$, the state $\tilde{n}_{(i,k)}(\mathbf{r}_\star, t_\star)$ will verify $\tilde{n}_{(i,1)}(\mathbf{r}_\star, t_\star) = n_i(\mathbf{r}_\star, t_\star)$. In other words, the exact dynamics of the initial model with non-local collisions is 'contained' in the dynamics of the new model with purely local collisions. This shows that a model with non-local collisions can be re-coded as a model with more channels per node, but with local collisions. □

A.2 The so-called 'face-centered hyper-cubic' four-dimensional lattice should be named '2D-face-centered body-centered hyper-cubic' lattice (see section 3.7)

Proof

The four-dimensional FCHC lattice is the set of points of $\mathbb{R}^4$ whose coordinates are signed integers with even sum (see Section 3.7). The basic *simple* hyper-cubic lattice on which the FCHC lattice is constructed is the set of all points with coordinates $(2n_1, 2n_2, 2n_3, 2n_4)$ where (n_1, n_2, n_3, n_4) are signed integers. Consider the elementary hyper-cube limited by 16 lattice points with Cartesian coordinates:

$$\begin{pmatrix}0\\0\\0\\0\end{pmatrix},\begin{pmatrix}2\\0\\0\\0\end{pmatrix},\begin{pmatrix}0\\2\\0\\0\end{pmatrix},\begin{pmatrix}0\\0\\2\\0\end{pmatrix},\begin{pmatrix}0\\0\\0\\2\end{pmatrix},\begin{pmatrix}2\\2\\0\\0\end{pmatrix},\begin{pmatrix}2\\0\\2\\0\end{pmatrix},\begin{pmatrix}2\\0\\0\\2\end{pmatrix},$$
$$\begin{pmatrix}2\\2\\2\\2\end{pmatrix},\begin{pmatrix}0\\2\\2\\2\end{pmatrix},\begin{pmatrix}2\\0\\2\\2\end{pmatrix},\begin{pmatrix}2\\2\\0\\2\end{pmatrix},\begin{pmatrix}2\\2\\2\\0\end{pmatrix},\begin{pmatrix}0\\0\\2\\2\end{pmatrix},\begin{pmatrix}0\\2\\0\\2\end{pmatrix},\begin{pmatrix}0\\2\\2\\0\end{pmatrix}. \tag{A.5}$$

The center of this hyper-cube is the point $(1, 1, 1, 1)$, which clearly belongs to the FCHC lattice since the sum of its coordinates is even. So, the FCHC lattice is 'body-centered'.

This hyper-cube has 8 (hyper-)faces which are obtained by setting successively each of the 4 coordinates equal to either 0 or 2. These hyper-faces are three-dimensional cubes. For instance, the hyper-face which verifies $x_4 = 0$ is limited by the 8 following lattice points:

$$\begin{pmatrix}0\\0\\0\\0\end{pmatrix},\begin{pmatrix}2\\0\\0\\0\end{pmatrix},\begin{pmatrix}0\\2\\0\\0\end{pmatrix},\begin{pmatrix}0\\0\\2\\0\end{pmatrix},\begin{pmatrix}2\\2\\0\\0\end{pmatrix},\begin{pmatrix}2\\0\\2\\0\end{pmatrix},\begin{pmatrix}2\\2\\2\\0\end{pmatrix},\begin{pmatrix}0\\2\\2\\0\end{pmatrix}. \tag{A.6}$$

The center of this hyper-face is the point $(1, 1, 1, 0)$ which does not belong to the FCHC lattice. Thanks to the invariances of the lattice, it is easy to convince oneself that none of the 8 hyper-faces are centered. So, the FCHC lattice is *not* 'face-centered', at least if 'face' is understood as three-dimensional 'hyper-face'.

The hyper-cube has 24 2D-faces which are obtained by setting successively each pair of coordinates equal to either $(0, 0)$, $(2, 0)$, $(0, 2)$ or $(2, 2)$. For instance, the 2D-face which verifies $(x_3, x_4) = (0, 0)$ is limited by the 4 following points:

$$\begin{pmatrix}0\\0\\0\\0\end{pmatrix},\begin{pmatrix}2\\0\\0\\0\end{pmatrix},\begin{pmatrix}0\\2\\0\\0\end{pmatrix},\begin{pmatrix}2\\2\\0\\0\end{pmatrix}. \tag{A.7}$$

The center of this 2D-face is the point $(1, 1, 0, 0)$ which clearly belongs to the FCHC lattice. Using again the invariances of the FCHC lattice, it is easy to show that all 24 2D-faces are centered. Thus the FCHC lattice is also '2D-face-centered'. □

A.3 The full statistics of a collection of Boolean variables contains more information than their individual statistics; when the variables are uncorrelated, the knowledge of either statistics yields the same amount of information (see Section 4.3.1)

The following proof is inspired by Fano (1961).

Proof

We compute the difference δ between the quantity of information contained in the knowledge of all the individual statistics and the knowledge of the full statistics (see Section 4.3.1 for notation):

$$\begin{aligned} \delta &= I^\star - I \\ &= \sum_{i=1}^{q}\left(p_i \log_2 p_i + \overline{p}_i \log_2 \overline{p}_i\right) - \sum_{Y\in\Gamma} \mathscr{P}(Y)\log_2 \mathscr{P}(Y), \end{aligned} \tag{A.8}$$

with $\overline{p}_i = (1 - p_i)$. We now use the fact that for a Bernoulli variable x_i, the probability p_i that x_i be equal to 1 is also the averaged value of x_i, namely:

$$p_i = \sum_{Y\in\Gamma} y_i \mathscr{P}(Y), \quad \text{and} \quad \overline{p}_i = \sum_{Y\in\Gamma} \overline{y}_i \mathscr{P}(Y).$$

We then rewrite δ as:

$$\begin{aligned} \delta &= \sum_{Y\in\Gamma} \mathscr{P}(Y) \sum_{i=1}^{q} (y_i \log_2 p_i + \overline{y}_i \log_2 \overline{p}_i) - \sum_{Y\in\Gamma} \mathscr{P}(Y)\log_2 \mathscr{P}(Y) \\ &= \sum_{Y\in\Gamma} \mathscr{P}(Y)\log_2 \frac{\prod_{i=1}^{b} p_i^{y_i} \overline{p}_i^{\overline{y}_i}}{\mathscr{P}(Y)}. \end{aligned}$$

Now considering the inequality $\ln a \leq a - 1$, which holds for any strictly positive value of a, the difference δ verifies the inequality:

$$\begin{aligned} \delta &\leq \sum_{Y\in\Gamma} \mathscr{P}(Y) \left(\frac{\prod_{i=1}^{b} p_i^{y_i} \overline{p}_i^{\overline{y}_i}}{\mathscr{P}(Y)} - 1 \right) \log_2 e \\ &\leq \sum_{Y\in\Gamma} \prod_{i=1}^{b} p_i^{y_i} \overline{p}_i^{\overline{y}_i} - \sum_{Y\in\Gamma} \mathscr{P}(Y). \end{aligned}$$

The normalization condition for $\mathscr{P}$ and p imposes that $\sum_{Y\in\Gamma} \mathscr{P}(Y)$ and $\sum_{Y\in\Gamma} \prod_{i=1}^{b} p_i^{y_i} \overline{p}_i^{\overline{y}_i}$ be equal to one, from which it follows that:

$$\delta \leq 0.$$

It is clear that $\delta = 0$ when $\mathscr{P}(Y) = \prod_{i=1}^{b} p_i^{y_i} \overline{p}_i^{\overline{y}_i}$, that is, when the individual Boolean variables x_i are *statistically independent*. □

A.4 The following global equilibrium conditions:

$$\prod_{\mathbf{r}_\star \in \mathscr{L}} \prod_{i=1}^{b} f_i^{S'_i} \bar{f}_i^{\bar{S}'_i} = \sum_{S \in \Gamma} \prod_{\mathbf{r}_\star \in \mathscr{L}} A(S \to S') \prod_{j=1}^{b} f_j^{S_j} \bar{f}_j^{\bar{S}_j} \quad \forall S' \in \Gamma \tag{A.9}$$

and

$$\prod_{j=1}^{b} f_j^{s'_j} \bar{f}_j^{\bar{s}'_j} = \sum_{s \in \gamma} A(s \to s') \prod_{j=1}^{b} f_j^{s_j} \bar{f}_j^{\bar{s}_j}, \quad \forall s' \in \gamma, \tag{A.10}$$

are strictly equivalent (see Section 4.4.1)

Proof

We first factorize the r.h.s. of (A.9) as follows: we single out a particular node $\mathbf{r}_0$ in $\mathscr{L}$, and denote by $\mathscr{L}_{(0)}$ the set of all nodes of $\mathscr{L}$ *except* $\mathbf{r}_0$. The r.h.s. of (A.9) can then be rewritten as:

$$\sum_{s \in \gamma} \sum_{\substack{S \in \Gamma, \\ S(\mathbf{r}_0) = s}} A(s \to S'(\mathbf{r}_0)) \prod_{j=1}^{b} f_j^{s_j} \bar{f}_j^{\bar{s}_j} \prod_{\mathbf{r}_\star \in \mathscr{L}_{(0)}} \underbrace{A(S \to S') \prod_{j=1}^{b} f_j^{S_j} \bar{f}_j^{\bar{S}_j}}_{S \text{ and } S' \text{ taken at node } \mathbf{r}_\star},$$

which is also equal to:

$$\sum_{s \in \gamma} A(s \to S'(\mathbf{r}_0)) \prod_{j=1}^{b} f_j^{s_j} \bar{f}_j^{\bar{s}_j} \left(\sum_{\substack{S \in \Gamma, \\ S(\mathbf{r}_0) = s}} \prod_{\mathbf{r}_\star \in \mathscr{L}_{(0)}} \underbrace{A(S \to S') \prod_{j=1}^{b} f_j^{S_j} \bar{f}_j^{\bar{S}_j}}_{S \text{ and } S' \text{ taken at node } \mathbf{r}_\star} \right),$$

We now denote by $\Gamma_{(0)}$ the phase space of the sub-lattice $\mathscr{L}_{(0)}$. The sum over all configurations S in Γ such that $S(\mathbf{r}_0) = s$ can be replaced by a sum over S in $\Gamma_{(0)}$. The above expression then becomes:

$$\sum_{s \in \gamma} A(s \to S'(\mathbf{r}_0)) \prod_{j=1}^{b} f_j^{s_j} \bar{f}_j^{\bar{s}_j} \left(\sum_{S \in \Gamma_{(0)}} \prod_{\mathbf{r}_\star \in \mathscr{L}_{(0)}} \underbrace{A(S \to S') \prod_{j=1}^{b} f_j^{S_j} \bar{f}_j^{\bar{S}_j}}_{S \text{ and } S' \text{ taken at node } \mathbf{r}_\star} \right),$$

The expression in parentheses has exactly the same form as the r.h.s. of (A.9), except that it concerns $\mathscr{L}_{(0)}$ rather than $\mathscr{L}$. The procedure can be repeated with a new particular node, and so on, leading to the following form for the r.h.s. of (A.9):

$$\prod_{\mathbf{r}_0 \in \mathscr{L}} \sum_{s \in \gamma} A(s \to S'(\mathbf{r}_0)) \prod_{j=1}^{b} f_j^{s_j} \bar{f}_j^{\bar{s}_j}.$$

Thus Equation (A.9) is equivalent to:

$$\prod_{\mathbf{r}_\star \in \mathscr{L}} \frac{\sum_{s\in\gamma} A(s\to S'(\mathbf{r}_\star)) \prod_{j=1}^{b} f_j^{s_j} \overline{f}_j^{\overline{s}_j}}{\prod_{i=1}^{b} f_i^{S'_i(\mathbf{r}_\star)} \overline{f}_i^{\overline{S}'_i(\mathbf{r}_\star)}} = 1, \quad \forall S' \in \Gamma. \tag{A.11}$$

Now consider an arbitrary Boolean state s' of the single-node phase space γ. Equation (A.11), which holds for all Boolean configurations S', also holds for *the* Boolean configuration S' for which $S'(\mathbf{r}_\star) = s'$, for all $\mathbf{r}_\star$ in $\mathscr{L}$. Thus (A.11) implies:

$$\left(\frac{\sum_{s\in\gamma} A(s\to s') \prod_{j=1}^{b} f_j^{s_j} \overline{f}_j^{\overline{s}_j}}{\prod_{i=1}^{b} f_i^{s'_i} \overline{f}_i^{\overline{s}'_i}} \right)^{\mathscr{N}} = 1, \quad \forall s' \in \gamma, \tag{A.12}$$

where $\mathscr{N}$ is the number of nodes in $\mathscr{L}$. The expression in parentheses is real and non-negative; so it must be equal to one to satisfy (A.12). This implies (A.10) immediately.

The reverse proof is clear: Equation (A.10), which is true for all s' in γ guarantees that (A.11) holds for all S' in Γ, from which it follows that (A.9) is true. □

A.5 The following equilibrium conditions:

$$\prod_{j=1}^{b} f_j^{s'_j} \overline{f}_j^{\overline{s}'_j} = \sum_{s\in\gamma} A(s\to s') \prod_{j=1}^{b} f_j^{s_j} \overline{f}_j^{\overline{s}_j}, \quad \forall s' \in \gamma, \tag{A.13}$$

$$\sum_{s,s'\in\gamma} (s'_i - s_i) A(s\to s') \prod_{j=1}^{b} f_j^{s_j} \overline{f}_j^{\overline{s}_j} = 0, \quad \forall i = 1,\ldots,b, \tag{A.14}$$

$$\sum_{i=1}^{b} \log\left(\frac{f_i}{1-f_i}\right)(s'_i - s_i) A(s\to s') = 0, \quad \forall s, s' \in \gamma, \tag{A.15}$$

are strictly equivalent when the semi-detailed balance condition is satisfied (see Section 4.4.1)

Proof

- **We prove that (A.13) ⇒ (A.14):**

We use the normalization condition (2.9) to write (A.13) under the following equivalent form:

$$\sum_{s\in\gamma} A(s'\to s) \prod_{j=1}^{b} f_j^{s'_j} \overline{f}_j^{\overline{s}'_j} = \sum_{s\in\gamma} A(s\to s') \prod_{j=1}^{b} f_j^{s_j} \overline{f}_j^{\overline{s}_j}, \quad \forall s' \in \gamma. \tag{A.16}$$

We multiply both sides by s'_i and sum over s' to deduce:

$$\sum_{s,s'\in\gamma} s'_i A(s'\to s)\prod_{j=1}^{b} f_j^{s'_j}\bar{f}_j^{\bar{s}'_j} = \sum_{s,s'\in\gamma} s'_i A(s\to s')\prod_{j=1}^{b} f_j^{s_j}\bar{f}_j^{\bar{s}_j}, \quad \forall i=1,\ldots,b. \tag{A.17}$$

On the l.h.s., both s and s' are dummy variables, which we can permute to obtain:

$$\sum_{s,s'\in\gamma} s_i A(s\to s')\prod_{j=1}^{b} f_j^{s_j}\bar{f}_j^{\bar{s}_j} = \sum_{s,s'\in\gamma} s'_i A(s\to s')\prod_{j=1}^{b} f_j^{s_j}\bar{f}_j^{\bar{s}_j}, \quad \forall i=1,\ldots,b. \tag{A.18}$$

Equation (A.14) follows immediately.

Note that we have reduced a set of 2^b equations to a set of b equations. Yet we made no restrictive assumption about the lattice gas model, and the semi-detailed balance condition (see Section 2.3.1) has not been used.

• **We prove that (A.14) ⇒ (A.15):**

We divide both sides of (A.14) by $\prod_{k=1}^{b}\bar{f}_k$,[1] and introduce the notation $\tilde{f}_i = f_i/\bar{f}_i$ to obtain:

$$\sum_{s,s'\in\gamma}(s'_i - s_i)A(s\to s')\prod_{j=1}^{b}\tilde{f}_j^{s_j} = 0, \quad \forall i=1,\ldots,b. \tag{A.19}$$

We now use a trick introduced by Gatignol in the context of gas models with discrete velocities (Gatignol, 1975): we multiply both sides by $\log\tilde{f}_i$ and sum over i. Equation (A.19) becomes:

$$\sum_{i=1}^{b}\log\tilde{f}_i\sum_{s,s'\in\gamma}(s'_i - s_i)A(s\to s')\prod_{j=1}^{b}\tilde{f}_j^{s_j} = 0, \tag{A.20}$$

which also reads:

$$\sum_{s,s'\in\gamma} A(s\to s')\log\left(\frac{\prod_{j=1}^{b}\tilde{f}_j^{s'_j}}{\prod_{j=1}^{b}\tilde{f}_j^{s_j}}\right)\prod_{j=1}^{b}\tilde{f}_j^{s_j} = 0. \tag{A.21}$$

We now introduce the *semi-detailed balance* condition (see Section 2.3.1):

$$\sum_{s\in\gamma} A(s\to s') = \sum_{s\in\gamma} A(s'\to s), \quad \forall s'\in\gamma$$

and we multiply both sides by $\prod_{j=1}^{b}\tilde{f}_j^{s'_j}$ and sum over s':

$$\sum_{s,s'\in\gamma} A(s\to s')\prod_{j=1}^{b}\tilde{f}_j^{s'_j} = \sum_{s,s'\in\gamma} A(s'\to s)\prod_{j=1}^{b}\tilde{f}_j^{s'_j}.$$

[1] We exclude the pathological case $f_k = 1$.

Interchanging the dummy variables s and s' on the r.h.s. yields:

$$\sum_{s,s'\in\gamma} A(s{\to}s')\left(\prod_{j=1}^{b}\tilde{f}_j^{s'_j}-\prod_{j=1}^{b}\tilde{f}_j^{s_j}\right)=0.$$

By adding this identically null expression to the l.h.s. of (A.21) we obtain:

$$\sum_{s,s'\in\gamma} A(s{\to}s')\left\{\log\left(\frac{\prod\tilde{f}_j^{s'_j}}{\prod\tilde{f}_j^{s_j}}\right)\prod\tilde{f}_j^{s_j}-\prod\tilde{f}_j^{s'_j}+\prod\tilde{f}_j^{s_j}\right\}=0, \quad \text{(A.22)}$$

or

$$\sum_{s,s'\in\gamma} A(s{\to}s')\prod_{j=1}^{b}\tilde{f}_j^{s_j}\left\{\log\left(\frac{\prod\tilde{f}_j^{s'_j}}{\prod\tilde{f}_j^{s_j}}\right)-\frac{\prod\tilde{f}_j^{s'_j}}{\prod\tilde{f}_j^{s_j}}+1\right\}=0, \quad \text{(A.23)}$$

where all products are taken from $j=1$ to b. The expression between braces is always negative or zero. Indeed, denoting by x the argument of the logarithm, this expression has the form $\log x - x + 1$ which is zero for $x=1$ and whose first derivative ($=1/x-1$) is positive for $x<1$, null for $x=1$ and negative for $x>1$. Thus, $\log x - x + 1$ must be negative for any value of $x \neq 1$. As a consequence, (A.23) implies that the expression between braces must be zero whenever $A(s{\to}s')$ is non-zero. In other words, $\prod_{j=1}^{b}\tilde{f}_j^{s'_j}$ and $\prod_{j=1}^{b}\tilde{f}_j^{s_j}$ must be equal for any states s and s' with a non-zero collisional transition probability $A(s{\to}s')$, that is:

$$A(s{\to}s')\left(\log\prod_{i=1}^{b}\tilde{f}_i^{s'_i}-\log\prod_{i=1}^{b}\tilde{f}_i^{s_i}\right)=0, \quad \forall s,s'\in\gamma, \quad \text{(A.24)}$$

or equivalently:

$$\sum_{i=1}^{b} A(s{\to}s')(s'_i-s_i)\log\tilde{f}_i=0, \quad \forall s,s'\in\gamma. \quad \text{(A.25)}$$

With the definition of $\tilde{f}_i$, this result yields (A.15).

• **We prove that (A.15) ⇒ (A.13):**

Equation (A.15) implies that for all s and s' in γ, the products $\prod\tilde{f}_i^{s'_i}$ and $\prod\tilde{f}_i^{s_i}$ are equal when $A(s{\to}s')$ is non-zero. Therefore, we deduce that:

$$A(s{\to}s')\prod_{j=1}^{b} f_j^{s'_j}\bar{f}_j^{\bar{s}'_j}=A(s{\to}s')\prod_{j=1}^{b} f_j^{s_j}\bar{f}_j^{\bar{s}_j}, \quad \forall s,s'\in\gamma. \quad \text{(A.26)}$$

Summing both sides over all s in γ and using the *semi-detailed balance condition* (2.17) gives (A.13). □

A.6 **The following relation holds:**

$$\sum_{s,s'\in\gamma} s_i A(s\to s') \prod_{j=1}^{b} f_j^{s_j} \bar{f}_j^{\bar{s}_j} = f_i, \quad \forall i = 1,\ldots,b \tag{A.27}$$

(see Section 4.4.1)

Proof

We use the normalization condition (2.9) to write the l.h.s. of (A.27) under the following form:

$$\sum_{s\in\gamma} s_i \prod_{j=1}^{b} f_j^{s_j} \bar{f}_j^{\bar{s}_j}.$$

The sum over s can be restricted to the Boolean states s with $s_i = 1$. The l.h.s. of (A.27) then reads:

$$f_i \sum_{\substack{s\in\gamma,\\ s_i=1}} \prod_{\substack{j=1,\\ j\neq i}}^{b} f_j^{s_j} \bar{f}_j^{\bar{s}_j}.$$

The sum is obviously equal to one, since it is taken over all possible Boolean states of the $b-1$ remaining channels (channel i excluded). □

A.7 **The square matrix M whose elements are:**

$$M^{[\kappa,\kappa']} = \sum_{i=1}^{b} q_i^{[\kappa]} q_i^{[\kappa']} f_{1i}, \qquad \kappa = 1,\ldots,\delta, \qquad \kappa' = 1,\ldots,\delta, \tag{A.28}$$

is invertible (see Section 4.5)

Proof

We denote by E the canonical real vector space of dimension b, and by I the δ-dimensional vector subspace of collisional invariants. The set of vectors $(q^{[\kappa]}, \kappa = 1,\ldots,\delta)$ is a basis of I. The vectors in E will be denoted by Roman letters rather than bold faced symbols, which we use for vectors in the physical space $\mathbb{R}^D$.

We introduce the symmetric quadratic form acting on $E \times E$, which associates to any pair of vectors $(x, y) \in E \times E$, the real number:

$$x \cdot y \equiv \sum_{i=1}^{b} x_i y_i (-f_{1i}).$$

The f_{1i}s are given in terms of the unperturbed average populations f_{0i} by $f_{1i} = -f_{0i}(1-f_{0i})$ (see Section 4.5). Thus, the coefficients $(-f_{1i})$ of the quadratic form are all positive and at least one of them is non-zero, if we exclude the pathological situations where the average populations are all zero or all one (completely empty or completely full lattice). This symmetric quadratic form is therefore positive definite and defines an inner product on E. The matrix elements of M can be expressed in terms of this inner product as:

$$M^{[\kappa,\kappa']} = -q^{[\kappa]} \cdot q^{[\kappa']}, \quad \kappa, \kappa' = 1, \ldots, \delta.$$

We now consider in I, a basis $(p^{[\kappa]}, \kappa = 1, \ldots, \delta)$ of orthogonal unitary vectors (orthogonal and unitary with respect to the inner product defined above). In other words, these vectors are such that:

$$p^{[\kappa]} \cdot p^{[\kappa']} = \delta_{\kappa\kappa'}, \quad \kappa, \kappa' = 1, \ldots, \delta.$$

We introduce the real $\delta \times \delta$ square matrix R which transforms the $q^{[\kappa]}$s into the $p^{[\kappa]}$s. Its matrix elements $R^{[\kappa\kappa']}$ are such that:

$$p^{[\kappa]} = \sum_{\kappa'=1}^{\delta} R^{[\kappa\kappa']} q^{[\kappa']}.$$

Consider the matrix $-R\,M\,R^{\top}$, where $R^{\top}$ is the transpose of R. The matrix elements of $-R\,M\,R^{\top}$ are the inner products $p^{[\kappa]} \cdot p^{[\kappa']}$, which are equal to 1 when $\kappa = \kappa'$ and 0 otherwise (the matrix $-R\,M\,R^{\top}$ is just the opposite of the unit matrix). Since the $q^{[\kappa]}$s and the $p^{[\kappa]}$s are two bases of I, the matrix R has a non-zero determinant. Therefore M also has a non-zero determinant. □

A.8 Computation of the r.h.s. of (4.47) for the class of models defined in Section 4.5.2

Let us call $S^{[\kappa]}$ the r.h.s. of (4.47):

$$S^{[\kappa]} = -\frac{1}{2} \sum_{i=1}^{b} q_i^{[\kappa]} f_{2i} \left(\sum_{\kappa'=1}^{\delta} \lambda_{(1)}^{[\kappa']} q_i^{[\kappa']} \right)^2, \tag{A.29}$$

where the first order corrections $\lambda_{(1)}^{[\kappa]}$ to the Lagrange multipliers are, according to (4.71):

$$\begin{aligned} \lambda_{(1)}^{[0]} &= 0, \\ \lambda_{(1)}^{[\alpha]} &= -\frac{1}{d(1-d)} \frac{j_{(1)\alpha}}{\xi_2}, \quad \alpha = 1, \ldots, D, \\ \lambda_{(1)}^{[D+1]} &= -\frac{1}{d(1-d)} \frac{w_{(1)}}{\xi_4}. \end{aligned} \tag{A.30}$$

We recall the definitions of the characteristic coefficients ξ_2, ξ_4 and ξ_6 (see Equation (4.64)) which depend on the geometric structure of the lattice gas:

$$
\begin{aligned}
\sum_{i=1}^{b} c_{i\alpha} c_{i\beta} &= \xi_2 \delta_{\alpha\beta}, \\
\sum_{i=1}^{b} c_{i\alpha} c_{i\beta} c_{i\gamma} c_{i\delta} &= \frac{1}{D(D+2)} \left(4\xi_4 + \frac{D^2 \xi_2^2}{b} \right) \\
&\quad \times (\delta_{\alpha\beta}\delta_{\gamma\delta} + \delta_{\alpha\gamma}\delta_{\beta\delta} + \delta_{\alpha\delta}\delta_{\beta\gamma}), \\
\sum_{i=1}^{b} \frac{\mathbf{c}_i^2}{2} c_{i\alpha} c_{i\beta} c_{i\gamma} c_{i\delta} &= \frac{1}{D(D+2)} \left(4\xi_6 + \frac{6D\xi_2\xi_4}{b} + \frac{D^3 \xi_2^3}{2b^2} \right) \\
&\quad \times (\delta_{\alpha\beta}\delta_{\gamma\delta} + \delta_{\alpha\gamma}\delta_{\beta\delta} + \delta_{\alpha\delta}\delta_{\beta\gamma}).
\end{aligned}
\tag{A.31}
$$

We also recall that the collisional invariants $q^{[\kappa]}$ of the class of models under consideration are (see Equation (4.65)):

$$
\begin{aligned}
q^{[0]} &= \begin{pmatrix} 1 \\ \cdot \\ \cdot \\ \cdot \\ 1 \end{pmatrix}, \qquad q^{[\alpha]} = \begin{pmatrix} c_{1\alpha} \\ \cdot \\ \cdot \\ \cdot \\ c_{b\alpha} \end{pmatrix} \qquad (\alpha = 1, \ldots, D), \\
q^{[D+1]} &= \begin{pmatrix} e_1 \equiv \frac{1}{2}\mathbf{c}_1^2 - \frac{D\xi_2}{2b} \\ \cdot \\ \cdot \\ \cdot \\ e_b \equiv \frac{1}{2}\mathbf{c}_b^2 - \frac{D\xi_2}{2b} \end{pmatrix}.
\end{aligned}
\tag{A.32}
$$

The expansion coefficient f_{2i} is $d(1-d)(1-2d)$, according to (4.54). Using the definitions (A.31) and (A.32), one can prove the following useful identities:

$$
\begin{aligned}
&\sum_{i=1}^{b} e_i^2 = \xi_4, \qquad \sum_{i=1}^{b} c_{i\alpha} c_{i\beta} e_i = \frac{2\xi_4}{D} \delta_{\alpha\beta}, \\
&\sum_{i=1}^{b} e_i^3 = \xi_6, \qquad \sum_{i=1}^{b} c_{i\alpha} c_{i\beta} e_i^2 = \left(\frac{2\xi_6}{D} + \frac{\xi_2\xi_4}{b} \right) \delta_{\alpha\beta},
\end{aligned}
\tag{A.33}
$$

which we will need to compute the r.h.s. of (4.47).

Starting from (A.29), and taking into account (A.30), we can rewrite $S^{[\kappa]}$ as:

$$S^{[\kappa]} = -\frac{1-2d}{2d(1-d)} \sum_{i=1}^{b} q_i^{[\kappa]} \left(\frac{\mathbf{j}_{(1)} \cdot \mathbf{c}_i}{\xi_2} + \frac{w_{(1)}\, e_i}{\xi_4} \right)^2 .$$

The expression in parentheses can be written in terms of a sum over component indices, running from 1 to D. These summations are implicit: with Einstein's convention, $j_\beta j_\gamma c_{i\beta} c_{i\gamma}$ stands for $\sum_{\beta,\gamma} j_\beta j_\gamma c_{i\beta} c_{i\gamma}$. Thus we have:

$$S^{[\kappa]} = -\frac{1-2d}{2d(1-d)} \sum_{i=1}^{b} q_i^{[\kappa]} \times \left(\frac{1}{\xi_2^2} j_{(1)\beta} j_{(1)\gamma} c_{i\beta} c_{i\gamma} + \frac{2}{\xi_2 \xi_4} w_{(1)} j_{(1)\beta} c_{i\beta} e_i + \frac{1}{\xi_4^2} w_{(1)}^2 e_i^2 \right). \tag{A.34}$$

We first examine the case $\kappa = 0$. The summation over i can be performed using the explicit expression (A.31) for the crystallographic tensor $\sum c_{i\alpha} c_{i\beta}$. Because crystallographic tensors of odd order are null, the second term in (A.34) does not contribute, and the expression for $S^{[0]}$ reads:

$$S^{[0]} = -\frac{1-2d}{2d(1-d)} \left(\frac{1}{\xi_2} \mathbf{j}_{(1)}^2 + \frac{1}{\xi_4} w_{(1)}^2 \right). \tag{A.35}$$

For the case $\kappa = \alpha$, $\alpha = 1, \ldots, D$, the algebra is quite similar. The first and third terms in (A.34) do not contribute since for $q^{[\kappa]} = c_{i\alpha}$ they involve odd order crystallographic tensors. The second term is computed with the second identity in (A.33). This gives:

$$S^{[\alpha]} = -\frac{1-2d}{2d(1-d)} \left(\frac{4}{D\xi_2} w_{(1)} j_{(1)\alpha} \right). \tag{A.36}$$

The case $\kappa = D+1$ is also straightforward. The first and third identities in (A.33) are used, which yields:

$$S^{[D+1]} = -\frac{1-2d}{2d(1-d)} \left(\frac{2\xi_4}{D\xi_2^2} \mathbf{j}_{(1)}^2 + \frac{\xi_6}{\xi_4^2} w_{(1)}^2 \right). \tag{A.37}$$

The second order perturbations $\lambda_{(2)}^{[\kappa']}$ to the Lagrange multipliers must satisfy the relation (4.47) which reads:

$$\sum_{\kappa'=1}^{\delta} M^{[\kappa,\kappa']} \lambda_{(2)}^{[\kappa']} = S^{[\kappa]}, \tag{A.38}$$

where the r.h.s. terms $S^{[\kappa]}$ are given by (A.35), (A.36) and (A.37). Since the matrix $M^{[\kappa,\kappa']}$ is diagonal (see Section 4.5.2), the second order perturbations $\lambda^{[\kappa']}_{(2)}$ follow immediately:

$$
\begin{aligned}
\lambda^{[0]}_{(2)} &= \frac{1-2d}{2d^2(1-d)^2}\left(\frac{1}{b\xi_2}\mathbf{j}^2_{(1)} + \frac{1}{b\xi_4}w^2_{(1)}\right),\\
\lambda^{[\alpha]}_{(2)} &= \frac{1-2d}{2d^2(1-d)^2}\left(\frac{4}{D\xi_2^2}w_{(1)}j_{(1)\alpha}\right), \quad \alpha = 1,\dots,D,\\
\lambda^{[D+1]}_{(2)} &= \frac{1-2d}{2d^2(1-d)^2}\left(\frac{2}{D\xi_2^2}\mathbf{j}^2_{(1)} + \frac{\xi_6}{\xi_4^3}w^2_{(1)}\right).
\end{aligned}
\tag{A.39}
$$

A.9 Computation of the momentum and effective energy fluxes involved in the first order macrodynamic equations (see Section 5.3.1)

By definition (see Section 2.1.3), momentum and effective energy fluxes are:

$$
\begin{aligned}
\Phi^{[\alpha]}_{\beta} &\equiv \sum_{i=1}^{b} c_{i\alpha}c_{i\beta}f_i^{(0)}, \qquad \alpha\beta = 1,\dots,D,\\
\Phi^{[D+1]}_{\beta} &\equiv \sum_{i=1}^{b} e_i c_{i\beta}f_i^{(0)}, \qquad \beta = 1,\dots,D,
\end{aligned}
\tag{A.40}
$$

where the $f_i^{(0)}$s are the equilibrium mean populations derived in Section 4.5.2 for the nearly equally distributed states of single-species thermal models:

$$
\begin{aligned}
f_i^{(0)} &= \frac{p}{b}\\
&+ \frac{1}{\xi_2}j_\gamma c_{i\gamma} + \frac{1}{\xi_4}we_i\\
&+ \frac{b(b-2p)}{2p(b-p)}\left[\frac{1}{\xi_2^2}j_\gamma j_\delta\left(c_{i\gamma}c_{i\delta} - \frac{\mathbf{c}_i^2}{D}\delta_{\gamma\delta}\right)\right.\\
&+ \frac{2}{\xi_2\xi_4}wj_\gamma\left(e_i - \frac{2\xi_4}{D\xi_2}\right)c_{i\gamma}\\
&+ \left.\frac{1}{\xi_4^2}w^2\left(e_i^2 - \frac{\xi_6}{\xi_4}e_i - \frac{\xi_4}{b}\right)\right]\\
&+ \mathcal{O}(j^3) + \mathcal{O}(j^2w) + \mathcal{O}(\eta^3).
\end{aligned}
\tag{A.41}
$$

We recall that the expansion parameter η represents the order of magnitude of j and w, that is, the proximity of the considered equilibrium state to the 'ideal' equally distributed equilibrium state with zero momentum density and effective energy density.

The key-ingredients to compute the fluxes are the following relations:

- $\sum_{i=1}^{b} c_{i\alpha} c_{i\beta} = \xi_2 \delta_{\alpha\beta}$,
- $\sum_{i=1}^{b} c_{i\alpha} c_{i\beta} c_{i\gamma} c_{i\delta} = \frac{1}{D(D+2)} \left(4\xi_4 + \frac{D^2 \xi_2^2}{b}\right) \times \left(\delta_{\alpha\beta}\delta_{\gamma\delta} + \delta_{\alpha\gamma}\delta_{\beta\delta} + \delta_{\alpha\delta}\delta_{\beta\gamma}\right)$,
- $\sum_{i=1}^{b} c_{i\alpha} c_{i\beta} \mathbf{c}_i^2 = \frac{1}{D}\left(4\xi_4 + \frac{D^2 \xi_2^2}{b}\right)\delta_{\alpha\beta}$, (A.42)
- $\sum_{i=1}^{b} c_{i\alpha} c_{i\beta} e_i = \frac{2\xi_4}{D}\delta_{\alpha\beta}$,
- $\sum_{i=1}^{b} c_{i\alpha} c_{i\beta} e_i^2 = \left(\frac{2\xi_6}{D} + \frac{\xi_2\xi_4}{b}\right)\delta_{\alpha\beta}$.

These relations are direct consequences of the definition (4.64) of the geometrical parameters ξ_2, ξ_4 and ξ_6 (see Section 4.5.2). To these relations, we must also add the fact that the sum over i of any product containing an odd number of components of $\mathbf{c}_i$ is zero (see Section 2.3.5).

With these relations, the computation of the momentum flux tensor is relatively straightforward: all terms in (A.41) with odd powers of $\mathbf{j}$ vanish, and using the relations (A.42) yields the desired expression (5.30):

$$\Phi_\beta^{[\alpha]} = g(p) j_\alpha u_\beta + P(p, \mathbf{j}^2, w)\delta_{\alpha\beta} + \mathcal{O}(\eta^3), \qquad \alpha, \beta = 1, \ldots, D,$$

where

$$g(p) = \frac{b-2p}{b-p}\frac{b}{D(D+2)}\left(\frac{4\xi_4}{\xi_2^2} + \frac{D^2}{b}\right),$$

and

$$P(p, \mathbf{j}^2, w) = \frac{\xi_2}{b} p - \frac{1}{D}\frac{g(p)}{p}\mathbf{j}^2 + \frac{2}{D} w.$$

For the energy flux, all terms in (A.41) with even powers of $\mathbf{j}$ vanish, and one obtains (5.33):

$$\Phi_\beta^{[D+1]} = g'(p) w u_\beta + \frac{2\xi_4}{D\xi_2} j_\beta + \mathcal{O}(\eta^3), \qquad \beta = 1, \ldots, D,$$

with

$$g'(p) = \frac{b-2p}{b-p}\left(1 + \frac{2b\xi_6}{D\xi_2\xi_4} - \frac{4b\xi_4}{D^2\xi_2^2}\right).$$

A.10 Computation of the second order solvability conditions for the Chapman–Enskog expansion (see Section 5.4.1)

We recall the second order solvability conditions (5.57) for the Chapman–Enskog expansion:

$$
\begin{aligned}
0 &= \partial_{t_1} \sum_{i=1}^{b} q_i^{[\kappa]} f_i^{(1)} + \partial_{1\beta} \sum_{i=1}^{b} q_i^{[\kappa]} c_{i\beta} f_i^{(1)} \\
&+ \partial_{t_2} \sum_{i=1}^{b} q_i^{[\kappa]} f_i^{(0)} + \frac{1}{2} \partial_{1\beta} \partial_{1\gamma} \sum_{i=1}^{b} q_i^{[\kappa]} c_{i\beta} c_{i\gamma} f_i^{(0)} \qquad \text{(A.43)} \\
&+ \frac{1}{2} \partial_{t_1}^2 \sum_{i=1}^{b} q_i^{[\kappa]} f_i^{(0)} + \partial_{1\beta} \partial_{t_1} \sum_{i=1}^{b} q_i^{[\kappa]} c_{i\beta} f_i^{(0)},
\end{aligned}
$$

for all $\kappa = 1, \ldots, \delta$, that is, for all collisional invariants.

Before analyzing separately the cases of the mass, momentum, and effective energy invariants corresponding respectively to $\kappa = 0$, $\kappa = 1, \ldots, D$ and $\kappa = D+1$, one can simplify (A.43) without any loss of generality.

First, the term $\sum_i q_i^{[\kappa]} f_i^{(1)}$ vanishes for all κ, since it represents the contribution of the first order corrections $f_i^{(1)}$ to the collisional invariants' densities, and this contribution is zero by the definition of $f_i^{(1)}$ (see Equation (5.4) and the preceding discussion).

Second, one can use the first order solvability condition (5.25):

$$
\partial_{t_1} \sum_{i=1}^{b} q_i^{[\kappa]} f_i^{(0)} + \partial_{1\beta} \sum_{i=1}^{b} q_i^{[\kappa]} c_{i\beta} f_i^{(0)} = 0, \qquad \kappa = 1, \ldots, \delta,
$$

to simplify the last two terms in (A.43). The resulting simplified second order solvability condition reads:

$$
\begin{aligned}
0 &= \partial_{t_2} \rho^{[\kappa]} + \partial_{1\beta} \sum_{i=1}^{b} q_i^{[\kappa]} c_{i\beta} f_i^{(1)} \\
&+ \frac{1}{2} \partial_{1\beta} \partial_{t_1} \Phi_\beta^{[\kappa]} + \frac{1}{2} \partial_{1\beta} \partial_{1\gamma} \sum_{i=1}^{b} q_i^{[\kappa]} c_{i\beta} c_{i\gamma} f_i^{(0)}, \qquad \text{(A.44)}
\end{aligned}
$$

for all $\kappa = 1, \ldots, \delta$, that is, for all collisional invariants. In (A.44), $\rho^{[\kappa]}$ is the density $\sum_i q_i^{[\kappa]} f_i^{(0)}$ of the collisional invariant $q^{[\kappa]}$, and $\mathbf{\Phi}^{[\kappa]}$ is the corresponding flux $\sum_i q_i^{[\kappa]} \mathbf{c}_i f_i^{(0)}$.

We now address successively the cases of mass, momentum and effective energy invariants. The mass invariant leads to the simplest algebra. Indeed, the first order solvability condition (5.35) for the mass invariant makes the last two terms in (A.44) vanish. In addition, the contribution $\sum_i c_{i\beta} f_i^{(1)}$ of $f_i^{(1)}$ to the mass flux (i.e. momentum density) vanishes. This leads to the second order solvability condition (5.58) for the mass invariant:

$$\partial_{t_2} p = 0.$$

To handle the case of the momentum invariant, we need the expression (5.30) of the zeroth order momentum flux $\sum_i c_{i\alpha} c_{i\beta} f_i^{(0)}$, that we truncate at the second order in η:

$$\sum_{i=1}^{b} c_{i\alpha} c_{i\beta} f_i^{(0)} = \left(\frac{\xi_2}{b} p + \frac{2}{D} w \right) \delta_{\alpha\beta} + \mathcal{O}(\eta^2), \qquad \alpha, \beta = 1, \ldots, D.$$

We also need the expression (5.53) for the first order momentum flux $\sum_i c_{i\alpha} c_{i\beta} f_i^{(1)}$:

$$\sum_{i=1}^{b} c_{i\alpha} c_{i\beta} f_i^{(1)} = \nu^{(c)} \left(\partial_{1\alpha} j_\beta + \partial_{1\beta} j_\alpha - \frac{2}{D} \partial_{1\gamma} j_\gamma \delta_{\alpha\beta} \right).$$

Finally, we must compute $\sum_i c_{i\alpha} c_{i\beta} c_{i\gamma} f_i^{(0)}$ up to the first order in η. This requires the second identity in (A.42), that is, the very definition of ξ_4. The resulting expression is:

$$\sum_{i=1}^{b} c_{i\alpha} c_{i\beta} c_{i\gamma} f_i^{(0)} =$$

$$\frac{1}{D(D+2)} \left(\frac{4\xi_4}{\xi_2} + \frac{D^2 \xi_2}{b} \right) \left(j_\gamma \delta_{\alpha\beta} + j_\beta \delta_{\alpha\gamma} + j_\alpha \delta_{\beta\gamma} \right) + \mathcal{O}(\eta^2).$$

After incorporating these ingredients in (A.44), the second order solvability condition for the momentum invariant becomes:

$$\begin{aligned} 0 \;=\; & \partial_{t_2} j_\alpha - \partial_{1\beta} \left(\nu^{(c)} \left(\partial_{1\alpha} j_\beta + \partial_{1\beta} j_\alpha - \frac{2}{D} \partial_{1\gamma} j_\gamma \delta_{\alpha\beta} \right) \right) \\ & + \frac{1}{2} \partial_{1\beta} \left(\frac{\xi_2}{b} \partial_{t_1} p + \frac{2}{D} \partial_{t_1} w \right) \delta_{\alpha\beta} \\ & + \frac{1}{2} \partial_{1\beta} \left(\frac{1}{D(D+2)} \left(\frac{4\xi_4}{\xi_2} + \frac{D^2 \xi_2}{b} \right) \left(\partial_{1\alpha} j_\beta + \partial_{1\beta} j_\alpha + \partial_{1\gamma} j_\gamma \delta_{\alpha\beta} \right) \right) \\ & + \mathcal{O}(\eta^2). \end{aligned}$$

We then use the first order solvability conditions (5.35) and (5.37) for the mass and effective energy respectively, to re-express the time-derivatives of p and w in terms of space-derivatives. This yields:

$$\begin{aligned}
0 &= \partial_{t_2} j_\alpha - \partial_{1\beta}\left(\nu^{(c)} \left(\partial_{1\alpha} j_\beta + \partial_{1\beta} j_\alpha - \frac{2}{D} \partial_{1\gamma} j_\gamma \delta_{\alpha\beta} \right) \right) \\
&- \frac{1}{2} \partial_{1\beta} \left(\frac{1}{D^2} \left(\frac{4\xi_4}{\xi_2} + \frac{D^2 \xi_2}{b} \right) \partial_{1\gamma} j_\gamma \delta_{\alpha\beta} \right) \\
&+ \frac{1}{2} \partial_{1\beta} \left(\frac{1}{D(D+2)} \left(\frac{4\xi_4}{\xi_2} + \frac{D^2 \xi_2}{b} \right) \left(\partial_{1\alpha} j_\beta + \partial_{1\beta} j_\alpha + \partial_{1\gamma} j_\gamma \delta_{\alpha\beta} \right) \right) \\
&+ \mathcal{O}(\eta^2),
\end{aligned}$$

or in compact form:

$$\begin{aligned}
0 &= \partial_{t_2} j_\alpha - \partial_{1\beta}\left(\nu^{(c)} \left(\partial_{1\alpha} j_\beta + \partial_{1\beta} j_\alpha - \frac{2}{D} \partial_{1\gamma} j_\gamma \delta_{\alpha\beta} \right) \right) \\
&+ \partial_{1\beta} \left(\frac{1}{2D(D+2)} \left(\frac{4\xi_4}{\xi_2} + \frac{D^2 \xi_2}{b} \right) \left(\partial_{1\alpha} j_\beta + \partial_{1\beta} j_\alpha - \frac{2}{D} \partial_{1\gamma} j_\gamma \delta_{\alpha\beta} \right) \right) \\
&+ \mathcal{O}(\eta^2).
\end{aligned}$$

This gives the second order solvability condition (5.59) for the momentum invariant:

$$\partial_{t_2} j_\alpha + \partial_{1\beta} \left(-\nu \left(\partial_{1\alpha} j_\beta + \partial_{1\beta} j_\alpha - \frac{2}{D} \partial_{1\gamma} j_\gamma \delta_{\alpha\beta} \right) \right) = \mathcal{O}(\eta^2),$$

where the coefficient ν, which *a priori* depends on $(p, \mathbf{j}^2, w)$, is given by:

$$\nu = \nu^{(c)} - \frac{1}{2D(D+2)} \left(\frac{4\xi_4}{\xi_2} + \frac{D^2 \xi_2}{b} \right).$$

The case of the effective energy invariant is somewhat simpler. We need expression (5.33) of the zeroth order effective energy flux $\sum_i e_i c_{i\beta} f_i^{(0)}$, that we truncate at the second order in η:

$$\sum_{i=1}^{b} e_i c_{i\beta} f_i^{(0)} = \frac{2\xi_4}{D\xi_2} j_\beta + \mathcal{O}(\eta^2), \qquad \beta = 1, \ldots, D,$$

and we use expression (5.56) for the first order effective energy flux:

$$\sum_{i=1}^{b} e_i c_{i\beta} f_i^{(1)} = -\zeta^{(c)} \, \partial_{1\beta} w.$$

Finally we must compute $\sum_i e_i c_{i\beta} c_{i\gamma} f_i^{(0)}$ to first order in η, which requires the fourth and fifth identities in (A.42); the result reads:

$$\sum_{i=1}^{b} e_i c_{i\beta} c_{i\gamma} f_i^{(0)} = \left(\frac{2\xi_4}{Db} p + \left(\frac{2\xi_6}{D\xi_2} + \frac{\xi_2}{b} \right) w \right) \delta_{\beta\gamma} + \mathcal{O}(\eta^2)$$

When all these results are incorporated in (A.44), the second order solvability condition for the effective energy invariant becomes:

$$\begin{aligned} 0 &= \partial_{t_2} w - \partial_{1\beta} \left(\zeta^{(c)} \partial_{1\beta} w \right) \\ &+ \frac{1}{2} \partial_{1\beta} \left(\frac{2\xi_4}{D\xi_2} \partial_{t_1} j_\beta \right) \\ &+ \frac{1}{2} \partial_{1\beta} \left(\frac{2\xi_4}{Db} \partial_{1\beta} p + \left(\frac{2\xi_6}{D\xi_2} + \frac{\xi_2}{b} \right) \partial_{1\beta} w \right) \\ &+ \mathcal{O}(\eta^2). \end{aligned}$$

Using the first order solvability condition (5.36) for the momentum, we re-express the time-derivatives of j_β in terms of space-derivatives, and we obtain:

$$\begin{aligned} 0 &= \partial_{t_2} w - \partial_{1\beta} \left(\zeta^{(c)} \partial_{1\beta} w \right) \\ &- \frac{1}{2} \partial_{1\beta} \left(\frac{2\xi_4}{D\xi_2} \partial_{1\beta} \left(\frac{\xi_2}{b} p + \frac{2}{D} w \right) \right) \\ &+ \frac{1}{2} \partial_{1\beta} \left(\frac{2\xi_4}{Db} \partial_{1\beta} p + \left(\frac{2\xi_6}{D\xi_2} + \frac{\xi_2}{b} \right) \partial_{1\beta} w \right) \\ &+ \mathcal{O}(\eta^2), \end{aligned}$$

or in compact form:

$$\begin{aligned} 0 &= \partial_{t_2} w - \partial_{1\beta} \left(\zeta^{(c)} \partial_{1\beta} w \right) \\ &+ \frac{1}{2} \partial_{1\beta} \left(\left(\frac{2\xi_6}{D\xi_2} + \frac{\xi_2}{b} - \frac{4\xi_4}{D^2\xi_2} \right) \partial_{1\beta} w \right) \\ &+ \mathcal{O}(\eta^2). \end{aligned}$$

This leads to the second order solvability condition (5.61) for the effective energy invariant:

$$\partial_{t_2} w + \partial_{1\beta} \left(-\zeta \, \partial_{1\beta} w \right) = \mathcal{O}(\eta^2),$$

with:

$$\zeta = \zeta^{(c)} + \left(\frac{\xi_6}{D\xi_4} + \frac{\xi_2}{2b} - \frac{2\xi_4}{D^2\xi_2}\right).$$

This completes the computation of the second order solvability conditions.

A.11 Cumulants (Section 8.2)

Consider the definition of the commutative n-product:

$$\left\langle B^{(1)} \mid B^{(2)} \mid \ldots \mid B^{(n)} \right\rangle_{\text{LE}} (\mathbf{r}, t) \equiv \sum_{i=1}^{b} \kappa_i^{(n)}(\mathbf{r}, t)\, B_i^{(1)}\, B_i^{(2)} \,\ldots\, B_i^{(n)}, \quad \text{(A.45)}$$

with:

$$\kappa_i^{(n)}(\mathbf{r}, t) = (-1)^n \frac{\partial^n}{\partial \phi_0^n} \log\left[1 + \exp\left\{-\underline{\phi}(\mathbf{r}, t) * \underline{A}_i\right\}\right], \quad \text{(A.46)}$$

where $\underline{\phi}(\mathbf{r}, t) = \underline{\phi}(\langle \underline{A}(\mathbf{r}, t)\rangle)$ is given implicitly by relation (8.16), and ϕ_0 is the thermodynamic potential conjugate to the number of particles per node (proportional to the chemical potential).

We remark that $\kappa_i^{(n)}(\mathbf{r}, t)$ is the nth cumulant of the local-equilibrium distribution, and therefore the n-product defined by (A.45) is a *cumulant average*:

$$\left\langle B^{(1)} \mid B^{(2)} \mid \ldots \mid B^{(n)} \right\rangle_{\text{LE}} (\mathbf{r}, t) = \left\langle\!\left\langle B^{(1)} B^{(2)} \ldots B^{(n)} \right\rangle\!\right\rangle_{\text{LE}} (\mathbf{r}, t). \quad \text{(A.47)}$$

The first few cumulants (see also Chapter 4) are given explicitly by the expressions (Abramowitz and Stegun, 1965):

$$\begin{aligned}
\kappa_i^{(1)}(\mathbf{r}, t) &= f_i^{\text{LE}}(\mathbf{r}, t), \\
\kappa_i^{(2)}(\mathbf{r}, t) &= \left\langle \left(\delta n_i^{\text{LE}}(\mathbf{r}, t)\right)^2 \right\rangle = f_i^{\text{LE}}(\mathbf{r}, t) \left(1 - f_i^{\text{LE}}(\mathbf{r}, t)\right), \\
\kappa_i^{(3)}(\mathbf{r}, t) &= \left\langle \left(\delta n_i^{\text{LE}}(\mathbf{r}, t)\right)^3 \right\rangle \\
&= f_i^{\text{LE}}(\mathbf{r}, t) \left(1 - f_i^{\text{LE}}(\mathbf{r}, t)\right) \left(1 - 2 * f_i^{\text{LE}}(\mathbf{r}, t)\right), \\
\kappa_i^{(4)}(\mathbf{r}, t) &= \left\langle \left(\delta n_i^{\text{LE}}(\mathbf{r}, t)\right)^4 \right\rangle - 3 \left\langle \left(\delta n_i^{\text{LE}}(\mathbf{r}, t)\right)^2 \right\rangle^2 \\
&= f_i^{\text{LE}}(\mathbf{r}, t) \left(1 - f_i^{\text{LE}}(\mathbf{r}, t)\right) \left(1 - 6 f_i^{\text{LE}}(\mathbf{r}, t) + 6 \left(f_i^{\text{LE}}(\mathbf{r}, t)\right)^2\right).
\end{aligned} \quad \text{(A.48)}$$

Higher order cumulants can be obtained by evaluating the determinant

(Risken, 1984):

$$\kappa_i^{(n)}(\mathbf{r},t) = (-1)^{(n-1)}$$

$$\times \begin{vmatrix} f_i^{\mathrm{LE}}(\mathbf{r},t) & 1 & 0 & 0 & \cdots \\ f_i^{\mathrm{LE}}(\mathbf{r},t) & f_i^{\mathrm{LE}}(\mathbf{r},t) & 1 & 0 & \cdots \\ f_i^{\mathrm{LE}}(\mathbf{r},t) & f_i^{\mathrm{LE}}(\mathbf{r},t) & \binom{2}{1} f_i^{\mathrm{LE}}(\mathbf{r},t) & 1 & \cdots \\ f_i^{\mathrm{LE}}(\mathbf{r},t) & f_i^{\mathrm{LE}}(\mathbf{r},t) & \binom{3}{1} f_i^{\mathrm{LE}}(\mathbf{r},t) & \binom{3}{2} f_i^{\mathrm{LE}}(\mathbf{r},t) & \cdots \\ \cdots & \cdots & \cdots & \cdots & \cdots \end{vmatrix}_n , \qquad (A.49)$$

where we have used the fact that n_i is a Boolean variable, and therefore:

$$\langle (n_i(\mathbf{r},t))^n \rangle_{\mathrm{LE}} = \langle n_i(\mathbf{r},t) \rangle_{\mathrm{LE}} = f_i^{\mathrm{LE}}(\mathbf{r},t). \qquad (A.50)$$

Alternatively we can use the recursion formula:

$$\kappa_i^{(n+1)}(\mathbf{r},t) = \kappa_i^{(n)}(\mathbf{r},t) - \sum_{m=1}^{n} \binom{n-1}{m-1} \kappa_i^{(m)}(\mathbf{r},t)\, \kappa_i^{(n-m+1)}(\mathbf{r},t), \quad (A.51)$$

which is obtained from the observation that the generating function for the cumulants:

$$G(x) \equiv \log \langle \exp\{x n_i(\mathbf{r},t)\} \rangle_{\mathrm{LE}} = \sum_{n=1}^{\infty} \kappa_i^{(n)}(\mathbf{r},t) \frac{x^n}{n!}, \qquad (A.52)$$

satisfies the differential equation:

$$G''(x) = G'(x)\left(1 - G'(x)\right). \qquad (A.53)$$

It is easy to verify that the cumulants are connected by the relation:

$$\frac{\partial}{\partial \langle \underline{A}(\mathbf{r},t) \rangle} \kappa_i^{(n)}(\mathbf{r},t) = -\kappa_i^{(n+1)}(\mathbf{r},t) \left(\frac{\partial}{\partial \langle \underline{A}(\mathbf{r},t) \rangle} \underline{\phi}(\mathbf{r},t) \right) * \underline{A}_i. \qquad (A.54)$$

Consider this relation for $n = 1$:

$$\frac{\partial}{\partial \langle \underline{A}(\mathbf{r},t) \rangle} f_i^{\mathrm{LE}}(\mathbf{r},t) = -\kappa_i^{\mathrm{LE}}(\mathbf{r},t) \left(\frac{\partial}{\partial \langle \underline{A}(\mathbf{r},t) \rangle} \underline{\phi}(\mathbf{r},t) \right) * \underline{A}_i. \qquad (A.55)$$

If we multiply (A.55) by $\underline{A}_i$ on the right on both sides of the equation and then sum over the channels, we obtain the equality:

$$\frac{\partial}{\partial \langle \underline{A}(\mathbf{r},t) \rangle} \langle \underline{A}(\mathbf{r},t) \rangle = \underline{\underline{1}} = -\left(\frac{\partial}{\partial \langle \underline{A}(\mathbf{r},t) \rangle} \underline{\phi}(\mathbf{r},t) \right) * \langle \underline{A}|\underline{A} \rangle_{\mathrm{LE}}(\mathbf{r},t), \quad (A.56)$$

which implies:

$$\frac{\partial}{\partial \langle \underline{A}(\mathbf{r},t) \rangle} \underline{\phi}(\mathbf{r},t) = -\langle \underline{A}|\underline{A} \rangle_{\mathrm{LE}}^{-1}(\mathbf{r},t). \qquad (A.57)$$

Thus, Equation (A.54) becomes:

$$\frac{\partial}{\partial \langle \underline{A}(\mathbf{r},t)\rangle} \kappa_i^{(n)}(\mathbf{r},t) = \kappa_i^{(n+1)}(\mathbf{r},t)\, \underline{A}_i * \langle \underline{A}|\underline{A}\rangle_{\text{LE}}^{-1}(\mathbf{r},t). \tag{A.58}$$

Multiplying (A.58) by $B_i^{(1)}B_i^{(2)} \ldots B_i^{(n)}$ and summing over i yields:

$$\frac{\partial}{\partial \langle \underline{A}(\mathbf{r},t)\rangle} \langle B^{(1)} \mid B^{(2)} \mid \ldots \mid B^{(n)}\rangle_{\text{LE}}(\mathbf{r},t)$$
$$= \langle \underline{A}|\underline{A}\rangle_{\text{LE}}^{-1}(\mathbf{r},t) * \langle \underline{A} \mid B^{(1)} \mid B^{(2)} \mid \ldots \mid B^{(n)}\rangle_{\text{LE}}(\mathbf{r},t), \tag{A.59}$$

and, as a particular example of this general relation:

$$\frac{\partial}{\partial \langle \underline{A}(\mathbf{r},t)\rangle} \langle \underline{\mathbf{J}}(\mathbf{r},t)\rangle_{\text{LE}} = \langle \underline{A}|\underline{A}\rangle_{\text{LE}}^{-1}(\mathbf{r},t) * \langle \underline{A}|\underline{\mathbf{J}}\rangle_{\text{LE}}(\mathbf{r},t), \tag{A.60}$$

which proves Equation (8.33).

A.12 Matrices (Section 8.3)

Here we discuss the properties of the non-symmetric matrix $\mathscr{L}^{\text{LE}}(\mathbf{r},t)$, defined in Section 8.3 by:

$$\mathscr{L}_{ij}^{\text{LE}}(\mathbf{r},t) = \sum_{\{s\}\{s'\}} s_i' \frac{s_j - f_j^{\text{LE}}(\mathbf{r},t)}{\kappa_j^{\text{LE}}(\mathbf{r},t)} A(s\to s')\, F_{\{s\}}^{\text{LE}}(\mathbf{r},t), \tag{A.61}$$

where:

$$F_{\{s\}}^{\text{LE}}(\mathbf{r},t) = \prod_{k=1}^{b} \left(f_k^{\text{LE}}(\mathbf{r},t)\right)^{s_k} \left(1 - f_k^{\text{LE}}(\mathbf{r},t)\right)^{(1-s_k)}. \tag{A.62}$$

First, we prove that the conserved quantities $\{\underline{A}_i\}_{i=1}^{b}$ are left eigenvectors of $\mathscr{L}^{\text{LE}}(\mathbf{r},t)$, with eigenvalue one:

$$\sum_{i=1}^{b} \underline{A}_i\, \mathscr{L}_{ij}^{\text{LE}}(\mathbf{r},t) = \underline{A}_j, \tag{A.63}$$

The proof makes use of the fact that the average collision matrix connects only those configurations with the same value for the conserved quantities. Hence:

$$\sum_{i=1}^{b} \underline{A}_i\, \mathscr{L}_{ij}^{\text{LE}}(\mathbf{r},t)$$
$$= \sum_{\{s\}\{s'\}} \left(\sum_{i=1}^{b} \underline{A}_i s_i'\right) \frac{s_j - f_j^{\text{LE}}(\mathbf{r},t)}{\kappa_j^{\text{LE}}(\mathbf{r},t)} A(s\to s')\, F_{\{s\}}^{\text{LE}}(\mathbf{r},t)$$
$$= \sum_{\{s\}\{s'\}} \left(\sum_{i=1}^{b} \underline{A}_i s_i\right) \frac{s_j - f_j^{\text{LE}}(\mathbf{r},t)}{\kappa_j^{\text{LE}}(\mathbf{r},t)} A(s\to s')\, F_{\{s\}}^{\text{LE}}(\mathbf{r},t)$$

$$= \sum_{i \neq j} \underline{A}_i \sum_{\{s\}\{s'\}} s_i \frac{s_j - f_j^{\mathrm{LE}}(\mathbf{r},t)}{\kappa_j^{\mathrm{LE}}(\mathbf{r},t)} A(s \to s')\, F_{\{s\}}^{\mathrm{LE}}(\mathbf{r},t)$$

$$+ \underline{A}_j \frac{1}{f_j^{\mathrm{LE}}(\mathbf{r},t)} \sum_{\{s\}\{s'\}} s_j \, \langle \xi \rangle_{\{s\} \to \{\sigma\}} \, F_{\{s\}}^{\mathrm{LE}}(\mathbf{r},t)$$

$$= \underline{A}_j \, . \tag{A.64}$$

The right eigenvectors of $\mathscr{L}^{\mathrm{LE}}(\mathbf{r},t)$ with eigenvalue one are:

$$\sum_{j=1}^{b} \mathscr{L}_{ij}^{\mathrm{LE}}(\mathbf{r},t)\, \kappa_j^{\mathrm{LE}}(\mathbf{r},t) \underline{A}_j = \kappa_i^{\mathrm{LE}}(\mathbf{r},t) \underline{A}_i . \tag{A.65}$$

The proof of this relation is similar to the previous one:

$$\sum_{j=1}^{b} \mathscr{L}_{ij}^{\mathrm{LE}}(\mathbf{r},t)\, \kappa_j^{\mathrm{LE}}(\mathbf{r},t) \underline{A}_j$$

$$= \sum_{\{s\}\{s'\}} s_i' \left(\sum_{j=1}^{b} (s_j - f_j^{\mathrm{LE}}(\mathbf{r},t))\, \underline{A}_j \right) A(s \to s')\, F_{\{s\}}^{\mathrm{LE}}(\mathbf{r},t)$$

$$= \sum_{\{s\}\{s'\}} s_i' \left(\sum_{j=1}^{b} (s_j' - f_j^{\mathrm{LE}}(\mathbf{r},t))\, \underline{A}_j \right) \langle \xi \rangle_{\{s\} \to \{\sigma\}} \, F_{\{s\}}^{\mathrm{LE}}(\mathbf{r},t)$$

$$= \sum_{j \neq i} \underline{A}_j \sum_{\{s\}\{s'\}} s_i'(s_j' - f_j^{\mathrm{LE}}(\mathbf{r},t)) \, \langle \xi \rangle_{\{s\} \to \{\sigma\}} \, F_{\{s\}}^{\mathrm{LE}}(\mathbf{r},t)$$

$$+ \underline{A}_i \, (1 - f_i^{\mathrm{LE}}(\mathbf{r},t)) \sum_{\{s\}\{s'\}} s_i' \, A(s \to s')\, F_{\{s\}}^{\mathrm{LE}}(\mathbf{r},t)$$

$$= \kappa_i^{\mathrm{LE}}(\mathbf{r},t) \underline{A}_i \, . \tag{A.66}$$

The other right eigenvectors of $\mathscr{L}^{\mathrm{LE}}(\mathbf{r},t)$, which constitute the set of *kinetic* eigenvectors, are:

$$\sum_{j=1}^{b} \mathscr{L}_{ij}^{\mathrm{LE}}(\mathbf{r},t)\, \kappa_j^{\mathrm{LE}}(\mathbf{r},t)\, C_j^{(n)}$$

$$= \lambda^{(n)} \, \kappa_i^{\mathrm{LE}}(\mathbf{r},t)\, C_i^{(n)}; \quad n = 1, \ldots, (b - N_{\mathrm{hydro}}), \tag{A.67}$$

with eigenvalues $|\lambda^{(n)}| < 1$, for $n = 1, \ldots, (b - N_{\mathrm{hydro}})$, a property that follows from the fact that the equilibrium state is stable. The kinetic eigenvectors are orthogonal to the constants of motion, in the sense that:

$$\langle \underline{A} | C^{(n)} \rangle_{\mathrm{LE}} (\mathbf{r},t) = 0 \, ; \quad n = 1, \ldots, (b - N_{\mathrm{hydro}}), \tag{A.68}$$

where the scalar product has been defined in section A.11. This property can be derived from the following set of equalities for $n = 1, \ldots, (b - N_{\text{hydro}})$:

$$\sum_{i,j} \underline{A}_i \mathscr{L}_{ij}^{\text{LE}}(\mathbf{r},t) \kappa_j^{\text{LE}}(\mathbf{r},t)\; C_j^{(n)} = \sum_{j=1}^{b} \left(\sum_{i=1}^{b} \underline{A}_i \mathscr{L}_{ij}^{\text{LE}}(\mathbf{r},t) \right) \kappa_j^{\text{LE}}(\mathbf{r},t)\; C_j^{(n)}$$

$$= \sum_{j=1}^{b} \kappa_j^{\text{LE}}(\mathbf{r},t) \underline{A}_j C_j^{(n)} = \left\langle \underline{A} | C^{(n)} \right\rangle_{\text{LE}} (\mathbf{r},t), \tag{A.69}$$

$$\sum_{i,j} \underline{A}_i \mathscr{L}_{ij}^{\text{LE}}(\mathbf{r},t) \kappa_j^{\text{LE}}(\mathbf{r},t) C_j^{(n)} = \sum_{i=1}^{b} \underline{A}_i \left(\sum_{j=1}^{b} \mathscr{L}_{ij}^{\text{LE}}(\mathbf{r},t) \kappa_j^{\text{LE}}(\mathbf{r},t) C_j^{(n)} \right)$$

$$= \lambda^{(n)} \sum_{i=1}^{b} \kappa_i^{\text{LE}}(\mathbf{r},t) \underline{A}_i C_i^{(n)} = \lambda^{(n)} \left\langle \underline{A} | C^{(n)} \right\rangle_{\text{LE}} (\mathbf{r},t). \tag{A.70}$$

From (A.69) and (A.70), we obtain:

$$\left(\lambda^{(n)} - 1\right) \left\langle \underline{A} \,|\, C^{(n)} \right\rangle_{\text{LE}} (\mathbf{r},t) = 0; \qquad n = 1, \ldots, (b - N_{\text{hydro}}), \tag{A.71}$$

which implies (A.68), since $\lambda^{(n)} \neq 1$.

Using the properties given by Equations (A.63), (A.65) and (A.70), we can decompose $\mathscr{L}^{\text{LE}}(\mathbf{r},t)$ in the form given in (8.39):

$$\mathscr{L}_{ij}^{\text{LE}}(\mathbf{r},t) = \kappa_i^{\text{LE}}(\mathbf{r},t) \underline{A}_i * \langle \underline{A} | \underline{A} \rangle_{\text{LE}}^{-1} (\mathbf{r},t) * \underline{A}_j + \tilde{\mathscr{L}}_{ij}^{\text{LE}}(\mathbf{r},t); \tag{A.72}$$

$\tilde{\mathscr{L}}_{ij}^{\text{LE}}(\mathbf{r},t)$ has the same set of eigenvectors as $\mathscr{L}_{ij}^{\text{LE}}(\mathbf{r},t)$, with eigenvalues:

$$\sum_{j=1}^{b} \tilde{\mathscr{L}}_{ij}^{\text{LE}}(\mathbf{r},t)\; \kappa_j^{\text{LE}}(\mathbf{r},t)\; \underline{A}_j = 0, \tag{A.73}$$

$$\sum_{j=1}^{b} \tilde{\mathscr{L}}_{ij}^{\text{LE}}(\mathbf{r},t)\; \kappa_j^{\text{LE}}(\mathbf{r},t)\; C_j^{(n)} = \lambda^{(n)}\; \kappa_i^{\text{LE}}(\mathbf{r},t)\; C_i^{(n)}, \tag{A.74}$$

where $\left|\lambda^{(n)}\right| < 1$ for $n = 1, \ldots, (b - N_{\text{hydro}})$.

A.13 Projectors (Section 8.3)

We define the projection operator $\mathscr{P}$, the projector onto the set of constants of motion, by its action on an arbitrary vector $\{B_l(\mathbf{r},t) f_l^{\text{LE}}(\mathbf{r},t)\}_{l=1}^{b}$:

$$(\mathscr{P}B)_j\,(\mathbf{r},t)\; f_j^{\text{LE}}(\mathbf{r},t)$$

$$= \sum_{l=1}^{b} \kappa_j^{\text{LE}}(\mathbf{r},t)\; \underline{A}_j * \langle \underline{A} | \underline{A} \rangle_{\text{LE}}^{-1} (\mathbf{r},t) * \underline{A}_l B_l(\mathbf{r},t) \times f_l^{\text{LE}}(\mathbf{r},t)$$

$$= \kappa_j^{\text{LE}}(\mathbf{r},t)\; \underline{A}_j * \langle \underline{A} | \underline{A} \rangle_{\text{LE}}^{-1} (\mathbf{r},t) * \langle \underline{A}\, B(\mathbf{r},t) \rangle_{\text{LE}}\,. \tag{A.75}$$

In particular, we notice that:

$$\langle \underline{A}\,(\mathscr{P}B)\,(\mathbf{r},t)\rangle_{\mathrm{LE}} \;=\; \sum_{j=1}^{b} \underline{A}_j\,(\mathscr{P}\underline{B})_j\,(\mathbf{r},t)\; f_j^{\mathrm{LE}}(\mathbf{r},t) \;=\; \langle \underline{A}\,B(\mathbf{r},t)\rangle_{\mathrm{LE}}\,. \tag{A.76}$$

The complementary projector $\mathscr{Q} = 1 - \mathscr{P}$ projects onto the set orthogonal to the constants of motion, i.e.:

$$\begin{aligned}(\mathscr{Q}B)_j\,(\mathbf{r},t) f_j^{\mathrm{LE}}(\mathbf{r},t) &\\ = \sum_{l=1}^{b} \left(\delta_{jl} - \kappa_j^{\mathrm{LE}}(\mathbf{r},t)\,\underline{A}_j * \langle \underline{A}|\underline{A}\rangle_{\mathrm{LE}}^{-1}\,(\mathbf{r},t) * \underline{A}_l\right) B_l(\mathbf{r},t)\, f_l^{\mathrm{LE}}(\mathbf{r},t) &\\ = B_j(\mathbf{r},t) f_j^{\mathrm{LE}}(\mathbf{r},t) - \kappa_j^{\mathrm{LE}}(\mathbf{r},t)\,\underline{A}_j * \langle \underline{A}|\underline{A}\rangle_{\mathrm{LE}}^{-1}\,(\mathbf{r},t) * \langle \underline{A}\,B(\mathbf{r},t)\rangle_{\mathrm{LE}}\,, &\end{aligned} \tag{A.77}$$

such that:

$$\langle \underline{A}\,(\mathscr{Q}B)\,(\mathbf{r},t)\rangle_{\mathrm{LE}} \;=\; \sum_{j=1}^{b} \underline{A}_j\,(\mathscr{Q}B)_j\,(\mathbf{r},t) f_j^{\mathrm{LE}}(\mathbf{r},t) \;=\; 0, \tag{A.78}$$

that is, the quantity $(\mathscr{Q}B)\,(\mathbf{r},t)$ has no projections onto the set of constants of motion. The quantity that appears on the r.h.s. of Equation (8.50) is precisely of this form, with $B_l(\mathbf{r},t) f_l^{\mathrm{LE}}(\mathbf{r},t) \;=\; \mathbf{c}_l \cdot \nabla_{\mathbf{r}_1}\, f_l^{\mathrm{LE}}(\mathbf{r},t)$.

A.14 Time correlation functions (Section 8.4)

Within the Boltzmann approximation the evolution of the equilibrium two-point correlation function (see also Section 6.1) is given by:

$$\begin{aligned}\langle \delta n_i(\mathbf{r},\tau)\delta n_j(\mathbf{r}',0)\rangle_{(\mathrm{eq})}^{(B)} &\\ = \sum_{k=1}^{b} \exp\{-\mathbf{c}_i \cdot \nabla_{\mathbf{r}}\}\; \mathscr{L}_{ik}^{\mathrm{eq}}\, \langle \delta n_k(\mathbf{r},\tau-1)\delta n_j(\mathbf{r}',0)\rangle_{(\mathrm{eq})}^{(B)}\,, &\end{aligned} \tag{A.79}$$

with:

$$\mathscr{L}_{ij}^{\mathrm{eq}} = \sum_{\{s\}\{s'\}} s_i' \,\frac{s_j - f_j^{(\mathrm{eq})}}{\kappa_j^{(\mathrm{eq})}}\, \langle \xi \rangle_{\{s\}\to\{\sigma\}} \prod_{k=1}^{b} \left(f_k^{(\mathrm{eq})}\right)^{s_k} \left(1 - f_k^{(\mathrm{eq})}\right)^{(1-s_k)}. \tag{A.80}$$

Equation (A.79) is iterated to yield:

$$\begin{aligned}\langle \delta n_i(\mathbf{r},\tau)\delta n_j(\mathbf{r}',0)\rangle_{(\mathrm{eq})}^{(B)} &\\ = \sum_{k,l} \exp\{-\mathbf{c}_i \cdot \nabla_{\mathbf{r}}\} \left\{\left[\mathscr{T}^{\mathrm{eq}}\right]^{(\tau-1)}\right\}_{ik} \mathscr{L}_{kl}^{\mathrm{eq}}\, \langle \delta n_l(\mathbf{r},0)\delta n_j(\mathbf{r}',0)\rangle_{(\mathrm{eq})}^{(B)} &\\ = \sum_{k,l} \exp\{-\mathbf{c}_i \cdot \nabla_{\mathbf{r}}\} \left\{\left[\mathscr{T}^{\mathrm{eq}}\right]^{(\tau-1)}\right\}_{ik} \mathscr{L}_{kl}^{\mathrm{eq}}\, \langle \delta n_l(\mathbf{r},0)\delta n_j(\mathbf{r}',0)\rangle_{(\mathrm{eq})}\,, &\end{aligned} \tag{A.81}$$

with:

$$\mathscr{T}_{ij}^{\rm eq} = \mathscr{L}_{ij}^{\rm eq} \exp\{-\mathbf{c}_j \cdot \nabla_{\mathbf{r}}\}. \tag{A.82}$$

The equal time correlation function is:

$$\langle \delta n_l(\mathbf{r}, 0)\, \delta n_j(\mathbf{r}', 0)\rangle_{\rm (eq)} = \kappa_j^{\rm (eq)}\, \delta_{jl}\, \delta(\mathbf{r}, \mathbf{r}'), \tag{A.83}$$

where $\kappa_j^{\rm (eq)}$ denotes the second cumulant $\kappa_j^{(2)}$ (see Section 4.7). Using Equations (A.81) and (A.83), we have:

$$\begin{aligned}
&\frac{1}{V} \sum_{\mathbf{r}\,\mathbf{r}'} \langle \delta n_i(\mathbf{r}, \tau) \delta n_j(\mathbf{r}', 0)\rangle_{\rm (eq)}^{(B)} \\
&= \frac{1}{V} \sum_{\mathbf{r}\,\mathbf{r}'} \exp\{-\mathbf{c}_i \cdot \nabla_{\mathbf{r}}\} \left\{ \left[\mathscr{T}^{\rm eq}\right]^{(\tau-1)} \right\}_{ik} \mathscr{L}_{kl}^{\rm eq} \kappa_j^{\rm (eq)} \delta_{jl} \delta(\mathbf{r}, \mathbf{r}') \\
&= \frac{1}{V} \sum_{\mathbf{r}} \exp\{-\mathbf{c}_i \cdot \nabla_{\mathbf{r}}\} \left\{ \left[\mathscr{T}^{\rm eq}\right]^{(\tau-1)} \right\}_{ik} \mathscr{L}_{kj}^{\rm eq} \kappa_j^{\rm (eq)} \\
&= \left\{ \left[\mathscr{L}^{\rm eq}\right]^{\tau} \right\}_{ij} \kappa_j^{\rm (eq)}.
\end{aligned} \tag{A.84}$$

Therefore:

$$\begin{aligned}
&\sum_{\tau=0}^{\infty}{}' \frac{1}{V} \sum_{\mathbf{r}\,\mathbf{r}'} \underline{\hat{\mathbf{J}}}_i\, \underline{\hat{\mathbf{J}}}_j \langle \delta n_i(\mathbf{r}, \tau) \delta n_j(\mathbf{r}', 0)\rangle_{\rm (eq)}^{(B)} \\
&= \sum_{\tau=0}^{\infty}{}' \sum_{i,j} \underline{\hat{\mathbf{J}}}_i\, \underline{\hat{\mathbf{J}}}_j \left\{ \left[\mathscr{L}^{\rm eq}\right]^{\tau} \right\}_{ij} \kappa_j^{\rm (eq)} \\
&= \sum_{\tau=0}^{\infty}{}' \left\langle \underline{\hat{\mathbf{J}}} \mid \underline{\hat{\mathbf{J}}}(\tau) \right\rangle_{\rm (eq)}.
\end{aligned} \tag{A.85}$$

References

Abramowitz, M. & Stegun, I.A. 1965. *Handbook of Mathematical Functions.* Dover, New York.

Alder, B.J. & Wainwright, T.E. 1970. Decay of the velocity autocorrelation function, *Phys. Rev. A* **1**(1), 18–21.

Appert, C. & Zaleski, S. 1990. Lattice gas with a liquid-gas transition, *Phys. Rev. Lett.* **64**(1), 1–4.

Ashcroft, N.W. & Mermin, N.D. 1976. *Solid State Physics.* Holt-Saunders, Philadelphia.

Batchelor, G.K. 1967. *An Introduction to Fluid Dynamics.* Cambridge University Press, Cambridge.

Bell, H., Mœller-Wenghoffer, H., Kollmar, A., Stockmeyer. R., Springer, T. & Stiller, H. 1975. Neutron Brillouin scattering in fluid neon, *Phys. Rev. A* **11**(1), 316–327.

Bender, C.M. & Orszag, S.A. 1978. *Advanced Mathematical Methods for Scientists and Engineers.* McGraw-Hill, Philadelphia.

Bernardin, D. & Sero-Guillaume, O.E. 1990. Lattice gases mixtures models for mass diffusion, *Eur. J. Mech. B* **9**, 21.

Bonetti, M. & Boon, J.P. 1989. Chaotic dynamics in open flows : the excited jet, *Phys. Rev. A* **40**(6), 3322–3345.

Boon, J.P., editor 1992. *Lattice Gas Automata Theory, Implementation and Simulation (special issue of J. Stat. Phys.* **68***(3/4)).* Plenum Press, New York (Proceedings of a NATO Advanced Research Workshop, Nice (France), June 25–28, 1991.).

Boon, J.P., Dab, D., Kapral, R. & Lawniczak, A. 1996. Lattice gas automata for reactive systems, *Phys. Rep.* **273**, 55–147.

Boon, J.P. & Yip, S. 1980. *Molecular Hydrodynamics.* McGraw-Hill, New York (Reprinted by Dover, New York, 1991).

Brito, R., Bussemaker, H.J. & Ernst, M.H. 1992. A fluctuation formula for the non-Galilean factor in lattice gas automata, *J. Phys. A Math. Gen.* **25**, L949–L954.

Brito, R. & Ernst, M.H. 1992. Ring kinetic theory for tagged-particle problems in lattice gases, *Phys. Rev. A* **46**(2), 875–887.

Brito, R., Ernst, M.H. & Kirkpatrick, T.R. 1991. Staggered diffusivities in lattice gas cellular automata, *J. Stat. Phys.* **62**(1/2), 283–295.

Bussemaker, H., Ernst, M.H. & Dufty, J.W. 1995. Generalized Boltzmann equation for lattice gas automata, *J. Stat. Phys.* **78**(5/6), 1521–1554.

Cauchy, A. 1851. Rapport sur un mémoire présenté par A. Bravais, et intitulé 'Etudes sur la cristallographie', *C. R. Acad. Sci. Paris* **32**(8), 284–289.

Cercignani, C. 1994. On the thermodynamics of a discrete velocity gas, *Transp. Theor. Stat. Phys.* **23**(1-3), 1–8.

Chapman, S. & Cowling, T.G. 1970. *The Mathematical Theory of Non-uniform Gases.* Cambridge University Press, Cambridge.

Chen, S., Zhao, K.H. & Doolen, G.D. 1989. A lattice gas model with temperature, *Physica D* **37**(1/3), 42–59.

Clavin, P., d'Humières, D., Lallemand, P. & Pomeau, Y. 1986. Automates cellulaires pour les problèmes à frontières libres en hydrodynamique à 2 et 3 dimensions, *C. R. Acad. Sci. Paris, série II* **303**, 1169–1174.

Cornubert, R., d'Humières, D. & Levermore, D. 1991. A Knudsen layer theory for lattice gases, *Physica D* **47**, 241–259.

Dab, D., Lawniczak, A., Boon, J.P. & Kapral, R. 1990. Cellular automaton model for reactive systems, *Phys. Rev. Lett.* **64**(20), 2462–2465.

Das, S.P., Bussemaker, H.J. & Ernst, M.H. 1993. Generalized hydrodynamics and dispersion relations in lattice gases, *Phys. Rev. A* **48**(1), 245–255.

Doolen, G., editor 1990. *Lattice Gas Methods for Partial Differential Equations.* Addison-Wesley, New York (Reprints collection).

Drazin, P.G. & Reid, W.H. 1981. *Hydrodynamic Stability.* Cambridge University Press, Cambridge.

Dubrulle, B., Frisch, U., Hénon, M. & Rivet, J.P. 1990. Low viscosity lattice gases, *J. Stat. Phys.* **59**, 1187–1226.

Eliott, J.P. & Dawber, P.G. 1979. *Symmetry in Physics.* MacMillan Press Ltd., London.

Ernst, M.H. 1990a. Statistical mechanics of cellular automata fluids, in *Liquids, Freezing and the Glass Transition*, 1–94, eds. D. Levesque, J.P. Hansen & J. Zinn–Justin. Elsevier Science, Amsterdam.

Ernst, M.H. 1990b. Linear response theory for cellular automata fluids, in *Fundamental Problems in Statistical Mechanics VII*, ed. H. van Beijeren. North Holland Publ., Amsterdam.

Ernst, M.H. 1991. Temperature and heat conductivity in cellular automata fluids, in *Discrete Models of Fluid Dynamics*, 186–197, ed. A.S. Alves. World Scientific, Singapore.

Ernst, M.H. & Das, S.P. 1992. Thermal cellular automata fluids, *J. Stat. Phys.* **66**(1/2), 465–483.

Ernst, M.H. & Dufty, J.W. 1990. Hydrodynamics and time correlation functions for cellular automata, *J. Stat. Phys.* **58**(1/2), 57–86.

Ernst, M.H., Hauge, E.H. & van Leeuwen, J.M.J. 1971. Asymptotic time behavior of correlation functions, *Phys. Rev. A* **4**(5), 2055–2065.

Fano, R.M. 1961. *Transmission of Information.* Wiley, New York (p. 43).

Fleury, P.A. & Boon, J.P. 1969. Brillouin scattering in simple liquids : argon and neon, *Phys. Rev.* **186**, 244–254.

Forster, D. 1975. *Hydrodynamic Fluctuations, Broken Symmetry, and Correlation Functions.* Benjamin, Reading, Massachusetts.

Frenkel, D. & Ernst, M.H. 1989. Simulation of diffusion in a two-dimensional lattice-gas cellular automaton: a test of mode-coupling theory, *Phys. Rev. Lett.* **63**, 2165–2168.

Frisch, U., Hasslacher, B. & Pomeau, Y. 1986. Lattice gas automata for the Navier–Stokes equation, *Phys. Rev. Lett.* **56**, 1505–1508.

Frisch, U., d'Humières, D., Hasslacher, B., Lallemand, P., Pomeau, Y. & Rivet, J.P. 1987. Lattice gas hydrodynamics in two and three dimensions, *Complex Systems* **1**, 649–707.

Gatignol. R. 1975. *Théorie Cinétique des Gaz à Répartition Discrète de vitesses.* Springer-Verlag, Berlin.

Grosfils, P. 1994. Hydrodynamique statistique des gaz sur réseau, Ph.D., Université Libre de Bruxelles.

Grosfils, P., Boon, J.P. & Lallemand, P. 1992. Spontaneous fluctuation correlations in thermal lattice gas automata, *Phys. Rev. Lett.* **68**, 1077–1080.

Grosfils, P., Boon, J.P., Brito, R. & Ernst, M.H. 1993. Statistical hydrodynamics of lattice gas automata, *Phys. Rev. E* **48**, 2655–2668.

Hanon, D. & Boon, J.P. 1997. Diffusion and correlations in a lattice gas automata, *Phys. Rev. E* **56**(6), 6331–6339.

Hardy, J. & Pomeau, Y. 1972. Thermodynamics and hydrodynamics for a modeled fluid, *J. Math. Phys.* **13**, 1042–1051.

Hardy, J., Pomeau, Y. & de Pazzis, O. 1973. Time evolution of a two-dimensional model system: Invariant states and time correlation functions, *J. Math. Phys.* **14**, 1746–1759.

Hardy, J., de Pazzis, O. & Pomeau, Y. 1976. Molecular dynamics of a classical lattice gas: Transport properties and time correlation functions, *Phys. Rev. A* **13**, 1949–1961.

Hardy, J. & Pomeau, Y. 1977. Microscopic model for viscous flow in two dimensions, *Phys. Rev. A* **16**, 720–726.

Hénon, M. 1987a. Isometric collision rules for the 4-D FCHC lattice gas, *Complex Systems* **1**, 475–494.

Hénon, M. 1987b. Viscosity of a lattice gas, *Complex Systems* **1**, 763–789.

Hénon, M. 1989. On the relation between lattice gases and cellular automata, in *Proceedings of the Workshop on Discrete Kinetic Theory, Lattice Gas Dynamics and Foundations of Hydrodynamics, Torino (Italy), September 20–24, 1988*, 160–161, ed. R. Monaco. World Scientific, Singapore.

Hénon, M. 1992. Implementation of the FCHC lattice gas model on the connection machine, in *Proceedings of the NATO Advanced Research Workshop on Lattice Gas Automata Theory, Implementation, and Simulation, Nice (France), June 25–28, 1991, J. Stat. Phys.* **68**(3/4), 353–377, ed. J.P. Boon.

van der Hoef, M.A. & Frenkel, D. 1990. Long-time tails of the velocity autocorrelation function in two- and three-dimensional lattice-gas cellular automata, *Phys. Rev. A* **41**, 4277–4285.

van der Hoef, M.A. & Frenkel, D. 1991. Evidence for faster-than-t^{-1} decay of the velocity autocorrelation function in a *2D* fluid, *Phys. Rev. Lett.* **66**, 1591–1594.

Homsy, G.M. 1987. Viscous fingering in porous media, *Annu. Rev. Fluid Mech.* **19**, 271–311.

Hopcroft, J.E. & Ullman, J.D. 1979. *Introduction to Automata Theory, Languages and Computation.* Addison-Wesley, New York.

Huang, K. 1963. *Statistical Mechanics.* Wiley, New York.

Huerre, P. & Monkewitz, P.A. 1990. Local and global instabilities in spatially developing flows, *Annu. Rev. Fluid Mech.* **22**, 473–537.

d'Humières, D., Lallemand, P. & Frisch, U. 1986. Lattice gas models for 3-D hydrodynamics, *Europhys. Lett.* **2**, 291–297.

d'Humières, D., Qian, Y.H. & Lallemand, P. 1989. Invariants in lattice gas models, in *Proceedings of the Workshop on Discrete Kinetic Theory, Lattice Gas Dynamics and Foundations of Hydrodynamics, Torino (Italy), September 20-24, 1988*, 102–113, ed. R. Monaco. World Scientific, Singapore.

d'Humières, D., Lallemand, P. & Qian, Y.H. 1990. Finding the linear invariants of lattice gases, in *Computational Physics and Cellular Automata*, eds. A. Pires, D.P. Landau & H. Herrman. World Scientific, Singapore.

Kinchin, A.I. 1957. *Mathematical Foundations of Information Theory.* Dover, New York.

König, M., Eisenlohr, H. and Eckelmann, H. 1990. The fine structure in the Strouhal-Reynolds number relationship of the laminar wake of a circular cylinder, *Phys. Fluids A* **2**(9), 1607–1614.

Kubo. R. 1958. Some aspects of the statistical mechanical theory of irreversible processes, in *Lectures in Theoretical Physics, vol. 1*, 120–140, eds. W.E. Brittin & L.G. Dunham. Wiley, New York.

Lamb, H. 1945. *Hydrodynamics*, 6th edition. Dover, New York.

Landau, L.D. & Lifschitz, E.M. 1987. *Fluid Mechanics*, 2nd edition, vol. 6, in 'A Course of Theoretical Physics'. Pergamon Press, Oxford.

Lutsko, J.F., Boon, J.P. & Somers, J.A. 1992. Lattice gas automata simulations of viscous fingering in a porous medium, in *Numerical Methods for the Simulation of Multi-Phase and Complex Flow*, 124–135, ed. T.M.M. Verheggen. Springer-Verlag, Berlin.

Majda, A. 1984. *Compressible Fluid Flow and Systems of Conservation Laws in Several Space Variables*. Springer-Verlag, Berlin.

Mareschal, M. & Holian, B.L., editors 1992. *Microscopic Simulations of Complex Hydrodynamic Phenomena*. Plenum, New-York.

Molvig, K, Donis, P., Miller,R., Myczkowski, J. & Vichniac, G. 1989. Multi-species lattice-gas automata for realistic fluid dynamics, In *Cellular Automata and Modeling of Complex Systems*, 206–231, eds. P. Manneville, R. Bideaux, N. Boccara & G. Vichniac. Springer-Verlag, Berlin.

Mori, H. 1965a. Transport, collective motion, and Brownian motion, *Prog. Theor. Phys.* **33**, 423–455.

Mori, H. 1965b. A continued fraction representation of the time-correlation function, *Prog. Theor. Phys.* **34**, 399–416.

Naitoh, T., Ernst, M.H., van der Hoef, M.A. & Frenkel, D. 1991. Extended mode coupling and simulations in cellular-automata fluids, *Phys. Rev. A* **44**, 2484–2494.

McNamara, G.R. 1990. Diffusion in a lattice gas automaton, *Europhys. Lett.* **12**, 329.

McQuarrie, D.A. 1976. *Statistical Mechanics*. Harper and Row, New York.

Noullez, A. 1990. Automates de Gaz sur Réseau: Aspects Théoriques et Simulations, Ph.D., Université Libre de Bruxelles.

Papoulis. A. 1965. *Probability, Random Variables, and Stochastic Processes*. McGraw-Hill, New York.

Procaccia, I., Ronis, D. & Oppenheim, I. 1979. Statistical mechanics of stationary states. IV. Far-from-equilibrium stationary states and the regression of fluctuations, *Phys. Rev.* A **20**(6), 2533–2546.

Résibois, P. & de Leener, M. 1977. *Classical Kinetic Theory of Fluids*. John Wiley, New York.

Risken, H. 1984. *The Fokker-Planck Equation*. Springer-Verlag, Berlin.

Rivet, J.P. 1987. Simulation d'écoulements tri-dimensionnels par la méthode des gaz sur réseaux: premiers résultats, *C. R. Acad. Sci. Paris, série II* **305**, 751–756.

Rivet, J.P. 1987. Green–Kubo formalism for lattice gas hydrodynamics and Monte-Carlo evaluation of shear viscosities, *Complex Systems* **1**, 839–851.

Rivet, J.P. 1988. Hydrodynamique par la méthode des gaz sur réseaux, Ph.D., Université de Nice.

Rivet, J.P. 1993. Spontaneous symmetry-breaking in the 3-D wake of a long cylinder simulated by the lattice gas method and drag coefficient measurements, *Appl. Sci. Res.* **51**, 123–126.

Rothman, D.H. & Keller, J.M. 1988. Immiscible cellular automaton fluids, *J. Stat. Phys.* **52**, 1119–1127.

Rothman, D.H. & Zaleski, S. 1994. Lattice-gas models of phase separation: interfaces, phase transitions, and multi-phase flow, *Rev. Mod. Phys.* **66**, 1417–1479.

Rothman, D.H. & Zalseki, S. 1977. *Lattice Gas Cellular Automata*, Cambridge University Press, Cambridge.

Schmitz, R. & Dufty, J.W. 1990. Positivity of transport coefficients for cellular automata, *Phys. Rev. A* **41**(8), 4294–4297.

Shannon, C.E. & Weaver, W. 1969. *The Mathematical Theory of Communication*, fourth edition. The University of Illinois Press, Urbana.

Suárez, A. & Boon, J.P. 1997a. Non-linear lattice gas hydrodynamics, *J. Stat. Phys.* **87**(5/6), 1123–1130.

Suárez, A. & Boon, J.P. 1997b. Non-linear hydrodynamics of lattice gas automata with semi-detailed balance, *Int. J. Mod. Phys. C* **8**(4), 653–674.

Succi, S. 2000. *The Lattice Boltzmann Equation for Fluid Dynamics and Beyond.* Oxford University Press, Oxford (U.K.) (to be published).

Succi, S. Benzi, R. & Higuera, F. 1991. The lattice Boltzmann equation: A new tool for computational fluid mechanics, *Physica D* **47** (1/2), 219–230.

Tribel, O. & Boon, J.P. 1995. Lattice gas automata with interaction potential, *J. Stat. Phys.* **81**, 361–377.

Tribel, O. & Boon, J.P. 1997. Entropy and correlations in lattice gas automata without detailed balance, *Int. J. Mod. Phys. C* **8**(4), 641–652.

Tung, W.K. 1985. *Group Theory in Physics.* World Scientific, Singapore.

Williamson, C.H.K. 1989. Oblique and parallel modes of vortex shedding in the wake of a circular cylinder at low Reynolds numbers, *J. Fluid Mech.* **206**, 579–627.

Wolfram, S. 1986. *Theory and Applications of Cellular Automata.* World Scientific, Singapore.

Wolfram, S. 1986. Cellular automaton fluids 1: basic theory, *J. Stat. Phys.* **45**, 471–525.

Zanetti, G. 1989. Hydrodynamics of lattice gas automata, *Phys. Rev. A* **40**(3), 1539–1548.

Zwanzig, R. 1960. Statistical mechanics of irreversibility, in *Lectures on Theoretical Physics, vol.3*, 106–141, eds. W.E. Brittin, B.W. Downs & J. Downs. Wiley, New York.

Zwanzig, R. 1965. Time-correlation functions and transport coefficients in statistical mechanics, *Ann. Rev. Phys. Chem.* **16**, 67–102.

Author index

Subject index

Page numbers in bold type indicate main sections.